Strengthening Global Resilience to Natural Disasters

Qamar Ali
https://orcid.org/0000-0002-8178-7591
Virtual University of Pakistan, Pakistan

Muhammad Tariq Iqbal Khan
Government College University, Pakistan

IGI Global
Scientific Publishing
Publishing Tomorrow's Research Today

Vice President of Editorial	Melissa Wagner
Director of Acquisitions	Mikaela Felty
Director of Book Development	Jocelynn Hessler
Production Manager	Mike Brehm
Cover Design	Jose Rosado

Published in the United States of America by
IGI Global Scientific Publishing
701 East Chocolate Avenue
Hershey, PA, 17033, USA
Tel: 717-533-8845 | Fax: 717-533-7115
Website: https://www.igi-global.com E-mail: cust@igi-global.com

Library of Congress Cataloging-in-Publication Data

Names: Ali, Qamar, 1988- editor. | Iqbal Khan, Muhammad Tariq, 1988- editor.
Title: Strengthening global resilience to natural disasters / edited by Qamar Ali, Muhammad Tariq Iqbal Khan.
Description: Hershey, PA : IGI Global Scientific Publishing, [2026] | Includes bibliographical references and index. | Summary: "The objective of the proposed study is to evaluate the influence of natural catastrophes on important economic indicators and development metrics across a range of socioeconomic categories"-- Provided by publisher.
Identifiers: LCCN 2024055548 (print) | LCCN 2024055549 (ebook) | ISBN 9798369397459 (hardcover) | ISBN 9798369397466 (paperback) | ISBN 9798369397473 (ebook)
Subjects: LCSH: Natural disasters--Economic aspects.
Classification: LCC HC79.D45 S77 2025 (print) | LCC HC79.D45 (ebook) | DDC 363.34068/1--dc23/eng/20250131
LC record available at https://lccn.loc.gov/2024055548
LC ebook record available at https://lccn.loc.gov/2024055549
British Cataloguing in Publication Data
A Cataloguing in Publication record for this book is available from the British Library.

Table of Contents

Preface xiii

Chapter 1
Understanding the Science of Natural Disaster 1
Qamar Ali, Virtual University of Pakistan, Pakistan
Muhammad Tariq Iqbal Khan, Government College University, Faisalabad, Pakistan

Chapter 2
Climate Change and Natural Disasters 21
Deepika Paira, VEL TECH University, India
A. Rohini, CVR College of Engineering, India
Sudhanshu Chandra, Maulana Azad National Urdu University, India
Pamarthi Satyanarayana, VEL TECH University, India
Mohit Sharma, Maharshi Dayanand University, Rohtak, India
P. Selvakumar, Department of Science and Humanities, Nehru Institute of Technology, Coimbatore, India
Manjunath T. C., Rajarajeswari College of Engineering, India

Chapter 3
A Review of the United States of America's Disaster Management System 45
Murat Şengöz, Independent Researcher, Turkey

Chapter 4
Disaster Education and Awareness Programs 73
Y. Ebenezer, VEL TECH University, India
Kasagani Durga, NBM Law College, India
N. B. Chandrakala, Andhra University, India
G. Shashank, Chandigarh School of Business, Chandigarh Group of Colleges, Jhanjeri, India
Mohit Sharma, Maharshi Dayanand University, Rohtak, India
P. Selvakumar, Department of Science and Humanities, Nehru Institute of Technology, Coimbatore, India
T. C. Manjunath, Rajarajeswari College of Engineering, India

Chapter 5
Tips for Effectively Communicating Human Capital Analysis Analysis to Decision Makers 103
Shivansu Sachan, CSJM University, India
Ankit Kumar, CSJM University, India

Chapter 6
Embracing Local Knowledge: Integrating Traditional Practices in Disaster Risk Reduction..................... 133
Menard Musendekwa, Reformed Church University, Zimbabwe

Chapter 7
Building Climate Resilience in the Health Sector Through Community-Based Adaptation Strategies 153
Naboshree Bhattacharya, Amity University, Ranchi, India

Chapter 8
Fostering Economic Resilience in a World of Increasing Natural Disasters: A Multi-Dimensional Perspective..................... 183
Naboshree Bhattacharya, Amity University, Ranchi, India
Divya Bansal, Amity University, Noida, India

Chapter 9
Building Resilience: Addressing the Economic, Social, and Environmental Impacts of Natural Disasters in a Changing Climate..................... 231
Muhammad Aamir Mahmood, Virtual University of Pakistan, Pakistan

Chapter 10
Psychological and Social Impact of Natural Disasters: A Case Study of Wayanad District of India 243
Anitha Parveen J., Lovely Professional University, Punjab, India
Guddo, Lovely Professional University, Punjab, India

Chapter 11
Social, Psychological, Motivational, and Educational Effects of 6 February Earthquakes: A Qualitative Study on Teachers..................... 263
Ceyda Akilli, Fırat University, Turkey

Chapter 12
Preparedness on Seismic Risk: A Case Study of a Community in Oaxaca, Mexico 291
Tatiana Gouzeva, Instituto Politecnico Nacional, Mexico

Chapter 13
Resilience and Renewal: The Literary Reflections and Societal Impact of Disasters in England 313
Gülten Akgül, The Ministry of National Education, Turkey

Compilation of References 347

About the Contributors 403

Index 409

Detailed Table of Contents

Preface.. xiii

Chapter 1
Understanding the Science of Natural Disaster 1
Qamar Ali, Virtual University of Pakistan, Pakistan
Muhammad Tariq Iqbal Khan, Government College University, Faisalabad, Pakistan

This chapter examines the relationship between natural hazards and vulnerabilities such as economic, social, environmental, and physical factors, which heighten the global impact of natural disasters. Recent data show alarming trends with 3.6 million deaths, 7.7 billion people affected, and \$3.3 trillion in economic losses due to natural disasters. These events, including earthquakes, floods, and wildfires, have increased in frequency and intensity due to human activities such as rapid urbanization, environmental degradation, and climate change. Developing countries are particularly vulnerable due to their socio-economic conditions, exacerbating disaster impacts. This study advocates for sustainable urban planning and resilient human settlements as essential strategies to mitigate disaster effects and promote sustainable development. Recommendations include adopting integrated, eco-friendly development strategies that prioritize environmental conservation and human well-being to enhance resilience against future disasters.

Chapter 2
Climate Change and Natural Disasters 21
Deepika Paira, VEL TECH University, India
A. Rohini, CVR College of Engineering, India
Sudhanshu Chandra, Maulana Azad National Urdu University, India
Pamarthi Satyanarayana, VEL TECH University, India
Mohit Sharma, Maharshi Dayanand University, Rohtak, India
P. Selvakumar, Department of Science and Humanities, Nehru Institute of Technology, Coimbatore, India
Manjunath T. C., Rajarajeswari College of Engineering, India

The intersection between these two issues is pivotal for understanding the broader impacts events. As a result, the frequency, intensity, and distribution are significantly affected. influences natural disasters is through the intensification of extreme weather events. For instance, warmer ocean temperatures provide more energy for tropical storms, leading to stronger hurricanes and typhoons. These storms, in turn, bring about more severe storm surges, heavy rainfall, and flooding. Similarly, the increased evaporation rates due to higher temperatures can exacerbate drought conditions, leading to prolonged dry periods that severely impact agriculture, water supply, and ecosystem health. Wildfires also become the relationship between is not limited to the direct effects of changing weather patterns. There are also indirect impacts that exacerbate vulnerabilities in both human and natural systems. For example, and lakes. Communities around the world experience these impacts in varied ways, often reflecting existing social and economic inequalities.

Chapter 3
A Review of the United States of America's Disaster Management System 45
Murat Şengöz, Independent Researcher, Turkey

This article focuses on enhancing the joint operational capabilities of the military, police, fire department, and other relevant organizations, as well as ensuring effective command and control relationships in disaster management. Effective response and coordination in disaster situations require collaboration and integration among different organizations. The article examines the disaster management system and practices in the United States, focusing on the roles and responsibilities of the Federal Emergency Management Agency (FEMA) and other relevant agencies. It also analyzes how cooperation between military and civilian authorities is achieved and how existing legal regulations support and enhance these processes. Supported by case studies and literature reviews, this study highlights critical elements such as training, exercises, technological infrastructure, and legal regulations for developing joint operational capabilities. Finally, recommendations for improving coordination and command-control relationships in disaster management are presented.

Chapter 4
Disaster Education and Awareness Programs 73
Y. Ebenezer, VEL TECH University, India
Kasagani Durga, NBM Law College, India
N. B. Chandrakala, Andhra University, India
G. Shashank, Chandigarh School of Business, Chandigarh Group of Colleges, Jhanjeri, India
Mohit Sharma, Maharshi Dayanand University, Rohtak, India
P. Selvakumar, Department of Science and Humanities, Nehru Institute of Technology, Coimbatore, India
T. C. Manjunath, Rajarajeswari College of Engineering, India

Disaster education plays a pivotal role in building resilient communities by equipping individuals, organizations, and local governments with the knowledge and skills necessary to prepare for, respond to, and recover from various types of disasters. In an era marked by increasing frequency and intensity of natural and man-made disasters—ranging from hurricanes and earthquakes to floods and wildfires—the importance of disaster education cannot be overstated. Effective disaster education programs help mitigate risks, reduce vulnerabilities, and foster a culture of preparedness, ultimately enhancing community resilience. At its core, disaster education aims to raise awareness about the potential hazards and risks that communities face. It provides critical information on the nature of different types of disasters, their impacts, and the steps that individuals and communities can take to protect themselves. This foundational knowledge is essential for developing a proactive approach to disaster preparedness

Chapter 5
Tips for Effectively Communicating Human Capital Analysis Analysis to Decision Makers 103
Shivansu Sachan, CSJM University, India
Ankit Kumar, CSJM University, India

In data-work trade, Analysis for Human Capital Analysis play an impact in forming lot of choices for organizations. In integration of these probability into decision forms requires communication to limit the separate between complex information of knowledge and significant techniques. This theory assures that lucrative workforce-related information is interpreted into aware organizational decisions and provides essential phrases for effectively presenting analysis for Human Capital Analysis to decision formers. The first piece of advice highlights how crucial it is to modify analysis for Human Capital Analysis. Decision makers are likely to focus on analysis that directly help in the formation of important destinations. This involves identifying key performance indicators (KPIs) that align with decision makers and are situated within the context of overarching trade forms. Include a brief summary of key perspectives and their implications for business goals. Use visuals such as charts or graphs to clearly highlight trends and results.

Chapter 6
Embracing Local Knowledge: Integrating Traditional Practices in Disaster Risk Reduction........ 133
Menard Musendekwa, Reformed Church University, Zimbabwe

This chapter examines the pivotal role of local knowledge and traditional practices in enhancing disaster risk reduction (DRR) strategies. Communities across the globe often rely on indigenous wisdom and cultural heritage to address environmental challenges, mitigate hazards, and strengthen resilience to disasters. By emphasizing the value of local expertise, this chapter seeks to bridge the gap between traditional knowledge systems and formal disaster management frameworks. It advocates for a more holistic and inclusive approach to disaster preparedness, where community-based practices are integrated with modern DRR strategies to foster sustainable resilience.

Chapter 7
Building Climate Resilience in the Health Sector Through Community-Based Adaptation Strategies 153
Naboshree Bhattacharya, Amity University, Ranchi, India

The global health landscape is increasingly shaped by the complex impacts of climate change, posing risks to human health and health systems. Building resilience in the health sector is a critical priority. This study explores the transformative potential of community-based adaptation (CBA) strategies in strengthening health systems' capacity to withstand climate-related shocks. CBA empowers communities to identify their vulnerabilities and develop context-specific resilience strategies. The study presents a comprehensive "Triple-A" framework for operationalizing CBA. The study examines the institutional and policy landscape shaping CBA initiatives. It identifies priorities to accelerate CBA uptake and strengthened institutions for mainstreaming CBA into health planning. This study makes a compelling case for CBA's transformative potential in building climate-resilient health systems. It offers a hopeful vision for a more resilient, equitable, and sustainable future for global health.

Chapter 8
Fostering Economic Resilience in a World of Increasing Natural Disasters: A Multi-Dimensional Perspective........ 183
Naboshree Bhattacharya, Amity University, Ranchi, India
Divya Bansal, Amity University, Noida, India

The increasing frequency and intensity of natural disasters pose a significant threat to global economic resilience. The economic impacts of these disasters are far-reaching and complex, encompassing direct costs such as infrastructure damage and indirect consequences like supply chain disruptions. Developing countries and vulnerable communities often bear the brunt of the damage, lacking the resources and capacity to effectively prepare for and respond to disasters. Building economic resilience requires a proactive, multi-dimensional approach that integrates prevention, response, and recovery strategies. This chapter examines the complex relationship between natural disasters and economic resilience, drawing upon empirical evidence, theoretical frameworks, and practical insights. It explores advancements in risk assessment technologies, financial protection mechanisms, and post-disaster reconstruction models, emphasizing the role of international cooperation and public-private partnerships.

Chapter 9
Building Resilience: Addressing the Economic, Social, and Environmental Impacts of Natural Disasters in a Changing Climate 231
Muhammad Aamir Mahmood, Virtual University of Pakistan, Pakistan

With global climate changes, natural disasters occur more often, increase in severity and unpredictability, and represent an unprecedented challenge to global societies. In this chapter, we analyze different types of natural disasters, like earthquakes, hurricanes, floods and wildfires and their indisputable economic, social and environmental impacts. It looks at ways in which climate change is a primary cause behind the worsening of weather patterns and the rising vulnerability of urban and rural communities to these disasters. In addition, this chapter examines global initiatives to boost disaster resilience by reviewing climate adaptation strategies, policy frameworks, technological innovation, and a multi stakeholder approach involving governments, NGOs, private sector actors, and local communities. Case studies of countries like Japan, Bangladesh and the Caribbean show how successful resilience initiatives can heavily dampen the impact of natural calamities on societies.

Chapter 10
Psychological and Social Impact of Natural Disasters: A Case Study of Wayanad District of India 243
Anitha Parveen J., Lovely Professional University, Punjab, India
Guddo, Lovely Professional University, Punjab, India

Kerala's geographic vulnerability to natural disasters, particularly in Wayanad, has been identified by the Centre for Earth Science Studies as one of the most landslide-prone areas in the state, significantly impacting residents' mental health. This study explores the psychological issues arising from such disasters, including PTSD, depression, and anxiety, which often persist for years due to displacement, loss of social support, emotional instability, and chronic distress. Landslides, in particular, have devastating consequences, orphaning children, fracturing families, and disrupting communities. While immediate rescue efforts prioritize physical safety, addressing psychological needs is crucial for long-term recovery. The study highlights the urgent need for increased psychological support, awareness, and policy initiatives to address the mental health challenges faced by disaster survivors. By integrating mental health into disaster management strategies, stakeholders can promote sustainable recovery and well-being for affected individuals and communities in Wayanad.

Chapter 11
Social, Psychological, Motivational, and Educational Effects of 6 February Earthquakes: A Qualitative Study on Teachers.. 263
Ceyda Akilli, Fırat University, Turkey

As a result of two earthquakes with magnitudes of 7.7 Mw and 7.6 Mw in Kahramanmaraş on 6 February 2023, 11 cities in Turkey were severely damaged, and over 50,000 citizens lost their lives. This study aims to explore the social, psychological, and motivational effects of these earthquakes on teachers and their impact on educational activities. A qualitative phenomenological design was used, with 46 teachers from Elazığ, Malatya, and Diyarbakır, areas declared disaster zones, participating. Data was collected using a semi-structured interview form and analyzed through descriptive and content analysis. Results revealed that the earthquakes negatively impacted teachers socially, psychologically, and motivationally, leading to trauma-related disorders, phobias, and reduced professional motivation. Some female teachers, however, noted increased solidarity and empathy. The findings also showed disruptions in educational quality, decreased academic performance, increased absenteeism, and a negative shift in school climate.

Chapter 12
Preparedness on Seismic Risk: A Case Study of a Community in Oaxaca, Mexico 291
Tatiana Gouzeva, Instituto Politecnico Nacional, Mexico

The book chapter presents some results on children's seismic risk knowledge and perception for a sample size of n=522. The study has been conducted in Huajuapan de Leon, Oaxaca, Mexico. The municipality is located in a highly seismic region. The data was collected by employing a questionnaire-based survey. Some of the main findings are the following: a) earthquakes (72.4%) and floods (34.1%) were recognized as a threat, but not drought, however, the lack of water is one of the major concerns of the community, b) around 40% perceived that they lack the knowledge on seismic risk; 57% recognized they live in a highly seismic region, around 95% responded that they know what to do in case of an earthquake while in a classroom, c) most of the students perceived that earthquakes would happen sometime in the future, e.g., 48.3% considered an earthquake would occur "next year", and d) around 68% of the participants "sometimes" feel scared or upset when thinking of earthquakes.

Chapter 13
Resilience and Renewal: The Literary Reflections and Societal Impact of Disasters in England 313
Gülten Akgül, The Ministry of National Education, Turkey

This chapter explores the historical and literary significance of the 1665 Great Plague and the 1666 Great Fire of London, their profound impact on social, economic, and cultural transformations in England. Through a qualitative approach, the research combines the historical analysis of primary sources such as Samuel Pepys' diaries and Daniel Defoe's "A Journal of the Plague Year" with literary analysis. Findings highlight how the Great Plague disrupted social structures, while also sparking acts of solidarity within communities. The aftermath of the Great Fire facilitated urban rebuilding, leading to modern urban planning and improved sanitation systems. The chapter also examines the psychological resilience of individuals and communities during these crises, particularly through religious and community responses. The research reveals the long-term impact of these events on cultural memory and resilience-building. By examining the historical strategies like quarantine measures and fire brigades, the study offers valuable insights for modern disaster preparedness and management.

Compilation of References 347

About the Contributors 403

Index 409

Preface

DESCRIPTION OF THE TOPIC

Natural disasters are one of the most vivid dilemmas of the contemporary societies. These disasters have become deadly to human life, infrastructure and social-economic systems as they are triggered by environmental conditions in most cases enhanced to a deadly degree by anthropogenic alterations. The growth of frequency, intensity, and unpredictability of natural catastrophes, as experienced over past decades, in addition to the existing latent vulnerabilities faced by human beings has brought to light the pressing need of interdisciplinary as well as comprehensive disaster reduction (DRR), management, and resiliency.

The book, strengthening global resilience to natural disasters, provides comprehensive research on the complex impact of natural disasters and puts forward measures that will magnify resilience. The chapters touch on everything including climate change, community-based adaptation and disaster education both in theory and practical aspects. Natural disasters are no longer the focus of geoscientists alone; they touch on economists, sociologists, public health experts, urban planners, psychologists and literary experts. This compendium is a much-needed addition to that developing multidisciplinary discipline.

IN MODERN WORLD APPLICABILITY

The ripple effects of the natural disasters are widespread in the more connected and globalized world these days. Due to the socioeconomic disparities between countries, the vulnerable populations in the developing world pay a steep price. Such phenomena as Turkey earthquakes in 2023, continuous climate-related fires, and increased sea levels serve as evidence of the fact that many areas are still unprepared. The correlation of disasters has created even more significant meaning than the actual physical harm; it affects teaching, mental health, vulnerability to gender, economic capacity, and memory.

Since countries seek to achieve the UN Sustainable Development Goals (SDGs), particularly on climate and natural disasters, sustainable cities, and inequalities, there has never been a better time to learn about the impacts of natural disasters and how they can be mitigated. The book stands at this critical point of science, policy and action by the community.

TARGET AUDIENCE

There is a wide but specific audience of this book and this audience includes:

Policymakers and economists looking at the macroeconomic impacts of calamities.
Researchers and scientists on climate and the environment.
Emergency planners and disaster risk reduction experts.
Medical workers and epidemiologists in the field of public health.
Scholars and learners in academia such as the urban planning, sociology and geography departments.
Humanitarian aid agencies, NGOs and the government.
Researchers of gender studies pay attention to the vulnerability in case of disaster.

Chapters are also selected to bring value in both conceptual knowledge and practice, and consequently the targeted readers on one hand should be made-up of scholars in academia and professionals in other fields.

SUMMARY OF THE CHAPTERS

Chapter 1: Knowing the Science behind the Natural Disasters

In this introductory chapter, the author investigates the impact of natural disasters which is worsened by different forms of weakness. The authors use appalling statistics and recent instances to advocate for sustainable urban planning and suggest environmentally friendly practices to counter the threat to people in the future.

Chapter 2 Climate Change and Natural Disasters

This chapter will reside on how climate change affects natural disasters where extreme weather events are aggravated. It examines the direct and indirect effects on people and environments, water, and crops.

Chapter 3: An Overview of the Disaster Management system in the USA

Offering a thorough study of the FEMA and other predominant agencies within the U.S., this chapter focuses on the combined operating potentialities and the legal and organizational constructions that have allowed the collective response to the occurrence of catastrophes.

Chapter 4: Disaster Educational and Awareness Programs

In exploring how education can help in disaster preparedness, the chapter also emphasizes the need to nurture the role of education through the development of a form of learning programmes within the community.

Chapter 5: Advice With Communicating Human Capital Analysis

It is a special chapter which bridged between data-driven human capital analysis and disaster decision-making. FiPro gives instrumental and communication tools and strategies to make sure that critical knowledge can contribution policy and planning.

Chapter 6: The Power of Local Knowledge: Irradiation of Local Knowledge in DRR

With an approach towards indigenous and traditional practices, this chapter promotes encompassing the knowledge of the community into formal DRR frameworks to benefit resilience and relevance to culture.

Chapter 7: Promoting Climate Resilience of the Health Sector

In this case, the authors have given a Triple-A framework of community-based adaptation in health systems. The chapter lists institutional facilitators and obstacles to the inclusion of CBA within climate-resilient health planning.

Chapter 8: Promoting Resilience to Natural Disasters in Economies

The chapter presents the multi-dimensional model of economic resilience, with the focus on the risk evaluation, financial security as well as the necessity of the international cooperation amid the prospect of the growing risks of disasters.

Chapter 9: Building Resilience: Addressing the Economic, Social, and Environmental Impacts of Natural Disasters in a Changing Climate

This chapter, based on global case studies, examines the influence of disasters in the environmental, social and economical systems. It examines policies and partnerships with stakeholders paving ways to the success of adaptation strategies.

Chapter 10: Psychological and Social Effect of Natural Disasters

As the case study, Wayanad in India is used as an example in this chapter to discuss the psyche aspects of the disaster. It makes the case that psychosocial care should be included in disaster management schemes.

Chapter 11: Social, Psychological, and Educational Consequences of the 2023 Turkey Earthquake

Based on qualitative studies of the teachers, this chapter describes the personal and professional trauma caused by the earthquakes, such as organizational rupture in education, as well as in school settings.

Chapter 12: Seismic Risk consciousness on Oaxaca, Mexico

This is one of the empirical studies, which focuses on the perception and knowledge of children regarding the seismic risks at a high-risk area. It emphasizes the localized education in minimizing any vulnerability.

Chapter 13: Renewal and Resilience through the Literary Reflections

Against the historical backdrop of disasters in England, the chapter does look at how cultural resilience is captured and transmitted through literature to provide psychological and societal referrals to current disaster management.

CONCLUSION: CONTRIBUTION TO THE FIELD

Strengthening Global Resilience to Natural Disasters helps the changing field of disasters studies by offering an interdisciplinary and broad approach to the subject matter. In an insightful combination of empirical research, theories, case studies of communities, and historical perspective, the book informs knowledge on the resilience in scientific, social, psychological, and policy strategies perspectives.

It can be summed up as following contributions:

A vulnerability that is sensitive to gender.
Focus on local and common knowledge.
Economic and health resilience frameworks.
Literary reflections, education, and mental health in holistic preparation of disaster.

This book therefore, is not only informative but also enables action. It promotes the amalgamation of different knowledge systems, community involvement and cross-disciplinary approaches to create a more resilient safer world of tomorrow.

Chapter 1 Understanding the Science of Natural Disaster

Qamar Ali
https://orcid.org/0000-0002-8178-7591
Virtual University of Pakistan, Pakistan

Muhammad Tariq Iqbal Khan
Government College University, Faisalabad, Pakistan

ABSTRACT

This chapter examines the relationship between natural hazards and vulnerabilities such as economic, social, environmental, and physical factors, which heighten the global impact of natural disasters. Recent data show alarming trends with 3.6 million deaths, 7.7 billion people affected, and $3.3 trillion in economic losses due to natural disasters. These events, including earthquakes, floods, and wildfires, have increased in frequency and intensity due to human activities such as rapid urbanization, environmental degradation, and climate change. Developing countries are particularly vulnerable due to their socio-economic conditions, exacerbating disaster impacts. This study advocates for sustainable urban planning and resilient human settlements as essential strategies to mitigate disaster effects and promote sustainable development. Recommendations include adopting integrated, eco-friendly development strategies that prioritize environmental conservation and human well-being to enhance resilience against future disasters.

DOI: 10.4018/979-8-3693-9745-9.ch001

INTRODUCTION

Natural disasters have always occurred (Rossello et al., 2020) and have become an international problem in recent times (Guo et al., 2015). Natural disasters are when hazards interact with economic, social, environmental and physical vulnerabilities. Vulnerability in country to natural disasters is usually at the level of development and economic structure (Benali et al., 2018). The effects of natural disasters have increased universally today and so there is a constrict urgency to achieve sustainable cities and human settlements (Panwar & Sen, 2019). The estimated global natural disasters with 3.6 million deaths, 7.7 billion affected people, and $3.3 trillion economic losses are recorded at 13,386 (Fang et al., 2019). Among such natural disasters like earthquakes, droughts, floods, extreme temperatures, epidemics, wildfire, insects infection, storms, mass movements (dry), landslides, mass movements, have increased globally and in their intensity. (Fang et al., 2019; Panwar & Sen, 2019). The risk of a disaster was perceived due to natural hazards until the 1970s. After decades of research, social scientists revealed the social dimension of disaster risk and claimed that disasters can be linked with human activity (Peduzzi, 2019). The rapid increase in the population, development of coastal zones, deterioration in wetlands, and changing climatic conditions like global warming were responsible for the increase in the vulnerability of human beings (Wang, 2019).

The pursuit of over exploitation of the nonrenewable resources has been going on for ages. the development activity and the attempts for reduction of the gap between led to ecological) supply and demand for the betterment of human living standards although there is a hue and cry and environmental imbalance leading to disasters, I've read about it and that disasters are cyclical but disaster follow a cycle. An overexploitation may not be the only reason for disasters, as they are an in evident part of nature and human life. While being accelerated by it, it can be overturned by the growing coping capacity of the people. Sustainable development activity is one which limits the gap between development and else ecological degradation would overrun the former, to relevant environment.

Since to understand this balance, one must work with empathetic and scientific point of view together. Rather times towards disaster mitigation must bring about development and now only the All the developmental projects which were undertaken in that area would be ecofriendly and pro-environment.

Natural disasters and man made disasters (technology) are generally considered as disasters. Environmental activities linked provoke natural disasters such as earthquakes, floods and hurricanes. with the law of nature. While natural disasters are called the “Act of God”, man-made disasters occur due to human activities termed as “Act of Man”. In contrast, a hybrid-form of natural disasters occurs due to both environmental and human activities (Ramli et al., 2014). Natural disasters are catastrophic events due to natural hazards including weather-related (meteorological/hydrological—windstorms, tornados, floods, drought, and heatwaves), internal (under the Earth—earthquakes, tsunamis, and volcanic eruptions), external (topographical—landslides, avalanches, wildfire, and dry mass movement), and biological phenomena (epidemics, infections, and locust swarms) (Fang et al., 2019). Table 1.1 shows the statistics of different types of natural disasters in the world from 2009 to 2019.

Table 1. The scenario of natural disasters in the world

Natural Disaster (type)	Disaster frequency (No.)		Deaths (No.)		No. of affected (Million)		Economic loss (Billion USD)	
	2009-2018*	2019	2009-2018*	2019	2009-2018*	2019	2009-2018*	2019
Drought	17	16	2004	85	72.8	29.2	10.1	0.1
Earthquake	25	32	26931	262	3.7	1.5	42.2	2.4
Extreme Temperature	21	21	7277	2909	1.2	0.2	0.9	0
Flood	149	194	4913	5110	71.9	31.0	40.0	36.8
Landslide	19	25	1011	727	0.3	0.2	0.4	0.2
Mass movement (dry)	1	0	14	0	●<0.1	0	●<0.1	0
Storm	99	90	2835	2525	34.3	32.8	75.8	59.3
Volcanic activity	4	4	134	21	0.4	●<0.1	0.2	0
Wildfire	9	14	93	116	0.1	●<0.1	6.2	3.7

* *Average*

Source: Centre for Research on the Epidemiology of Disasters (CRED, 2019)

It became quite evident from the developed and underdeveloped countries (Benali et al., 2017, Sawada & Takasaki, 2017) that the consequences of natural disasters. In addition physical damage and death caused by natural disaster are smaller in developed countries. In other words, for example, 90% of disaster related deaths take place in economically developed countries (Zenklusen, 2007). Poor infrastructure, few resources and unstable institutions make developing countries vulnerable to natural disasters (Trinh et al., 2021). Natural disasters mostly affect developing countries because of poor environment, weak adaptation and poor socioeconomic status (Hamidi et al, 2020). Additionally, Asia is the continent which faced dozens of natural disasters, contributing to 44% and ending life those of 58% of disaster related deaths. (Taghizadeh-Hesary et al. 2019).

The Asian countries were affected by disasters like cyclones and tsunamis in China, the Philippines, and Vietnam; floods, landslides, and earthquakes in Pakistan, Bangladesh, India, and Nepal (Shah et al., 2019). The tsunami in December 2004 was a major natural disaster, which damaged more than 18 Southeast Asian countries. It was responsible for the loss of about 250,000 human lives in one day, more than one million homeless people, and damage to houses. It was responsible for the loss of millions of dollars (Cavallo et al., 2014). The earthquake in Pakistan of 7.6 magnitude on October 8, 2005, destroyed the Kashmir and Himalayan region and is accountable for about 87,350 human lives, 38,000 injuries, and 3.5 million homeless people (EERI, 2006). The Indian Ocean faced a cyclone 'Nargis' in 2008, which caused massive destruction in Myanmar and Ayeyarwady Delta region and was responsible for 84,500 human lives and 53,800 missing people. The economic cost of this disaster was about 10 billion USD (Mercene, 2017).

Climate change is not the cause of all disasters but is responsible for an increase in the intensity and frequency of most natural disasters (Weerasekara et al., 2021). It is an important challenge for human beings in the 21st century (Trinh et al., 2021). The impact of climate change is evident across countries (Ali et al., 2021a). The intensity of climate change will further increase in the 21st century (i.e., frequent heat waves, increase in wind speeds, and severity of droughts) (Weerasekara et al., 2021). A significant change in temperature, precipitations and other meteorological indicators underpin extreme weather events (Weerasekara et al., 2021). Moreover, environmental deterioration, land use patterns, population growth, unplanned development, and unnecessary use of fossil fuels are responsible for the severity of natural disasters (Lee & Tang, 2019). Thus, policymakers tried to mitigate the adverse impact of climate change on human beings (Trinh et al., 2021). Table 2 briefly explains the consequences of different types of natural disasters.

Table 2. Types of natural disasters and their effects

Natural Disaster	Explanation
Earthquakes	Earthquakes adversely affect both capital and labor, destroy infrastructure, buildings, and factories. Earthquakes may lead to industrial growth, when (a) capital-worker ratio decreases, (b) average product of capital rises, (c) output increases in the reconstruction phase, and (d) better-quality capital replaced the destroyed capital (Loayza et al., 2009).
Floods	Floods are responsible for the reduction in farming, transportation, and urban activities, which in turn reduce productivity. Intense floods are responsible for water-borne diseases that cause loss of total factor productivity. Moderate floods are linked with higher growth due to (a) floods can accelerate agricultural growth due to an increase in water and productivity, (b) floods can accelerate industrial growth through an increase in electricity generation and agricultural intermediate products, and (c) floods can positively affect the services sector by increasing the supply of inputs (Loayza et al., 2009).
Droughts	Droughts show adverse impacts on agricultural growth due to a shortage of irrigation water. It disturbs industrial growth by (a) reducing the supply of agricultural inputs, and (b) decreasing the supply of hydropower electricity. Its adverse impact is more on labor than capital, which implies the increase in k beyond the steady-state capital (k*) (Loayza et al., 2009).
Storms	Storms adversely affected agricultural production due to the destruction of seedlings and plants. Storms are linked with the destruction of physical capital, a vital input in the industrial sector. Its adverse impact is more on physical capital than incapacitating workers. However, moderate storms showed a favorable impact on industrial growth (Loayza et al., 2009).
Hurricanes, Cyclones, and Typhoons	A hurricane is a storm with violent winds (>74 mph). Technology helps to forecast storms and anticipate the destruction and damage. The world witnessed a substantial population growth and vulnerability of coastal areas, which in turn led to higher hurricane-related deaths, injuries, and economic costs. Hurricanes are called typhoons and cyclones in the Western Pacific Ocean and the Indian Ocean, respectively (Hidalgo & Baez, 2019).
Surge	An abnormal increase in seawater level is called a Surge. It occurs during a storm when the storm's winds pushing water inland. It is accountable for vital hurricane mortality. Its amplitude is location-specific (Hidalgo & Baez, 2019).
Winds	The winds are responsible for mortality in a hurricane. It is possible to forecast the damage due to the speed of winds. It leads to property destruction, wind pressure, and flying objects (Hidalgo & Baez, 2019).
Tornado	A tornado is an intense and destructive rotational column of air, from the base of a thunderstorm to the ground. It is not possible to forecast tornados. The craniocerebral injuries are accountable for sudden death or serious injury. Falling structures may lead to crush injuries, fractures, soft tissue injuries, abrasions, and contusions (Hidalgo & Baez, 2019).
Landslides	Landslides refer to the down-slope movement of rock and soil due to gravity. The speed of onset is rapid with major damage to systems and structures. Sometimes, it leads to heavy rains, flooding, and movements of debris (Hidalgo & Baez, 2019).
Tsunami	It is a long high sea wave that occurs due to an earthquake or submarine landslide. Its intensity depends on the depth of the water where the seismic issue occurred with a variable speed of onset. It causes immediate trauma, drowning, saltwater contamination, disturbance in water supplies, and damage to crops and infrastructure (Hidalgo & Baez, 2019).

continued on following page

Table 2. Continued

Natural Disaster	Explanation
Fires	It is possible to mark the areas prone to bushfires, as they tend to occur seasonally. Its speed depends upon several factors i.e. wind speed, temperature, and weather conditions. Natural forest fires mainly occur due to lightning strikes in windy and dry situations. It disturbed air quality in the surrounding area that causes respiratory issues (Hidalgo & Baez, 2019).
Heatwaves	Heatwaves are a silent disaster, as the human loss due to heatwaves is more than the earthquakes and other catastrophes events. Every person is susceptible to heat illness without precautions (Hidalgo & Baez, 2019).
Volcanoes	Volcano causes significant deterioration of nearby buildings. The environment becomes a threat to human beings due to the presence of airborne pollutants, which causes respiratory problems (Hidalgo & Baez, 2019).

However, disaster risk is not yet included in macroeconomic (i.e., economic variables) framework in existing literature (Akao & Sakamoto, 2018). Increasingly, private and public sectors have demanded ways to reduce losses associated with natural disasters due to awareness of socio-economic consequences of those disasters (Marin & Modica, 2017). The empirical question however, is whether natural disasters have any economic impact (Klomp & Valckx, 2014). Climatic changes and the vulnerabilities of modern societies are expected to increase losses stemming from natural disasters. (Panwar & Sen, 2019).

CONSEQUENCES OF NATURAL DISASTERS

Direct damages (human deaths, property loss and injuries), and indirect damages (potential wages & capital loss) are accountable to natural disasters (Panwar & Sen, 2019). During disasters, government expenditure for the rebuilding of destroyed infrastructures (e.g., roads, buildings, and bridges) increases government spending, budget adjustments and economic contraction (Shi & Sun, 2021). But the economic cost of disasters is slightly more in less industrialized economies (Albuquerque & Rajhi, 2019. More research is needed to understand the nexus between natural disasters and socio economic conditions, to formulate a sustainable and comprehensive policy. To attain sustainable development, it augurs well for developed and developing economies to have a comprehensive policy (Haibo et al., 2019). Cost of natural disasters is dependent on: frequency, intensity, magnitude, risk, duration, exposure and vulnerability. (Keerthiratne, 2017; Keerthiratne & Tol, 2018).

According to the collateral damage concept, destructive events are accountable for irreversible damage to human capital, which in turn leads to poverty and poor standards of living. The increase in human capital is beneficial to enhance productivity and drive economic growth. Natural disasters become a hurdle to achieving human development, especially in developing economies. The destructive events hamper health, education, nutrition, and several income-earning activities (Padli et al., 2018). Natural disasters like earthquakes, floods, and hurricanes significantly damage the power infrastructure. Energy plays a key role in industrial development in the 21st century. About 1.2 billion people do not have electricity access and 2.9 billion people use polluting fuels for cooking (Poudyal et al., 2019). Fig. 1.1 shows several consequences of natural disasters. It is hard to stop the existence of natural disasters (Chun et al., 2017). However, there is a potential for the system and people to avoid adverse effects. The ability to withstand and recover from a natural disaster depends upon different socio-economic indicators (Noy & Yonson, 2018).

Figure 1. Effects of natural disasters

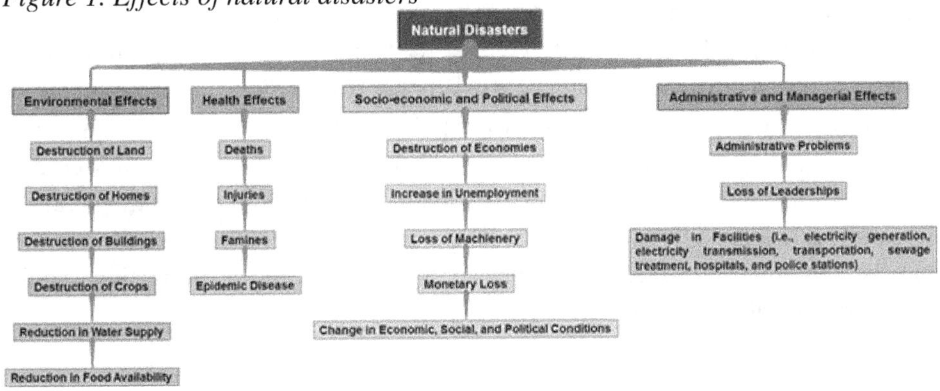

Reconstructed from Kapur (2016)

Due to space and time, it is difficult to study the impact of natural disasters on each sector of the economy. Thus, this study assesses the impact of natural disasters on economic growth, foreign reserves accumulation, agricultural development, and natural resources across all income groups.

DISASTER RESILIENCE

Policies can prevent (a) the size of a disaster, (b) the likelihood of that disaster to occur, and (c) the vulnerability of the population (Rose, 2004). Thanks to resilience-building (Chun et al., 2017), losses can be mitigated. Resilience means being able to respond both during and after natural disasters (Song et al., 2017). Natural disaster phenomenon exists everywhere in all geographical region only with a difference in the intensity. Yet, the mitigation policies can reduce post disaster loss because the occurrence of the natural hazard is difficult to avoid (Taghizadeh Hesary et al., 2021). On this basis, the focus of this study is on disaster resilience as a tool to handle natural disasters. Related terminologies, as referred to by UNISDR, are shown in Fig. 1.2. (Peduzzi, 2019).

The spiritual dimension of building resilience is often ignored, as religious values and norms lead to a safe and happy life. The guidance of Almighty Allah is essential for human living and will result in the prevention of a disaster. According to Islam, the occurrence of anything in this universe is due to the will of Almighty Allah, which is Most Merciful. It is mentioned in the Holy Quran that "Whatever of good reaches you, is from Allah, but whatever of evil befalls you, is from yourself, and We have sent You (Muhammad Peace Be Upon Him) as a Messenger to mankind, and Allah is Sufficient as a Witness" (Chapter 4, 79). According to The Quran, different incidents of disasters occurred in the world such as the people of Prophet Noah (A.S.), the tribe of Midian/the people of Prophet Shuaib (A.S.), the tribe of Sadom/people of Prophet Lut (A.S.), the tribe of Thamud/people of Prophet Saleh (A.S.), and the tribe of 'Ad/people of Prophet Hud (A.S.). These events clearly explained that it was not Allah Who wronged them but they wronged themselves (Chapter 39, 40). It is a fact that Allah never harms human beings but they do bad deeds. The disasters are viewed as an assessment of our faith in piety. Allah praised those who remain patient in front of hardships, as they will be rewarded in this world and the hereafter. However, proper measures of readiness and mitigation should come first (Ramli et al., 2014).

Figure 2. Related terminologies, defined by the UNISDR

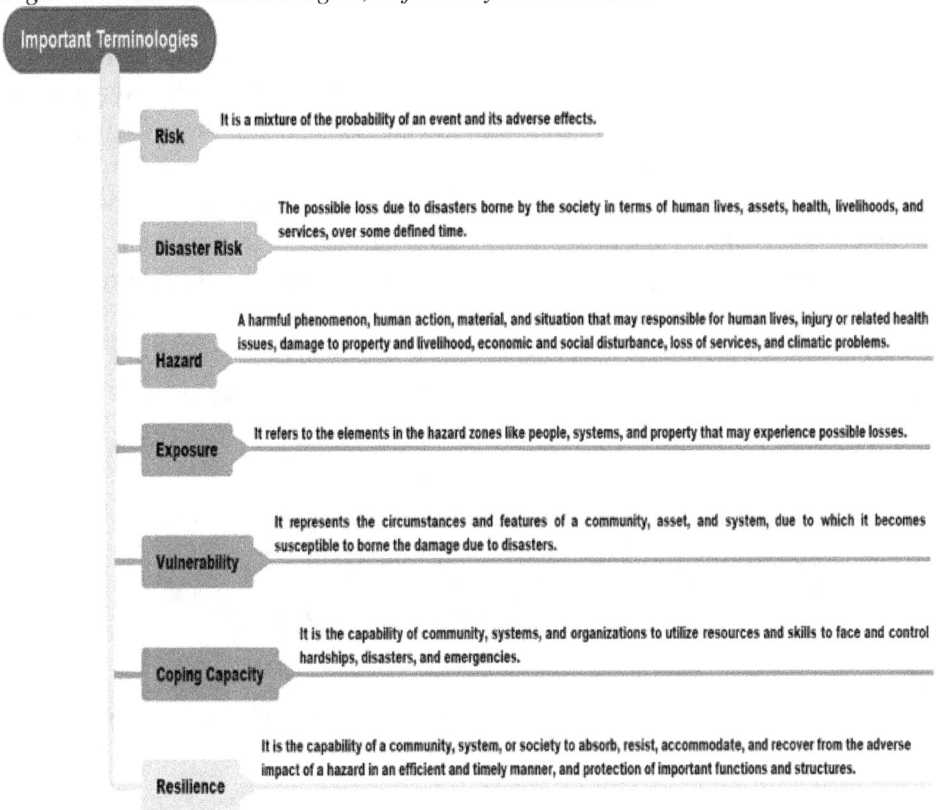

Reconstructed from (Peduzzi, 2019)

THE SENDAI FRAMEWORK FOR DRR 2015-2030: A WAY FORWARD

The severity of natural disasters has increased since the beginning of the 20th century (Hamidi et al., 2020), prompting the UN to conceptualize the 2015-2030 Sendai Framework for Disaster Risk Reduction (DRR) at an international conference in Sendai, Japan (UNISDR, 2015). The four established priorities in the framework include (1) understanding the risk of disasters, (2) strengthening governance to control disaster risk, (3) investing in DRR, and (4) increasing disaster readiness for a timely response, and "build back better" in recovery, reconstruction, and rehabilitation (UNISDR, 2019).

Thus, risk-informed development aims to reduce risk, avoid risk creation, and build resilience (Opitz-Stapleton et al., 2019). It explains the opportunities to improve disaster preparedness in society and speed up the restoration process (Marzi et al., 2019). Risk-informed development (Fig. 1.3) is a tool that prevents hazardous risk and strengthens resilience. This means that sustainable development strategies will fail unless they are risk-informed (UNDP, 2020). Sustainable development has strong linkages with economic progress, clean environment, and social advancement. It ensures environmental and economic stability in the long run by fulfilling present demands without disturbing the demands of future generations (Haibo et al., 2019). Fig. 1.4 shows efforts towards risk-informed sustainable development.

Figure 3. Aims of risk-informed sustainable development

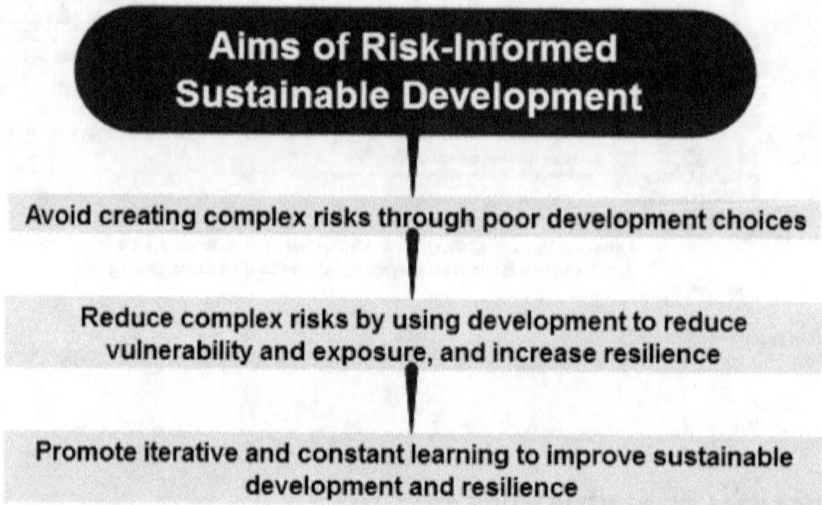

Reconstructed from Opitz-Stapleton et al. (2019)

Figure 4. A path of risk-informed sustainable development

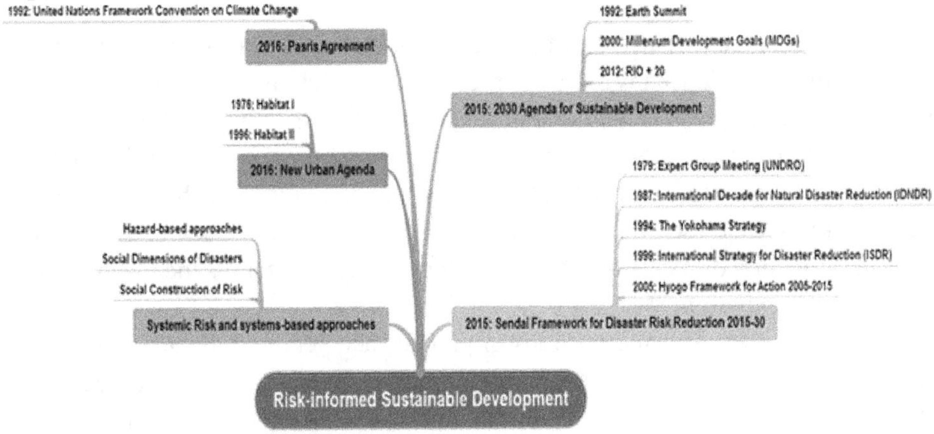

Reconstructed from UNDRR (2019)

The DRR targets are associated with 10 Sustainable Development Goals (SDGs) namely poverty reduction (SDG-1), zero-hunger (SDG-2), proper health and well-being (SDG-3), improved educational quality (SDG-4), clean water, and proper sanitation (SDG-6), sustainable industry, innovation and modern infrastructures (SDG-9), sustainable cities, communities and human settlements (SDG-11), climate change mitigation (SDG-13), life below water (SDG-14) and on land (SDG-15). Thus, DRR policies effectively decline disaster risks and enhance disaster resilience——which are core strategies to achieve the SDGs (Noy & Yonson, 2018; Tselios & Tompkins, 2017).

China is characterized by frequent disasters, proper rebuilding, and innovation abilities. Due to these characteristics, it may be possible that natural disasters positively influenced long-run economic growth. Therefore, Guo et al. (2015) used the data of 577 disasters in China from 1985 to 2011 to evaluate the economic costs of natural disasters. They confirmed an increase in economic growth through human capital after the occurrence of meteorological disasters. This study does not confirm any significant link between natural disasters and economic growth. Re-investment in human capital is required to manage recoveries and sustainable development.

Pakistan experienced a substantial increase in floods of different intensities. Only 16 floods hit Pakistan from 1973 to 1993, but the number of floods was 54 in the next 20 years. Keeping in view the problem of floods in Pakistan, Sardar et al. (2016) revealed the impact of flood-related damage on GDP growth from 1972 to 2013. The GDP growth and disaster management were found beneficial for the mitigation of floods-related hazards. The flood-related hazards adversely affect the GDP growth in Pakistan. The poverty rate increased in the flood-prone areas, having a large family size and fewer skilled workers.

The intensity and occurrence of natural disasters have been increased with time. Such circumstances led many researchers to reveal the vicious cycles between natural disasters and any subset of vulnerability, poverty, deforestation, and urbanization. Sodhi (2016) used the annual data of 179 countries from 1963 to 2012 and mentioned the adverse effects of natural disasters on economic growth and per capita income. It also increased the vulnerability of the population. They confirmed the surge in the natural disasters in the world, after normalizing for the increasing global income and population.

Natural disasters affected millions of people each year and disturb investment, output, debt, fiscal account, poverty, and external accounts across all income groups. In the case of HICs and MICs, Benali et al. (2017) revealed the dynamic link among natural disasters, government revenue, government expenditures, inflation rate, government debt, and economic growth from 1990 to 2013 in different countries. There exist a unidirectional causal association from natural disasters to government debt and government expenditures.

Pakistan is located in an extreme risk zone and faced huge economic loss due to earthquakes, floods, land sliding, temperature rising, and cyclones after the 1980s. Due to this fact, Baig et al. (2018) explained the causality between economic growth and natural disasters in Pakistan. The results confirmed an inverse link between economic growth and natural disaster loss. The causality analysis established a one-way causality from natural disasters to economic growth in the short and long run.

Natural disasters were accountable for complicated impacts on energy utilization in the short and long run. The infrastructure damage due to natural disasters showed a direct impact on energy utilization. Therefore, Doytch and Klein (2018) examined the effect of infrastructure deterioration due to natural disasters (geophysical, meteorological) on the different types of energy utilization in the panel of 80 countries using data from 1961 to 2011. A positive impact on the utilization of renewable energy was observed after 5 years of the incidence of a geophysical disaster in HICs. The positive impact was also recorded on industrial and residential energy use in LICs and MICs, respectively.

In Sri Lanka, natural disasters caused 27 million deaths from 1985 to 2013. The loss of human beings was higher due to droughts (47%) followed by floods (45%), extreme wind (6%), and epidemics (2%). Therefore, Keerthiratne and Tol (2018) explored the link between income inequality and natural disasters in Sri Lanka using household income and expenditure surveys from 1990 to 2013. The results revealed that the parallel occurrence of natural disasters and their lags are linked to decreased inequality in household income. However, natural disasters do not affect the inequality of household expenditure. Natural disasters increased the inequality of seasonal agricultural income and decreased the inequality of non-agricultural and non-seasonal agricultural income.

The Caribbean countries, situated in the North Atlantic Ocean hurricane belt, are known as the most disaster-prone territories in the world. They had low investment, large debts, high trade, reliance on Foreign Direct Investment (FDI) and foreign aid, and volatility in consumption. Due to this, Caribbean territories were vulnerable to climatic shocks. Mohan et al. (2018) explained the effect of natural disasters on the economy using panel data of 21 Caribbean countries from 1970 to 2011. They used four components of GDP like private consumption, government expenditures, investment, import, and export in the empirical analysis. Results revealed diversified effects of disasters on various parts of GDP such as private consumption, government expenditures, investment, import, and export. The differences in directions and timing of these shocks highlighted the difficulty to estimate the clear and cumulative effect of natural disasters and hurricanes on GDP.

In the recent era, economists showed interest to study the effect of natural hazards on economic growth. Due to frequent and large-scale disasters, socio-economic losses were observed in an economy. Therefore, Mukherjee and Hastak (2018) studied the economic consequence of disasters. A regional/country-specific variation was observed in terms of disaster-related indicators (fatalities and homelessness), and economic indicators (real GDP, trade openness, population, the share of investment in purchasing power parity, and share of government in purchasing power parity). The flood was found the most disturbing disaster in terms of economic growth at the region/country level.

The political and institutional indicators (better institutions, higher trade, higher government spending, higher income, and higher literacy) played a crucial role to mitigate the economic loss of disaster shock. Albuquerque and Rajhi (2019) revealed the impact of state fragility and natural disasters on financial and economic conditions in a panel of 66 developing countries. Natural disasters and state fragility showed a significant adverse effect on financial and economic indicators in MICs and LICs.

Entrepreneurship is a fundamental part of economic growth due to its potential for employment. However, natural disasters disturbed job creation because the owner faced difficulty in continuing the normal operations. Considering the importance of new business, Boudreaux et al. (2019) examined the impact of natural disasters on new business density in 79 countries from 2006 to 2016. Results confirmed an adverse impact of natural disasters on business start-up activities in the short run. The decrease in business start-up activities adversely affected economic growth and job creation. The adverse effect of natural disasters on business was continued for 1-2 years. Considering the type of natural disasters, it was observed that natural disasters affected business start-up activities in the MICs and LICs while geological natural disasters affected business start-up activities in the HICs.

Night light data is more reliable than regional income data to assess economic growth. The statistical department in developing nations is not capable of correctly estimating the national income due to measurement errors. However, night light data is physical data, which is available for all countries. Therefore, Fabian et al. (2019) explained the effect of earthquakes on night-time lights at a sub-national level from 1992 to 2013. The panel fixed effects regressions confirmed that earthquakes significantly decreased the light levels and light growth rates. The adverse effects continue for about 5 years.

The Philippines was exposed to natural disasters, having a 27.69% score in terms of the World Risk Index. On average, natural disasters were blamed for more than 1,000 human lives per annum. Thus, Lee and Tang (2019) examined the causality among financial development, poverty, economic growth, and natural disasters from 1970 to 2014 in the Philippines. The results established one-way causality from natural disasters to economic growth and financial development and a two-way causal relationship between economic growth, poverty, and financial strengthening. Therefore, the risk management practices and DRR policies would be beneficial to counter the adverse effects of natural disasters.

Brazil is vulnerable to climate disasters global warming. The Inter-Governmental Panel on Climate Change (IPCC) mentioned that a temperature rise is associated with intense droughts. Considering Brazil, Oliveira (2019) reported that natural disasters were accountable for the reduction in GDP growth. The environmental disturbance mostly deteriorated the services and agricultural sectors. However, the industrial sector was not disturbed due to environmental issues. The economy showed sensitivity subject to the large natural disasters, which further lead to the situation of emergency. Disaster-related micro-insurance could mitigate the negative effects of droughts and floods on the economy.

The effects of natural disasters on economic indicators remained not clear because literature showed positive, negative, and no effects of natural disasters on the economy. Re-examining this debate, Panwar and Sen (2019) described the economic and human life losses due to natural disasters (earthquakes, floods, storms, droughts) in 102 countries, which were further categorized into 73 developing and 29 developed countries. The economic impact of natural disasters was depended upon the type and intensity of natural disasters. The cost of natural disasters was more in developing economies.

The increase in the vulnerability of people to natural disasters was reported due to climate change, growing population, coastal zones development, and wetland draining. Natural disasters were responsible for significant human and economic losses in the United States. Wang (2019) showed the effect of natural disasters (tornado, hurricane, and earthquake) on population density from 1960 to 2000. Results confirmed that there was no significant effect of natural disasters on long-run growth. However, the negligible indirect effect of natural disasters on population density was reported across counties.

REFERENCES

Akao, K. I., & Sakamoto, H. (2018). A theory of disasters and long-run growth. *Journal of Economic Dynamics & Control*, *95*, 89–109. DOI: 10.1016/j.jedc.2018.08.006

Albuquerque, P. H., & Rajhi, T. (2019). Banking stability, natural disasters, and state fragility: Evidence from developing countries. *Research in International Business and Finance*, *50*, 430–443. DOI: 10.1016/j.ribaf.2019.06.001

Ali, S., Liu, Y., Ishaq, M., Shah, T., Ilyas, A., & Din, I. U. (2021). Climate change and its impact on the yield of major food crops: Evidence from Pakistan. *Foods*, *10*(6), 1285. DOI: 10.3390/foods10061285 PMID: 34199791

Baig, N., Khan, S., & Khan, M. A. (2018). Natural disasters and economic growth: Evidence from Pakistan. *Pakistan Journal of Applied Economics*, *28*(2), 189–207.

Benali, N., Abdelkafi, I., & Feki, R. (2017). The impact of natural disasters on public debt in middle-income countries. *International Journal of Economics and Finance*, *9*(5), 179–186. DOI: 10.5539/ijef.v9n5p179

Benali, N., Abdelkafi, I., & Feki, R. (2018). Natural-disaster shocks and government's behavior: Evidence from middle-income countries. *International Journal of Disaster Risk Reduction*, *27*, 1–6. DOI: 10.1016/j.ijdrr.2016.12.014

Boudreaux, C. J., Escaleras, M., & Skidmore, M. (2019). Natural disasters and entrepreneurship: Evidence from global data. *Journal of Natural Disaster Science*, *40*(1), 15–27. DOI: 10.2328/jnds.40.15

Cavallo, E., Galiani, S., Noy, I., & Pantano, J. (2014). Catastrophic natural disasters and economic growth. *The Review of Economics and Statistics*, *95*(5), 1549–1561. DOI: 10.1162/REST_a_00413

Centre for Research on the Epidemiology of Disasters (CRED). (2019). EM-DAT: The International Disaster Database. Université Catholique de Louvain. Retrieved from https://www.emdat.be

Chun, H., Chi, S., & Hwang, B. (2017). A spatial disaster assessment model of social resilience based on geographically weighted regression. *Sustainability (Basel)*, *9*(12), 2222. DOI: 10.3390/su9122222

Doytch, N., & Klein, Y. (2018). Natural disasters and energy consumption: A panel analysis. *Energy Economics*, *76*, 1–13. DOI: 10.1016/j.eneco.2018.09.014

Earthquake Engineering Research Institute (EERI). (2006). The Kashmir Earthquake of October 8, 2005: Impacts in Pakistan. EERI Special Earthquake Report. https://www.eeri.org/lfe/pdf/kashmir_eeri_report_2006.pdf

Fabian, M., Lessmann, C., & Sofke, T. (2019). Natural disasters and regional development: Evidence from earthquakes and night-time lights. *Journal of Economic Geography*, *19*(4), 797–819. DOI: 10.1093/jeg/lby036

Fang, J., Lau, C. K. M., Lu, Z., Wu, W., & Zhu, L. (2019). Natural disasters, climate change, and their impact on inclusive wealth in G20 countries. *Environmental Science and Pollution Research International*, *26*(2), 1455–1463. DOI: 10.1007/s11356-018-3634-2 PMID: 30426378

Fang, J., Lau, C. K. M., Lu, Z., Wu, W., & Zhu, L. (2019). Natural disasters, climate change, and their impact on inclusive wealth in G20 countries. *Environmental Science and Pollution Research International*, *26*(2), 1455–1463. DOI: 10.1007/s11356-018-3634-2 PMID: 30426378

Guo, J., Liu, H., Wu, X., Gu, J., Song, S., & Tang, Y. (2015). Natural disasters, economic growth and Sustainable development in China-An empirical study using provincial panel data. *Sustainability (Basel)*, *7*(12), 16783–16800. DOI: 10.3390/su71215847

Haibo, C., Zhang, K., & Su, Q. (2019). Sustainable development and disaster risk reduction: A case study of China. *Sustainability*, *11*(22), 6478. DOI: 10.3390/su11226478

Hamidi, A. R., Wang, J., & Guo, S. (2020). Natural disasters and economic growth: Empirical evidence from Asia. *Natural Hazards*, *101*(3), 707–727. DOI: 10.1007/s11069-020-03886-7

Hidalgo, D., & Baez, A. A. (2019). Natural disasters: Preparedness and response in the Americas. *Prehospital and Disaster Medicine*, *34*(S1), s10–s11. DOI: 10.1017/S1049023X19000283

Kapur, B. K. (2016). Economic consequences of natural disasters: A review. *Journal of Economic Surveys*, *30*(4), 664–687. DOI: 10.1111/joes.12123

Keerthiratne, S. (2017). Economic impact of natural disasters. University of Oxford, Department of Economics Working Paper. https://ora.ox.ac.uk/objects/uuid:12345678-1234-5678-1234-567890123456

Keerthiratne, S., & Tol, R. S. J. (2018). Impact of natural disasters on income inequality in Sri Lanka. *World Development*, *105*, 217–230. DOI: 10.1016/j.worlddev.2018.01.001

Klomp, J., & Valckx, K. (2014). Natural disasters and economic growth: A meta-analysis. *Global Environmental Change*, *26*, 183–195. DOI: 10.1016/j.gloenvcha.2014.02.006

Lee, C. C., & Tang, H. (2019). The impact of natural disasters on financial development: Evidence from the Philippines. *The Singapore Economic Review*, *64*(5), 1207–1229. DOI: 10.1142/S0217590818500118

Loayza, N. V., Olaberria, E., Rigolini, J., & Christiaensen, L. (2009). Natural disasters and growth: Going beyond the averages. World Bank Policy Research Working Paper 4980. DOI: 10.1596/1813-9450-4980

Marin, G., & Modica, M. (2017). Socio-economic exposure to natural disasters. *Environmental Impact Assessment Review*, *64*, 57–66. DOI: 10.1016/j.eiar.2017.03.002

Mercene, F. (2017). *Cyclone Nargis: A case study of disaster response*. Manila Bulletin.

Mohan, P. S., Strobl, E., & Watson, P. (2018). The short-term economic impact of tropical cyclones: Evidence from the Caribbean. *Natural Hazards*, *91*(1), 87–104. DOI: 10.1007/s11069-017-3113-9

Mukherjee, S., & Hastak, M. (2018). Economic consequences of natural disasters: A regional perspective. *International Journal of Disaster Risk Reduction*, *31*, 1216–1225. DOI: 10.1016/j.ijdrr.2018.03.017

Noy, I., & Yonson, R. (2018). Economic vulnerability and resilience to natural hazards: A survey of concepts and measurements. *Sustainability (Basel)*, *10*(8), 2850. DOI: 10.3390/su10082850

Oliveira, V. H. (2019). Natural disasters and economic growth in Brazil: A sectoral analysis. *Environment and Development Economics*, *24*(5), 497–517. DOI: 10.1017/S1355770X19000217

Opitz-Stapleton, S., Nadin, R., Watson, C., & Kellett, J. (2019). Risk-informed development: From crisis to resilience. Overseas Development Institute (ODI) Report. https://odi.org/en/publications/risk-informed-development-from-crisis-to-resilience

Padli, J., Habibullah, M. S., & Baharom, A. H. (2018). The impact of human development on natural disaster fatalities and damage: Panel data evidence. *Ekonomska Istrazivanja*, *31*(1), 1557–1573. DOI: 10.1080/1331677X.2018.1504689

Panwar, V., & Sen, S. (2019). Economic impact of natural disasters: An empirical re-examination. *Margin*, *13*(1), 109–139. DOI: 10.1177/0973801018800087

Peduzzi, P. (2019). The disaster risk, global change, and sustainability nexus. *Sustainability (Basel), 11*(4), 957. DOI: 10.3390/su11040957

Poudyal, R., Loskot, P., Nepal, R., Parajuli, R., & Khadka, S. K. (2019). Mitigating the current energy crisis in Nepal with renewable energy sources. *Renewable & Sustainable Energy Reviews, 116*, 109388. DOI: 10.1016/j.rser.2019.109388

Ramli, A., Mokhtar, M., & Aziz, B. A. (2014). Revisiting the concept of development, disaster and safety management: The Quranic perspective. *International Journal of Disaster Risk Reduction, 9*, 26–37. DOI: 10.1016/j.ijdrr.2014.03.006

Rose, A. (2004). Defining and measuring economic resilience to disasters. *Disaster Prevention and Management, 13*(4), 307–314. DOI: 10.1108/09653560410556528

Rossello, J., Becken, S., & Santana-Gallego, M. (2020). The effects of natural disasters on international tourism: A global analysis. *Tourism Management, 79*, 104080. DOI: 10.1016/j.tourman.2020.104080 PMID: 32287755

Sardar, A., Javed, S. A., & Amir-ud-Din, R. (2016). Natural disasters and economic growth in Pakistan: Evidence from flood-prone areas. *Pakistan Economic and Social Review, 54*(2), 199–222. https://www.jstor.org/stable/26616704

Sawada, Y., & Takasaki, Y. (2017). Natural disaster, poverty, and development: An introduction. *World Development, 94*, 2–15. DOI: 10.1016/j.worlddev.2016.12.035

Shah, A. A., Ye, J., Abid, M., & Ullah, R. (2019). Determinants of flood risk mitigation strategies at household level: A case of Khyber Pakhtunkhwa (KP) province, Pakistan. *Natural Hazards, 95*(1-2), 401–423. DOI: 10.1007/s11069-018-3508-2

Shi, Y., & Sun, J. (2021). The economic impacts of natural disasters: A review of models and empirical studies. *Natural Hazards Review, 22*(3), 04021015. DOI: 10.1061/(ASCE)NH.1527-6996.0000473

Sodhi, M. S. (2016). Natural disasters, the economy and population vulnerability: Lessons from global data. *International Journal of Production Economics, 176*, 69–78. DOI: 10.1016/j.ijpe.2016.03.008

Song, J., Huang, B., & Li, R. (2017). Measuring recovery to build up metrics of flood resilience based on pollutant discharge data: A case study in East China. *Water (Basel), 9*(8), 619. DOI: 10.3390/w9080619

Taghizadeh-Hesary, F., Sarker, T., Yoshino, N., Mortha, A., & Vo, X. V. (2021). Quality infrastructure and natural disaster resiliency: A panel analysis of Asia and the Pacific. *Economic Analysis and Policy, 69*, 394–406. DOI: 10.1016/j.eap.2020.12.021

Taghizadeh-Hesary, F., Yoshino, N., & Mortha, A. (2019). Assessing the macroeconomic impacts of natural disasters in Asia and the Pacific. ADBI Working Papers 966. Asian Development Bank Institute. https://www.adb.org/publications/assessing-macroeconomic-impacts-natural-disasters-asia-pacific

Trinh, T. A., Feeny, S., & Posso, A. (2021). The impact of natural disasters on migration: Evidence from Vietnam. *Economic Analysis and Policy*, *70*, 445–463. DOI: 10.1016/j.eap.2021.03.009

Tselios, V., & Tompkins, E. L. (2017). Local government, political decentralisation and resilience to natural hazard-induced disasters. *Environmental Hazards*, *16*(3), 228–252. DOI: 10.1080/17477891.2016.1277967

United Nations Development Programme (UNDP). (2020). Human Development Report 2020: The Next Frontier – Human Development and the Anthropocene. https://hdr.undp.org/en/2020-report

United Nations International Strategy for Disaster Reduction (UNISDR). (2015). Sendai Framework for Disaster Risk Reduction 2015-2030. United Nations. https://www.undrr.org/publication/sendai-framework-disaster-risk-reduction-2015-2030

United Nations International Strategy for Disaster Reduction (UNISDR). (2019). Global Assessment Report on Disaster Risk Reduction 2019. United Nations. https://gar.undrr.org

United Nations Office for Disaster Risk Reduction (UNDRR). (2019). Global Assessment Report on Disaster Risk Reduction 2019. United Nations. https://gar.undrr.org

Wang, J., & Dong, K. (2019). What drives environmental degradation? Evidence from 14 Sub Saharan African countries. *The Science of the Total Environment*, *656*, 165–173. DOI: 10.1016/j.scitotenv.2018.11.354 PMID: 30504018

Weerasekara, S., Wilson, C., Lee, B., & Hoang, V. N. (2021). Climate change and natural disasters: Evidence from the 2011 Thailand floods. *Environmental Development*, *39*, 100648. DOI: 10.1016/j.envdev.2021.100648

Zenklusen, O. (2007). *Natural disasters and economic development: A neoclassical review of theoretical perspectives and empirical evidence*. Ph.D. thesis, The University of St. Gallen, Switzerland.

Chapter 2
Climate Change and Natural Disasters

Deepika Paira
VEL TECH University, India

A. Rohini
https://orcid.org/0000-0002-9779-6572
CVR College of Engineering, India

Sudhanshu Chandra
https://orcid.org/0009-0006-6694-1032
Maulana Azad National Urdu University, India

Pamarthi Satyanarayana
https://orcid.org/0009-0008-4041-3674
VEL TECH University, India

Mohit Sharma
https://orcid.org/0009-0007-2280-8077
Maharshi Dayanand University, Rohtak, India

P. Selvakumar
https://orcid.org/0000-0002-3650-4548
Department of Science and Humanities, Nehru Institute of Technology, Coimbatore, India

Manjunath T. C.
https://orcid.org/0000-0003-2545-9160
Rajarajeswari College of Engineering, India

ABSTRACT

The intersection between these two issues is pivotal for understanding the broader impacts events. As a result, the frequency, intensity, and distribution are significantly affected. influences natural disasters is through the intensification of extreme weather events. For instance, warmer ocean temperatures provide more energy for tropical storms, leading to stronger hurricanes and typhoons. These storms, in turn, bring about more severe storm surges, heavy rainfall, and flooding. Similarly, the increased evaporation rates due to higher temperatures can exacerbate drought conditions, leading to prolonged dry periods that severely impact agriculture, water supply, and ecosystem health. Wildfires also become the relationship between is not limited to the direct effects of changing weather patterns. There are also indirect impacts

DOI: 10.4018/979-8-3693-9745-9.ch002

that exacerbate vulnerabilities in both human and natural systems. For example, and lakes. Communities around the world experience these impacts in varied ways, often reflecting existing social and economic inequalities.

INTRODUCTION TO CLIMATE CHANGE AND NATURAL DISASTERS: UNDERSTANDING THE INTERSECTION

The intersection between these two issues is pivotal for understanding the broader impacts events. As a result, the frequency, intensity, and distribution are significantly affected. influences natural disasters is through the intensification of extreme weather events. For instance, warmer ocean temperatures provide more energy for tropical storms, leading to stronger hurricanes and typhoons. These storms, in turn, bring about more severe storm surges, heavy rainfall, and flooding. Similarly, the increased evaporation rates due to higher temperatures can exacerbate drought conditions, leading to prolonged dry periods that severely impact agriculture, water supply, and ecosystem health. Wildfires also become the relationship between is not limited to the direct effects of changing weather patterns. There are also indirect impacts that exacerbate vulnerabilities in both human and natural systems. For example, and lakes. Communities around the world experience these impacts in varied ways, often reflecting existing social and economic inequalities. Vulnerable populations, particularly those in developing regions or living in informal settlements, exacerbated by climate change. Limited resources, inadequate infrastructure, and reduced adaptive capacity mean that these communities from extreme weather events. Conversely, wealthier nations, while better equipped to handle and recover from such events, still face significant economic and environmental costs associated with climate-induced disasters. Building resilience in communities and ecosystems is also crucial. This involves strengthening absorb and recover from-based preparedness programs, disaster risk reduction initiatives, and the integration of climate considerations into planning and development processes. Engaging local communities in these efforts ensures that the solutions are contextually appropriate and effectively of those most affected. In conclusion, the intersection of climate change and natural disasters highlights the complex and interconnected nature of contemporary environmental challenges. Understanding this relationship is essential towards solutions that address both the causes and consequences of these intertwined issues. Through comprehensive action and informed decision-making,

THE SCIENCE OF CLIMATE CHANGE: CAUSES, IMPACTS, AND CONSEQUENCES

field that seeks to understand the mechanisms driving global temperature shifts, the resulting fertilizer use. Industrial processes, such as cement production, further add to the GHG burden. Collectively, these activities enhance the greenhouse effect, leading to a cascade of climatic changes. The impacts experiencing a rise of approximately 1.2 degrees Celsius above pre-industrial levels. This warming has led to more frequent and severe heatwaves, affecting both human health and natural ecosystems. (Allen et al., 2014) . The thermal expansion of seawater and the addition of meltwater from glaciers increase the volume of the world's oceans, threatening coastal communities with flooding and erosion. Sea level rise also disrupts marine ecosystems and affects coastal infrastructure and economies. Changes in intense and frequent rainfall, leading to increased flooding and soil erosion, while others face prolonged droughts that strain water resources and agriculture. The alteration of precipitation patterns disrupts water cycles and can lead to reduced water availability, impacting drinking water supplies, agriculture, and energy production. Extreme weather events, such as The effects of climate change extend beyond immediate weather-related impacts. Ecosystems are being altered as species migrate to find suitable. Human societies are also facing significant challenges Health risks are increasing, with more frequent due to changing habitats of insects such as mosquitoes (Bender et al., 2010). Economic impacts are felt through measures. Social and economic inequalities are exacerbated as vulnerable populations, particularly in developing regions, face greater risks and fewer resources to cope with climate-induced stresses. Addressing the causes, impacts, and consequences further temperature rise. This involves changes that are already occurring or are anticipated, protecting vulnerable communities through targeted interventions. In summary, encompasses understanding the causes, impacts, and broader consequences of rising global temperatures and altered climate patterns. Human activities are the primary drivers of these changes, leading to a range of environmental and societal impacts. Through informed action and concerted efforts

Rising Temperatures and Extreme Weather Events: The New Normal

reshaping natural systems and human societies in profound ways. The steady increase in global temperatures, driven predominantly, has set the stage for a new era of climatic extremes. This warming trend is not merely a gradual shift but has accelerated the warming of the planet also has a profound impact on precipitation flooding (Bloodhart et al., 2015). This can overwhelm drainage systems, cause extensive property damage, and disrupt transportation and emergency services. Conversely, other areas suffer from prolonged droughts due to reduced precipitation and higher evaporation rates. Drought conditions stress water resources, negatively impact agriculture, and heighten the risk of wildfires. Both flooding and drought conditions highlight the growing volatility in weather patterns, making it increasingly challenging to manage water resources and agricultural production. Typhoons and other tropical cyclones are also on the rise due to warmer ocean temperatures. These storms are becoming more powerful, with increased wind speeds and higher rainfall rates. The enhanced energy available from warmer oceans fuels these storms, leading to more devastating storm surges, flooding, and wind damage. Coastal regions, particularly those with low elevation, face heightened risks of erosion and inundation, which can displace communities and cause significant economic losses. The destruction wrought reaching impacts, disrupting economies, damaging infrastructure, and challenging disaster response systems. Wildfires are another extreme weather event exacerbated by rising temperatures. Higher temperatures, combined with prolonged drought conditions, create ideal conditions for the ignition and spread of wildfires. These fires affecting respiratory health in communities across large regions. Species migration patterns are shifting, and some are moving toward higher altitudes disruptions in ecological balance and affect food chains, with potential consequences for both wildlife and human populations (Botzen et al., 2021). For instance, impact food availability for species that rely on these cues. Human societies are increasingly facing the multifaceted challenges posed by this new climate reality. Infrastructure designed for past climatic conditions may not be adequate to handle the stresses imposed by more extreme weather. Urban areas are particularly vulnerable due to their concentration of people and assets, as well as their potential exposure to heat islands—Economically, the costs associated with extreme weather events are rising. Repairing and rebuilding infrastructure, responding to natural disasters, and managing agricultural losses place a significant financial burden on governments, businesses, and individuals. The insurance industry also faces increased risks and costs due Building resilience in communities and ecosystems is also essential. This involves enhancing the capacity to recover from extreme weather events and reduce vulnerability. Community-based planning, improved disaster response mechanisms, and investments in climate-resilient infrastructure are critical components of building resilience (Census Bureau, 2019a). Additionally, engaging in international cooperation and knowledge-sharing can help countries learn from each other and implement effective strategies to address these challenges. In conclusion, the increase in extreme weather events represent a significant shift in our climate reality. The impacts are far-reaching, affecting health, infrastructure, ecosystems, and economies. Understanding and responding to this new normal as well as efforts to build resilience in communities and systems. By addressing these challenges proactively, it is possible to manage the risks

Sea-Level Rise and Coastal Vulnerability: Threats to Human Settlements

human settlements worldwide. This phenomenon, driven primarily by climate change, results from two main processes (Davis & Gertler, 2015). The consequences are increasingly evident, with rising seas encroaching on coastlines, eroding shores, and amplifying. Coastal areas, which are often densely populated and economically vital, are particularly vulnerable. In densely populated regions, such as the Nile (Census Bureau, 2019b), and the Mekong, sea-level rise threatens to displace millions of people. As saltwater intrudes into freshwater systems, agricultural productivity declines, which exacerbates food insecurity and economic instability. The social infrastructure in these areas, including roads, buildings, and sewage systems, is frequently ill-equipped to handle the increased stress of flooding and erosion, leading to costly damage and repair needs. Moreover, the economic implications extend beyond immediate damages. Coastal real estate values can plummet as areas become less desirable or uninhabitable, leading to a decrease in property taxes and potential losses in local and regional economies. Insurance companies may face increased claims, which could drive up premiums and further strain homeowners and businesses alike (Gibson & Mullins, 2020). Governments and municipalities must then allocate substantial resources to disaster response, adaptation, and mitigation efforts, placing additional strain on public finances (CRED, 2021). In response to these challenges, adaptation strategies are crucial. These include both implementing early warning systems. Integrated coastal zone management approaches that consider the interplay of natural systems and human activities are essential for building resilience. For instance, restoring wetlands and mangroves. However, adaptation alone may not be sufficient to fully address the scale of the threat. Proactive and ambitious equally critical to limiting future sea-level rise. International cooperation and robust climate policies are needed to achieve significant reductions in global warming and to support vulnerable regions in their adaptation efforts. from developed nations which are often the most affected yet least equipped to deal with these changes, is a key aspect of global climate action. Furthermore, raising public awareness about the risks and potential responses to sea-level rise can drive community engagement and support for necessary measures. Educational initiatives and participatory planning processes can empower individuals and communities to take proactive (Davenport et al., 2021) steps in safeguarding their environments and futures. In summary, sea-level rise represents a multifaceted threat to human settlements, affecting not just the physical landscape but also the social, economic, and environmental fabric of coastal regions (Feldman et al., 2012). Addressing this issue requires a comprehensive approach that combines effective adaptation strategies with ambitious climate change mitigation efforts, supported by international collaboration and community involvement. future.

Water Scarcity and Drought: Climate Change's Impact on Global Hydration

Water scarcity and drought are becoming increasingly pressing issues as climate change intensifies, profoundly impacting global hydration and the health of ecosystems and human societies. These challenges arise from a complex interplay of, all of which exacerbate the already uneven distribution of freshwater resources. Extended periods of reduced rainfall deplete surface water sources like rivers, lakes, and reservoirs, and diminish groundwater recharge, which is critical for sustaining water supplies during dry periods. In many arid and semi-arid regions, where water is already scarce, these changes can lead to acute processes.

Agriculture, a major consumer of water, is particularly vulnerable to these changes. Crop yields can suffer significantly from prolonged droughts, leading to reduced food production and increased prices. This exacerbates for both subsistence and economic activity. Additionally, drought can diminish the availability of water for irrigation, further straining agricultural productivity and potentially leading affected areas. The impacts of water scarcity extend beyond agriculture. Urban areas are not immune; cities can face severe water shortages as demand outstrips supply, leading to rationing, increased water prices, and potential public health crises. Moreover, as competition for limited water resources intensifies, conflicts can arise over access and rights, both within and between countries. Environmental ecosystems are also significantly impacted by water scarcity. Rivers and wetlands that rely on consistent water flows can suffer from reduced biodiversity and ecosystem degradation. Aquatic species dependent on stable water levels face habitat loss and altered reproductive cycles, disrupt food chains (Hennighausen & Suter, 2020). The degradation of wetlands, which serve as critical buffers against floods and help in water purification, can also have far-reaching environmental consequences. Policy and governance play pivotal roles in managing water resources. Comprehensive water management policies that . International cooperation is also vital, especially in regions where water sources cross national boundaries. Shared water resources require collaborative management to prevent conflicts and ensure that all parties benefit from sustainable practices (Herrnstadt & Muehlegger, 2014). Climate adaptation strategies must be integrated into water management plans. implementing policies to support communities that are most vulnerable to water scarcity. efforts. Mitigating climate change itself is a critical component of addressing water scarcity.

Reducing help stabilize temperatures and precipitation patterns, potentially reducing the international climate agreements and national cutting promoting development are integral to long-term water security. In conclusion, water scarcity and drought, exacerbated by climate change, pose significant threats to global hydration, agriculture, urban areas, and ecosystems. Tackling these issues requires a comprehensive approach that integrates effective water management, robust policy frameworks, international cooperation, and climate change mitigation efforts. ensuring a more resilient and sustainable water future for all.

Climate Migration and Displacement: Human Rights and Social Implications

Climate migration and displacement are emerging as critical issues of human rights and social, and prolonged droughts are increasingly displacing populations from their homes, creating new patterns of migration that challenge existing legal, social, and political frameworks. These climate-induced displacements pose significant human rights concerns and have profound social implications for both the displaced individuals and the communities receiving them. Human rights issues related to climate migration is the right to safety and security. As communities face increased vulnerability to climate-related disasters—such as floods, hurricanes, and wildfires (Kahn & Kotchen, 2011). This forced migration often occurs without adequate preparation or support, leaving individuals and families exposed to risks such as inadequate shelter, insufficient access to basic services, and heightened exposure to exploitation and abuse. For many, especially in developing regions, this displacement can lead to precarious living conditions in limited. climate migration can exacerbate existing inequalities and vulnerabilities. Marginalized groups, including women, children, elderly people, and indigenous communities, are disproportionately affected by both climate change and displacement. These groups often face higher risks due to socio-economic factors, including limited access to resources, political disenfranchisement, and systemic discrimination. For instance, women and children in displaced communities are particularly Addressing these disparities requires targeted support and protection measures to ensure that all individuals, particularly the most vulnerable, have their rights upheld and their needs addressed. The social implications of climate migration extend beyond the immediate experiences of displaced individuals. Receiving communities may face significant pressures as they absorb new populations. These pressures can strain local conflicts between newcomers and existing residents. Integrating migrants into new communities requires careful planning and investment in infrastructure, social services, and community-building (Kahneman & Frederick, 2002) efforts to foster social cohesion and mitigate potential conflicts. Successful integration relies on creating inclusive policies that promote social and economic opportunities for migrants, while also addressing the needs of host communities. At the same time, climate migration poses challenges for national and international legal frameworks. Current international laws and conventions, such as the Refugee Convention (Kelly & Goulden, 2008) primarily address persecution-based displacement and do not adequately cover climate-induced migration. This gap in legal protection leaves many climate migrants without formal status or rights, making them particularly vulnerable to exploitation and marginalization. There is a growing call migrants, including mechanisms for temporary or permanent protection and pathways for legal migration. Furthermore, supporting sustainable development can reduce the risk of displacement by making communities more resilient to climate impacts. For instance, improving early warning systems and building flood defenses need for forced migration. In conclusion, climate migration and displacement are pressing human rights and social safeguarding the rights of displaced individuals, supporting receiving communities, and updating legal frameworks to protect climate migrants. By integrating climate adaptation and mitigation strategies and fostering international cooperation, affected by climate change are met.

Natural Disasters and Climate Change: The Role of Human Activity

Natural increasingly intertwined, activity playing a pivotal role in both exacerbating the frequency shaping our responses to them. Historically, hurricanes, floods, droughts, and wildfires have occurred due to natural variability in climate and geological processes. However, contemporary patterns reveal a troubling correlation between human-induced climate change intensity and occurrence of these has altered atmospheric and oceanic conditions in ways that exacerbate natural disasters (Koubi et al., 2016). For instance, higher global temperatures contribute to more intense and frequent heatwaves, which can lead to severe drought conditions and amplify the risk of wildfires. Deforestation and land use changes also play a critical role in influencing natural disasters. The clearing. Forests act as natural barriers that stabilize soil and regulate water flow; their loss thus heightens the risk of landslides and erosion. In addition, urbanization often involves the construction of infrastructure that alters natural water flow patterns, further contributing to flood risks. The human impact on natural disaster patterns extends beyond the immediate physical environment. Socio-economic factors, such as population density, infrastructure quality, and economic resilience, influence how communities experience and recover from Addressing the role of human activity in natural disasters improving energy efficiency, and implementing reforestation and afforestation projects. By curbing emissions, we can help stabilize the climate and reduce the Resilience-building involves strengthening the capacity of investing in education and awareness programs to enhance community preparedness, improving early warning systems to provide timely information, and fostering social networks and support systems that can aid in recovery efforts. Resilient communities are better equipped to adapt to changing conditions and rebound from disasters, reducing long-term impacts on lives and livelihoods. Global and local governance natural disasters. International agreements and frameworks. In conclusion, natural disasters influenced by human activity. The exacerbation of natural disasters through natural disasters and safeguarding communities from their devastating effects.

Early Warning Systems and Emergency Preparedness: Saving Lives and Livelihoods

natural disasters and climate-related events, and they play a vital role in saving lives and protecting livelihoods. As the frequency and intensity of such events increase due to climate change, the importance of these systems becomes ever more apparent. Early warning systems, which provide timely and actionable information about impending hazards, are essential. These systems encompass a range of technologies and methods designed to detect and predict hazards such as storms, floods, heatwaves, and earthquakes, and to communicate risk information to at-risk populations. hazard monitoring, risk assessment, communication, and response coordination. Hazard monitoring involves the continuous observation of environmental conditions through various technologies (Krishnamurthy, 2012), such as satellites, weather radars, and seismic sensors. These tools collect data that are analyzed to predict the occurrence and potential impact of disasters. For example, meteorological satellites can track the formation and movement of tropical storms, while flood gauges can measure water levels in rivers and reservoirs. Risk assessment evaluates the potential impact of these hazards on communities density, infrastructure, and vulnerability. This assessment helps prioritize areas and populations at greatest risk and informs the design of targeted warning messages and preparedness plans. For instance, in flood-prone areas, risk assessments can determine which regions are most likely to experience severe flooding and guide the development of evacuation routes and emergency shelters. Communication is a critical element of early warning systems. Effective communication ensures that warnings are disseminated rapidly and clearly to those at risk. Modern communication technologies, including mobile alerts, social media, and mass media, play a significant role in reaching diverse audiences (Marchiori et al., 2012). However, ensuring that warnings reach all segments of the population, including those with limited access to technology or who speak different languages, requires a multifaceted approach. Community-based communication strategies, such as local warning networks and public information campaigns, are essential for ensuring that warnings are understood and acted upon. Response coordination involves organizing and mobilizing resources and personnel to address the immediate needs of affected communities. This includes activating emergency response plans, deploying rescue teams, and providing essential services such as medical care, food, and shelter. Effective response coordination disaster. Emergency preparedness complements early warning systems such as food, water, and medications. Educational initiatives that raise awareness about potential hazards and appropriate response actions help build a culture of preparedness and resilience (Martínez-Zarzoso, 2020) ensures that preparedness measures are tailored to their specific needs and contexts. can play a crucial role in disseminating information, organizing drills, and supporting vulnerable populations. By fostering local ownership and involvement, recover from disasters. into broader disaster risk reduction and climate adaptation strategies enhances their effectiveness. For example, incorporating early warning information into urban planning and infrastructure development can help reduce vulnerability to hazards. Additionally, integrating climate change projections into risk assessments and preparedness planning ensures that communities are prepared for future risks as well as current ones. Global and regional cooperation is also vital for strengthening early warning systems and preparedness efforts. Provide support and coordination for early warning and preparedness initiatives. Regional networks, such as the Pacific Tsunami Warning Center and the European Severe Storms Laboratory, contribute to cross-border collaboration and information sharing. In conclusion, hazards. By enhancing hazard monitoring, risk assessment, communication, and response co-ordination, and by fostering community involvement and integrating these efforts into broader risk reduction strategies, we can significantly improve our ability to save lives and protect livelihoods. As climate change continues to heighten the risks of disasters, investing in these systems and preparedness measures is critical for building resilient communities and safeguarding human and economic well-being.

Climate Change Adaptation and Resilience: Strategies for Vulnerable Communities

Vulnerable communities facing increasing risks from climate-related impacts. As global adverse effects, including extreme heat, flooding, droughts, and sea-level rise. Effective adaptation and resilience strategies aim to reduce the vulnerability of these populations, climate impacts, and ultimately promote sustainable development. Adaptation strategies involve modifying practices, policies, and infrastructure to better handle. For vulnerable communities, these strategies must be tailored to local conditions and needs, considering factors such as geography, socio-economic status, and existing vulnerabilities. A comprehensive adaptation approach often includes several key components: infrastructure improvements, ecosystem-based solutions, and community engagement. Infrastructure improvements are essential for building resilience against climate impacts. For example, upgrading drainage systems can help prevent flooding in areas prone to heavy rainfall. In coastal regions, constructing sea walls and restoring natural barriers like mangroves and wetlands can protect against storm surges and erosion, contributing to broader socio-economic development. Ecosystem-based solutions leverage the natural environment to enhance resilience. For instance, wetlands act as natural buffers against floods, while forests stabilize soils and regulate water flow. By maintaining or restoring these ecosystems, communities can benefit from their natural protective functions and improve their overall resilience. Integrating ecosystem management into adaptation planning also promotes biodiversity, which supports ecosystem health and improved water quality and carbon sequestration (Maas et al., 2020). Communities often possess valuable traditional knowledge and insights that can inform adaptation strategies. Participatory approaches, such as community-based planning and consultation, help identify priority needs and preferences, enhance local ownership of adaptation initiatives, and foster social cohesion. Empowering communities through education and capacity-building activities also strengthens their ability to respond to climate impacts and advocate for their needs. One effective adaptation strategy is to focus on building social resilience. This involves strengthening social networks and support systems within communities, which can enhance collective coping mechanisms and recovery efforts. Programs that address social inequalities and improve access to resources, education, and healthcare contribute to greater resilience by reducing vulnerabilities and enabling more equitable responses to climate impacts. Social safety nets, such as insurance schemes and emergency assistance programs, also play a critical role in supporting individuals and households during and after climate-related events. Integrating climate adaptation into broader development planning communities. This means embedding climate considerations into sectors such as agriculture, water management, and urban planning. For example, promoting climate-smart agricultural help farmers adapt to changing conditions and ensure food security. Similarly, incorporating climate risk assessments into urban planning can guide the development of resilient infrastructure and land-use practices (Mann & Gleick, 2015). Financial resources and technical support are crucial for implementing adaptation strategies effectively. Vulnerable communities often face significant challenges in accessing funding and technical expertise for adaptation projects. International climate finance provide essential, and the private sector can facilitate the flow of resources and knowledge to where they are most needed. Monitoring and evaluation are important for assessing the effectiveness of adaptation measures and ensuring continuous improvement. Developing metrics and indicators to track progress, assess outcomes, and identify challenges helps refine adaptation strategies and inform future planning. Engaging communities in monitoring activities also promotes transparency and accountability, ensuring that adaptation efforts remain relevant and effective. In conclusion, risks. By focusing on infrastructure improvements, ecosystem-based solutions, and community engagement, and by integrating adaptation into broader development planning, these strategies help reduce vulnerability, enhance coping capacity, and promote sustainable development. Financial resources, technical support, and robust monitoring and evaluation mechanisms we can build resilience and secure a more sustainable future for communities facing the challenges of a changing climate.

The Economic Costs of Climate Change and Natural Disasters: Assessing the Damage

The economic costs are profound and multifaceted, encompassing direct damages, indirect effects, and long-term economic implications. As climate change accelerates, the financial burdens on economies become increasingly severe. Assessing these costs is critical for understanding the full direct damages from natural disasters include the immediate financial losses associated with the destruction of infrastructure, property, and agricultural assets. For example, hurricanes real estate, while wildfires can destroy forests, homes, and entire communities (Zaval et al., 2014). The costs of repairing and rebuilding damaged infrastructure and replacing lost assets are substantial and can strain public and private finances. In the wake of a disaster, the need for emergency response, relief operations, and temporary shelters further compounds these direct costs. Indirect economic effects of natural disasters are less immediately apparent but can be equally damaging. These include disruptions to economic activities, Tourism-dependent regions may experience reduced visitor numbers due to damaged infrastructure or negative perceptions of safety. The loss of productivity and income due to these disruptions can result in reduced economic output and slower recovery for affected regions. Long-term economic implications include the persistent effects on growth and development. Frequent or severe disasters can erode the economic base of communities, leading to decreased investment and slower economic growth. For developing countries, which often have fewer resources to invest in resilience and recovery, the economic impacts can be particularly severe. The loss of infrastructure and assets can undermine development gains, perpetuating cycles of poverty and vulnerability. Additionally, can place additional financial pressures on governments and businesses. Quantifying the economic costs involves a range of methodologies and data sources (Lang & Ryder, 2016). Economic impact assessments typically consider direct damages, indirect economic losses, and long-term costs. Tools understand the scale of the economic challenges and prioritize investments in resilience and adaptation. The economic cost regions, particularly low-lying coastal areas and developing countries, often face higher costs due to, which can lead to significant economic losses and displacement of populations. In contrast, wealthier countries may have more resources to invest in adaptation and recovery but still face substantial financial impacts and financial mechanisms, including insurance and climate finance, can provide crucial support for managing the economic impacts of disasters. Insurance products, such as catastrophe bonds and parametric insurance, can help spread the financial risks of disasters and provide timely funding for recovery efforts. International climate finance Moreover, essential for improving our understanding of the economic impacts of climate change and natural disasters. Enhanced data on damages, losses, and recovery costs can inform more accurate risk assessments and guide effective policy responses., sector, and academic institutions can help advance knowledge and develop innovative solutions for managing economic risks.

Climate Justice and Equity: Addressing Disproportionate Impacts

Climate justice and equity are critical considerations in addressing the disproportionate marginalized and vulnerable communities, exacerbating existing social, economic, and environmental inequalities. This disparity calls for a focus on fairness and justice in climate policy and action, ensuring that those who contribute least to climate change and are most affected by its consequences are given priority in both mitigation and adaptation efforts (Yale PCCC, 2019). The disproportionate change highlight the need for climate justice, which seeks to address these inequities and ensure that the benefits of climate action are shared fairly. Climate justice involves recognizing the historical and ongoing contributions of marginalized communities to climate change and acknowledging the inequities in both responsibility and capacity. Equity in approach not only leads to more effective and just policies but also empowers communities to take ownership of their adaptation and mitigation efforts. Empowering local organizations and leaders who understand the specific challenges and opportunities within their Adaptation and mitigation strategies must be designed with an emphasis on equity (The World Bank, 2022), ensuring that interventions do not perpetuate or exacerbate existing inequalities. For instance, urban greening initiatives and renewable energy projects should be implemented in ways that benefit low-income communities and avoid displacing or marginalizing local populations. Similarly, climate adaptation programs should address the specific needs of vulnerable groups or offering financial assistance to low-income households to retrofit their homes for increased resilience. Addressing climate justice also involves recognizing and rectifying the socio-economic impacts of climate policies can have economic implications for communities dependent on fossil fuel industries. Just transition policies are essential to ensure that workers and communities affected by these transitions are provided with support, retraining opportunities, and alternative livelihoods. This approach helps mitigate the negative. In conclusion, addressing climate justice and equity is vital for ensuring that the impacts of climate change are managed in a fair and inclusive manner. By acknowledging the disproportionate impacts on marginalized communities, incorporating their perspectives into decision-making processes, and implementing equitable adaptation and mitigation strategies, we can work towards a more just and resilient response to climate change

International Cooperation and Climate Governance: Global Responses

requiring collective action and coordination among countries, organizations, and stakeholders. transboundary issue that transcends national borders, and its impacts are felt worldwide, necessitating a unified global response to effectively mitigate and adapt to its effects. The complexity and scale of the climate crisis call for a robust framework of international cooperation and governance to ensure effective and equitable solutions. Effective climate governance also involves enhance ambition over time. Regular reporting by countries on their NDCs and the implementation of climate policies is essential for tracking progress and identifying areas for improvement. Transparent and accurate data on emissions, climate impacts, and financial flows are critical for informed decision-making and maintaining trust among stakeholders. Regional and bilateral cooperation complement global efforts by addressing specific climate challenges and opportunities at a more localized level., demonstrate how collaborative efforts can enhance climate action and resilience within specific geographic areas. These regional frameworks a significant role in addressing shared climate challenges, such as transboundary pollution or shared water resources. Civil society (Woolridge, 2010) including non-governmental organizations (NGOs), research institutions, and advocacy groups, plays a vital role in international climate governance. These organizations contribute to climate action through advocacy, research, and grassroots initiatives. They help to amplify the voices of marginalized communities, drive public awareness, and hold governments and corporations accountable for their climate commitments. Collaborative partnerships between civil society and governments or businesses can lead to innovative solutions and more inclusive climate policies. The success of international climate cooperation hinges on addressing issues of equity and justice. Ensuring that climate actions and policies are fair and equitable, particularly for the most vulnerable and least developed countries (Westerling & Bryant, 2008) is essential for achieving global climate goals. This includes addressing the historical responsibility of developed nations for greenhouse gas emissions and ensuring that climate finance and support are allocated effectively to those in need. (Trenberth et al., 2014). Equity considerations also involve recognizing and supporting the rights of indigenous peoples, who often face disproportionate impacts and play crucial roles in climate adaptation and conservation. In conclusion, international cooperation and climate governance are critical for tackling the global climate crisis and achieving meaningful. Through global agreements like the Paris Agreement, financial mechanisms, and technology transfer initiatives, countries can work together to address climate challenges and support vulnerable communities. Effective governance requires robust monitoring, transparent reporting, and inclusive participation from all stakeholders, including civil society. By prioritizing equity and justice, and fostering collaborative efforts at regional and bilateral levels,

Case Studies: Successful Climate Change Mitigation and Adaptation Efforts

Case studies of successful provide valuable insights into effective strategies. These examples demonstrate how targeted actions, innovative solutions, and collaborative approaches can lead to meaningful progress in reducing greenhouse gas emissions and broader climate goals. One notable case study (Van der Linden, 2015) in Denmark's experience highlights the importance of government support, technological innovation, and public-private partnerships in driving a successful transition to renewable energy. Another successful mitigation effort is the implementation of energy efficiency measures in Japan. Following the Fukushima nuclear disaster in 2011, Japan faced increased energy demands and a pressing need to reduce its carbon footprint. In response, the Japanese government launched the "Top Runner Program," which sets ambitious energy efficiency standards for appliances and vehicles. In terms of adaptation, the city of Rotterdam in the Netherlands offers a compelling case study. The city has invested in "climate-proof" infrastructure, such as elevated buildings, flood-resistant barriers, and green roofs that absorb rainwater and reduce runoff. Rotterdam's approach also involves integrating climate adaptation into urban development plans and engaging citizens in resilience-building activities. This proactive and integrated approach to adaptation demonstrates how cities can effectively manage climate risks and enhance their resilience to future challenges. The small island nation of Maldives provides an example of successful adaptation in the face of elevation and susceptibility to coastal erosion, has pursued a multifaceted adaptation strategy that includes building sea walls, restoring mangroves, and developing floating infrastructure. adaptation projects that enhance local resilience and promote sustainable practices. The Maldives' efforts underscore the importance of integrating traditional knowledge with modern technology and the need for international support and cooperation in addressing the unique challenges faced by island nations. Improving soil health through conservation tillage (NOAA,

2019), diversifying crops to reduce vulnerability to pests and diseases, and adopting water-efficient irrigation systems not only increase agricultural productivity but also contribute to carbon sequestration and reduced greenhouse gas emissions. Kenya's experience highlights the role of local knowledge, capacity-building, and community engagement in successfully adapting agricultural practices to changing climatic conditions. The reforestation efforts in Costa Rica represent a successful case study in both climate mitigation and adaptation. Costa Rica has implemented a national reforestation program that aims to restore degraded lands and expand forest cover (Turco et al., 2014). This initiative has led to significant increases in forested areas, which sequester carbon dioxide, enhance biodiversity, and improve watershed management. The program also provides economic incentives to landowners and local communities, demonstrating the potential for integrating environmental and socio-economic benefits in climate action. adaptation goals. Efforts. Strong government policies and frameworks, technological innovation, community involvement, and international cooperation are all essential components of successful climate strategies. Each case study demonstrates the importance of tailoring solutions to local contexts and needs, leveraging diverse resources and expertise, and ensuring that climate actions are integrated into broader development goals. Furthermore, these examples underscore the need for ongoing evaluation and learning to refine and improve climate strategies. Monitoring and assessing the impacts of climate initiatives, sharing best practices, and adapting approaches based on new information and changing conditions In conclusion, successful climate change mitigation and adaptation efforts offer valuable lessons for addressing the global climate challenge. (Le Houérou, 1996) By examining case studies such as Denmark's wind energy transition, Japan's energy efficiency program, Rotterdam's climate adaptation strategy, the Maldives' coastal protection measures, Kenya's climate-smart agriculture, and Costa Rica's reforestation efforts, we can identify effective strategies, best practices, and key factors that contribute to successful climate action. These insights can guide future efforts and inform policies and practices for all.

The Role of Technology in Climate Change Mitigation and Disaster Response

Technology is transformative, improving our ability to respond to and recover from disasters. By integrating cutting-edge technologies into climate strategies, we can achieve more effective and scalable solutions to these global issues. In agriculture, precision farming technologies help to mitigate climate change by optimizing resource use and reducing crop health, and weather patterns. This information allows for precise application of water, fertilizers, and pesticides, minimizing environmental impacts and enhancing crop yields. Climate-smart agriculture practices, supported by these technologies(Owen et al., 2012), contribute to increased resilience and reduced greenhouse gas emissions in the agricultural sector. On the disaster response front, technology significantly enhances our ability to predict, prepare for, and manage natural disasters. Early warning systems are one of the most critical technological advancements in disaster management. These systems use a combination of satellite data, weather models, and sensors to provide timely alerts for events such as hurricanes, tsunamis, and earthquakes. For example, the European Space Agency's Sentinel satellites and NOAA's weather satellites monitor atmospheric conditions and track storm developments, providing crucial information for forecasting and early warning. sharing and coordination among emergency responders and affected communities (Owen & Wesselbaum, 2020). Social media platforms provide a means for individuals to report emergencies, share updates, and request assistance. Data identify patterns, predict impacts, and optimize response strategies. The program also provides economic incentives to landowners and local communities, demonstrating the potential for integrating environmental and socio-economic benefits in climate action. Strong government policies and frameworks, technological innovation, community involvement, and international cooperation are all essential components of successful climate strategies. Each case study demonstrates the importance of tailoring solutions to local contexts and needs, leveraging diverse resources and expertise, and ensuring that climate actions are integrated into broader development goals. Furthermore, these examples underscore the need for ongoing evaluation and learning to refine and improve climate strategies.

Drones, or are difficult to reach by traditional means (Owen & Wesselbaum, 2020). They provide real-time aerial imagery and data that enhance situational awareness and support decision-making in the aftermath of a disaster. Moreover, advancements in increasingly being applied to disaster response and climate adaptation. AI algorithms can analyze satellite images and climate models to predict extreme weather events, assess damage, and optimize resource allocation (Sims, 2003). Machine learning techniques are used to improve forecasting accuracy, analyze disaster impacts, and develop adaptive strategies based on historical data and real-time information. The integration of technology in climate change mitigation and disaster response also involves . Ensuring that technological solutions are accessible to all communities, including those in low-income or remote areas, is essential for maximizing their effectiveness and inclusivity. Additionally, that technologies are used responsibly and ethically are critical considerations in the deployment of climate and disaster technologies. Change mitigation and enhancing disaster response efforts. From renewable energy innovations and carbon capture technologies to early warning systems (Pachauri et al., 2014). GIS, drones, and AI, technological advancements provide critical tools for addressing climate challenges and improving our ability to manage and recover from natural disasters. By harnessing these technologies and addressing associated challenges, we can build a more resilient and sustainable future,

Building a Climate-Resilient Future through Collective Action

Building a climate-resilient future necessitates a concerted effort through collective action, integrating diverse stakeholders—from governments and businesses to communities and individuals—in a unified approach multifaceted challenges posed by climate change. Collective action amplifies the impact of individual efforts, leveraging the strengths and resources of various sectors to create comprehensive, scalable solutions that enhance resilience and sustainability. systemic nature of climate risks, drive innovation, and implement strategies that ensure long-term environmental, social, and economic stability. Governmental action forms the backbone of climate resilience, with policies and regulations providing a framework for coordinated efforts across different sectors. National governments play a crucial role in setting ambitious climate goals, enacting legislation, and allocating resources to support mitigation and adaptation initiatives (Pahl et al., 2014). For example, countries that commit to reducing greenhouse gas emissions under. Companies that adopt climate-friendly practices, such as reducing their carbon footprint, improving energy criteria guide businesses in aligning their operations with climate goals. Furthermore, private sector investments in green technologies, renewable energy, and climate adaptation solutions can accelerate progress and create economic opportunities (Pielke et al., 2008). Communities and local organizations are on the front lines of climate impacts and are essential to building resilience at the grassroots level. Local adaptation and empowering individuals to take proactive steps in safeguarding their environment. Community-led initiatives, such as tree planting campaigns, waste reduction programs, and climate education, exemplify how collective action at the local level contributes to broader resilience efforts. Individuals also have a critical role in climate resilience through their everyday actions and choices. Grassroots movements and citizen advocacy drive political will and influence policy changes, demonstrating the power of individual and community voices in shaping climate policy. Public awareness campaigns and educational programs help individuals understand their role in climate action and motivate them to engage in sustainable practices (Piguet, 2022). Collective action in building climate resilience also involves fostering collaboration across sectors and disciplines. Integrating scientific research, technology development, and policy-making ensures that climate solutions are evidence-based and effective. Partnerships between academic institutions, research organizations, and government agencies facilitate the development of innovative strategies and technologies that address climate challenges. Collaborative efforts, such as cross-sectoral task forces and multi-stakeholder platforms, enable diverse perspectives to contribute to comprehensive and inclusive climate solutions. A key component of building a climate-resilient future is addressing issues of equity and justice (Schultz & Elliott, 2013). Climate resilience efforts must consider the needs and vulnerabilities of marginalized and underserved communities, ensuring that they are not disproportionately affected by climate impacts. Equitable solutions involve providing support and resources to those who are most at risk and ensuring inequalities and build resilience in a manner that benefits all members of society. Monitoring, evaluation, and adaptive management are essential for ensuring the effectiveness of collective climate resilience efforts. Regular assessment of progress, impacts, and challenges allows for the adjustment of strategies and the identification of areas for improvement. In conclusion, building a climate-resilient future requires a multifaceted approach grounded in collective action. By leveraging the strengths of The path to a climate-resilient future is one of shared responsibility and collective effort, where each contribution strengthens our ability to navigate and thrive in a changing world.

CONCLUSION

Integrated approaches to mitigate risks, enhance resilience, and adapt to the evolving challenges of a warming planet. Climate change acts as a force multiplier, exacerbating the frequency, intensity, and unpredictability of natural disasters, ranging from more severe hurricanes and intensified heatwaves to unprecedented floods and prolonged droughts. The impacts of these disasters are far-reaching, affecting ecosystems, economies, and human lives, with particularly severe consequences for vulnerable communities that are least equipped to cope with such shocks. analytics, provide timely alerts that enable communities to prepare for and respond to impending threats more effectively. Collaboration across sectors and scales is critical to the success of these efforts. Governments, international organizations. Similarly, local and regional initiatives are essential for tailoring responses to specific vulnerabilities and risks, ensuring that solutions are contextually appropriate and effective -making processes is essential for creating inclusive and effective climate action. By addressing systemic inequalities and prioritizing the needs of those most affected, we can build a more equitable and resilient society. In cannot be overlooked in the quest to manage climate change and natural disasters. Advances in climate modeling, satellite monitoring, and early warning systems provide critical tools for predicting and responding to disasters. Technological solutions, such as green infrastructure, climate-smart agriculture, and renewable energy technologies, offer pathways to reduce emissions and enhance resilience. Leveraging these innovations effectively requires investment in research, development, and deployment, as well as ensuring that technologies are accessible and beneficial to all communities. Furthermore, public awareness and education are vital components of disaster preparedness and climate action. change and natural disasters, as well as promoting knowledge of preventive measures and adaptive strategies, empowers individuals and communities to take proactive steps. Educational initiatives can foster a culture of resilience and preparedness, enabling people to make informed decisions and participate actively in climate action efforts. In sum, natural disasters highlights the imperative for a holistic and coordinated approach to both mitigation and adaptation. By addressing the root causes of climate change, investing in resilience, fostering collaboration, and prioritizing equity, we can reduce the risk and impact of natural disasters while building a sustainable and resilient future. The challenges are significant, but the opportunities for transformative change are vast.

REFERENCES

Allen, M. R., Barros, V. R., Broome, J., Cramer, W., Christ, R., Church, J. A., Clarke, L., . . . Dubash, N. K. (2014). Climate change 2014: Synthesis report. Contribution of Working Groups I, II and III to the fifth assessment report of the Intergovernmental Panel on Climate Change. Intergovernmental Panel on Climate Change. https://www.ipcc.ch/report/ar5/syr/

Bender, M. A., Knutson, T. R., Tuleya, R. E., Sirutis, J. J., Vecchi, G. A., Garner, S. T., & Held, I. M. (2010). Modeled impact of anthropogenic warming on the frequency of intense Atlantic hurricanes. *Science*, *327*(5964), 454–458. DOI: 10.1126/science.1180568

Bloodhart, B., Maibach, E., Myers, T., & Zhao, X. (2015). Local climate experts: The influence of local TV weather information on climate change perceptions. *PLoS One*, *10*(11), e0141526. Advance online publication. DOI: 10.1371/journal.pone.0141526

Botzen, W. J. W., Duijndam, S. J., & van Beukering, P. (2021). Lessons for climate policy from behavioral biases towards COVID-19 and climate change risks. *World Development*, *137*, 105214. Advance online publication. DOI: 10.1016/j.worlddev.2020.105214

Census Bureau. (2019a). Household income [Data set]. United States Census Bureau. https://www.census.gov/topics/income-poverty/income/data/tables.html

Census Bureau. (2019b). County population by characteristics 2010-2019 [Data set]. United States Census Bureau. https://www.census.gov/data/tables/time-series/demo/popest/2010s-counties-detail.html

Centre for Research on the Epidemiology of Disasters. (2021). EM-DAT: The international disaster database. Université Catholique de Louvain. https://www.emdat.be/database

Davenport, F. V., Burke, M., & Diffenbaugh, N. S. (2021). Contribution of historical precipitation change to US flood damages. *Proceedings of the National Academy of Sciences of the United States of America*, *118*(4), e2017524118. Advance online publication. DOI: 10.1073/pnas.2017524118

Davis, L. W., & Gertler, P. J. (2015). Contribution of air conditioning adoption to future energy use under global warming. *Proceedings of the National Academy of Sciences of the United States of America*, *112*(19), 5962–5967. DOI: 10.1073/pnas.1423558112

Federal Emergency Management Agency. (2019). FEMA data feeds. https://www.fema.gov/data-feeds

Feldman, L., Maibach, E. W., Roser-Renouf, C., & Leiserowitz, A. (2012). Climate on cable: The nature and impact of global warming coverage on Fox News, CNN, and MSNBC. *The International Journal of Press/Politics*, *17*(1), 3–31. DOI: 10.1177/1940161211425410

Gibson, M., & Mullins, J. T. (2020). Climate risk and beliefs in New York floodplains. *Journal of the Association of Environmental and Resource Economists*, *7*(6), 1069–1111. DOI: 10.1086/710666

Hennighausen, H., & Suter, J. F. (2020). Flood risk perception in the housing market and the impact of a major flood event. *Land Economics*, *96*(3), 366–383. DOI: 10.3368/le.96.3.366

Herrnstadt, E., & Muehlegger, E. (2014). Weather, salience of climate change and congressional voting. *Journal of Environmental Economics and Management*, *68*(3), 435–448. DOI: 10.1016/j.jeem.2014.08.002

Kahn, M. E., & Kotchen, M. J. (2011). Business cycle effects on concern about climate change: The chilling effect of recession. *Climate Change Economics (Singapore)*, *2*(3), 257–273. DOI: 10.1142/S2010007811000292

Kahneman, D., & Frederick, S. (2002). Representativeness revisited: Attribute substitution in intuitive judgment. In Gilovich, T., Griffin, D., & Kahneman, D. (Eds.), *Heuristics and biases: The psychology of intuitive judgment* (pp. 49–81). Cambridge University Press. DOI: 10.1017/CBO9780511808098.004

Kelly, A. E., & Goulden, M. L. (2008). Rapid shifts in plant distribution with recent climate change. *Proceedings of the National Academy of Sciences of the United States of America*, *105*(33), 11823–11826. DOI: 10.1073/pnas.0802891105

Koubi, V., Spilker, G., Schaffer, L., & Bernauer, T. (2016). Environmental stressors and migration: Evidence from Vietnam. *World Development*, *79*, 197–210. DOI: 10.1016/j.worlddev.2015.11.016

Krishnamurthy, P. K. (2012). Disaster-induced migration: Assessing the impact of extreme weather events on livelihoods. *Environmental Hazards*, *11*(2), 96–111. DOI: 10.1080/17477891.2011.609879

Lang, C., & Ryder, J. D. (2016). The effect of tropical cyclones on climate change engagement. *Climatic Change*, *135*(3–4), 625–638. DOI: 10.1007/s10584-015-1590-0

Le Houérou, H. N. (1996). Climate change, drought and desertification. *Journal of Arid Environments, 34*(2), 133–185. DOI: 10.1006/jare.1996.0099

Maas, A., Wardropper, C., Roesch-McNally, G., & Abatzoglou, J. (2020). A (mis) alignment of farmer experience and perceptions of climate change in the US inland Pacific Northwest. *Climatic Change, 162*(3), 1011–1029. DOI: 10.1007/s10584-020-02713-6

Mann, M. E., & Gleick, P. H. (2015). Climate change and California drought in the 21st century. *Proceedings of the National Academy of Sciences of the United States of America, 112*(13), 3858–3859. DOI: 10.1073/pnas.1503667112

Marchiori, L., Maystadt, J.-F., & Schumacher, I. (2012). The impact of weather anomalies on migration in sub-Saharan Africa. *Journal of Environmental Economics and Management, 63*(3), 355–374. DOI: 10.1016/j.jeem.2012.02.001

Martínez-Zarzoso, I. (2020). Searching for grouped patterns of heterogeneity in the climate–migration link. *Weather, Climate, and Society, 12*(4), 695–710. DOI: 10.1175/WCAS-D-19-0116.1

National Oceanic and Atmospheric Administration. (2019). Storm events database. https://www.ncdc.noaa.gov/stormevents/

Owen, A. L., Conover, E., Videras, J., & Wu, S. (2012). Heat waves, droughts, and preferences for environmental policy. *Journal of Policy Analysis and Management, 31*(3), 556–577. DOI: 10.1002/pam.21627

Owen, P., & Wesselbaum, D. (2020). On thresholds in the climate–migration relationship. *International Review of Applied Economics, 34*(3), 400–412. DOI: 10.1080/02692171.2020.1749242

Pachauri, R. K., Allen, M. R., Barros, V. R., Broome, J., Cramer, W., Christ, R., . . . Dubash, N. K. (2014). Climate change 2014: Synthesis report. Contribution of Working Groups I, II and III to the fifth assessment report of the Intergovernmental Panel on Climate Change. Intergovernmental Panel on Climate Change. https://www.ipcc.ch/report/ar5/syr/

Pahl, S., Sheppard, S., Boomsma, C., & Groves, C. (2014). Perceptions of time in relation to climate change. *Wiley Interdisciplinary Reviews: Climate Change, 5*(3), 375–388. DOI: 10.1002/wcc.272

Pielke, R. A.Jr, Gratz, J., Landsea, C. W., Collins, D., Saunders, M. A., & Musulin, R. (2008). Normalized hurricane damage in the United States: 1900–2005. *Natural Hazards Review, 9*(1), 29–42. DOI: 10.1061/(ASCE)1527-6988(2008)9:1(29)

Piguet, E. (2022). Linking climate change, environmental degradation, and migration: An update after 10 years. *Wiley Interdisciplinary Reviews: Climate Change*, *13*(1), e746. Advance online publication. DOI: 10.1002/wcc.746

Schultz, J., & Elliott, J. R. (2013). Natural disasters and local demographic change in the United States. *Population and Environment*, *34*(3), 293–312. DOI: 10.1007/s11111-012-0171-7

Schutte, S., Vestby, J., Carling, J., & Buhaug, H. (2021). Climatic conditions are weak predictors of asylum migration. *Nature Communications*, *12*, 2067. Advance online publication. DOI: 10.1038/s41467-021-22255-4

Sims, C. A. (2003). Implications of rational inattention. *Journal of Monetary Economics*, *50*(3), 665–690. DOI: 10.1016/S0304-3932(03)00029-1

The PRS Group. (2019). The International Country Risk Guide. https://www.prsgroup.com/explore-our-products/icrg/

The World Bank. (2022). World development indicators [Data set]. https://databank.worldbank.org/source/world-development-indicators

Trenberth, K. E., Dai, A., van der Schrier, G., Jones, P. D., Barichivich, J., Briffa, K. R., & Sheffield, J. (2014). Global warming and changes in drought. *Nature Climate Change*, *4*(1), 17–22. DOI: 10.1038/nclimate2067

Turco, M., Llasat, M. C., von Hardenberg, J., & Provenzale, A. (2014). Climate change impacts on wildfires in a Mediterranean environment. *Climatic Change*, *125*(3–4), 369–380. DOI: 10.1007/s10584-014-1183-3

van der Linden, S. (2015). The social-psychological determinants of climate change risk perceptions: Towards a comprehensive model. *Journal of Environmental Psychology*, *41*, 112–124. DOI: 10.1016/j.jenvp.2014.11.012

Westerling, A. L., & Bryant, B. P. (2008). Climate change and wildfire in California. *Climatic Change*, *87*(Suppl 1), 231–249. DOI: 10.1007/s10584-007-9363-z

Wooldridge, J. M. (2010). *Econometric analysis of cross section and panel data* (2nd ed.). MIT Press.

Yale Program on Climate Change Communication. (2019). Climate change opinion maps. https://climatecommunication.yale.edu/visualizations-data/ycom-us-2018/

Zaval, L., Keenan, E. A., Johnson, E. J., & Weber, E. U. (2014). How warm days increase belief in global warming. *Nature Climate Change*, *4*(2), 143–147. DOI: 10.1038/nclimate2093

Chapter 3
A Review of the United States of America's Disaster Management System

Murat Şengöz
https://orcid.org/0000-0001-6597-0161
Independent Researcher, Turkey

ABSTRACT

This article focuses on enhancing the joint operational capabilities of the military, police, fire department, and other relevant organizations, as well as ensuring effective command and control relationships in disaster management. Effective response and coordination in disaster situations require collaboration and integration among different organizations. The article examines the disaster management system and practices in the United States, focusing on the roles and responsibilities of the Federal Emergency Management Agency (FEMA) and other relevant agencies. It also analyzes how cooperation between military and civilian authorities is achieved and how existing legal regulations support and enhance these processes. Supported by case studies and literature reviews, this study highlights critical elements such as training, exercises, technological infrastructure, and legal regulations for developing joint operational capabilities. Finally, recommendations for improving coordination and command-control relationships in disaster management are presented.

DOI: 10.4018/979-8-3693-9745-9.ch003

1. INTRODUCTION

Disasters, whether natural or human-induced, have significant and often devastating impacts on communities and societies, requiring swift and coordinated interventions to mitigate damage and aid in recovery. These events, ranging from hurricanes and wildfires to industrial accidents and pandemics, can overwhelm local, regional, and even national systems. The aftermath of disasters typically requires an integrated approach from various sectors, including governmental bodies, emergency services, non-governmental organizations, and private enterprises, all of whom must collaborate effectively to ensure the success of disaster response efforts. The complexity of managing these crises demands seamless coordination, communication, and operational integration between different responding agencies (Haddow, Bullock, & Coppola, 2017).

Disaster management is inherently interdisciplinary and involves the combined efforts of military, police, fire departments, medical teams, and other emergency responders. These agencies must work together, often under extreme pressure, to ensure a timely and effective response. One of the critical elements for success in disaster management is enhancing the joint operational capabilities of these agencies, allowing them to function cohesively during complex, chaotic situations. This entails the development of shared protocols, joint training programs, and effective communication strategies that ensure smooth interaction among agencies and the reduction of response times (Waugh & Tierney, 2007). Failure to integrate and harmonize the operational capabilities of these diverse agencies can result in inefficient disaster responses, exacerbating the loss of life and property (Kapucu, 2008).

In this context, the disaster management system in the United States offers a wealth of insights and practices that can serve as a model for improving interagency coordination and the establishment of efficient command-and-control relationships. The Federal Emergency Management Agency (FEMA), the nation's lead disaster response agency, plays a crucial role in directing disaster management processes and coordinating efforts between federal, state, and local agencies. FEMA's operations are supported by a range of local and state-level emergency management bodies, each tasked with implementing disaster response strategies within their respective jurisdictions (FEMA, 2018). Through its partnership with various organizations, including military and civilian institutions, FEMA has played a pivotal role in shaping the disaster management landscape, ensuring that disaster response is not only coordinated but also strategically organized to reduce harm to affected populations.

A fundamental element in the success of disaster management lies in the capacity of military, police, fire departments, and other agencies to engage in joint operations. The development of joint operational capabilities ensures that these agencies can efficiently share resources, coordinate activities, and respond as a unified force during disaster events. As seen in the aftermath of major disasters like Hurricane Katrina and the 2020 California Wildfires, the need for joint operational capabilities becomes all the more apparent, with different organizations often having to collaborate in high-stress environments, where rapid decisions must be made, and response efforts must be optimized. Joint operational capability refers to the capacity of different responding agencies to collaborate seamlessly, allowing them to address challenges in real-time while minimizing delays and maximizing resource utilization (Comfort, 2007).

As emphasized in the literature, enhancing joint operational capabilities requires a multifaceted approach, involving the training and preparation of personnel, the development of shared technical infrastructures, and clear communication protocols. Training programs are particularly vital in ensuring that personnel from different agencies are familiar with each other's roles, operational procedures, and expectations in the context of disaster management. Joint exercises, such as those regularly conducted by FEMA, simulate real-world disaster scenarios and allow participating agencies to practice coordinated responses, enabling them to assess strengths and weaknesses in their operations. Such training programs and exercises help in fostering a sense of trust and familiarity among different organizations, ensuring that they can work together effectively when disasters occur (Kapucu, 2008).

Moreover, technological infrastructure plays an indispensable role in modern disaster management. The development and integration of cutting-edge communication systems, Geographic Information Systems (GIS), and data-sharing platforms have enabled disaster response agencies to streamline coordination efforts. Advanced technologies provide real-time situational awareness, improve the accuracy of decision-making, and facilitate the rapid dissemination of information across different agencies. In addition, they assist in tracking the location of personnel, equipment, and supplies, ensuring that resources are allocated efficiently during emergency operations (Rodríguez, Donner, & Trainor, 2018). For example, FEMA's integration of the National Response Framework (NRF) with digital tools has significantly improved the coordination between state, federal, and local responders, enabling more precise and timely interventions. The continuous upgrading and testing of technological systems are essential to ensuring that they can withstand the challenges presented by large-scale disasters.

Another critical factor in enhancing the effectiveness of disaster management is the establishment of efficient and clearly defined command-and-control relationships. The command-and-control structure in disaster management determines the decision-making processes, the allocation of resources, and the delegation of tasks during an emergency. A well-structured command system ensures that all responding agencies know their roles and responsibilities and are able to act in a coordinated fashion. Effective command-and-control relationships also facilitate quick decision-making, which is vital when time is of the essence. According to Alexander (2002), the absence of a clear command structure can lead to confusion, delays, and conflicting actions, which can worsen the impacts of disasters.

In the U.S., command-and-control relationships are firmly established through a hierarchical system, with FEMA providing overarching leadership and coordination. In addition to federal leadership, state and local governments maintain considerable autonomy in managing disaster responses within their jurisdictions. However, these agencies operate under the guidance of national frameworks, ensuring that all response efforts are aligned and that resources are deployed where they are needed most. The Stafford Act, which authorizes federal assistance for disaster response, and the Homeland Security Act, which created the Department of Homeland Security (DHS), provide the legal framework for coordination across different levels of government and between military and civilian authorities (White, 2012). These legal regulations further enhance the operational framework by outlining the roles, responsibilities, and procedures that must be followed during disaster responses.

This study aims to explore the key elements involved in establishing effective joint operational capabilities and command-and-control relationships in disaster management. Through a comprehensive literature review and case study analysis, this research will examine how training programs, technological innovations, and legal regulations can strengthen disaster management efforts. Additionally, this study will assess how lessons learned from past events like Hurricane Katrina and the 2020 California Wildfires can inform future disaster response strategies and improve the operational preparedness of agencies involved in disaster management.

2. DISASTER MANAGEMENT SYSTEM AND DEVELOPMENT OF JOINT OPERATIONAL CAPABILITIES

The United States boasts a comprehensive and multilayered disaster management system, designed to ensure swift, efficient, and coordinated responses to both natural and human-made disasters. This integrated system involves numerous federal, state, and local agencies, each contributing its unique expertise to manage disaster response. Central to this system is the Federal Emergency Management Agency (FEMA), which acts as the primary authority responsible for overseeing disaster management efforts at the national level (FEMA, 2018). FEMA's broad strategic framework encompasses disaster preparedness, response, and recovery, ensuring a holistic approach to crisis management. In an effort to improve the efficiency of these processes, FEMA coordinates closely with various federal agencies, state and local governments, non-governmental organizations (NGOs), and private-sector entities to enhance interagency cooperation and collaboration (Haddow, Bullock, & Coppola, 2017).

The foundation of the U.S. disaster management system rests on effective coordination and shared responsibility among numerous actors. FEMA is the linchpin of this structure, providing guidance, funding, and coordination for disaster response and recovery efforts. However, FEMA alone is not responsible for managing disasters. Rather, it serves as the facilitator, orchestrating the actions of a diverse array of local, state, and federal agencies, each tasked with fulfilling specific roles during emergency events. These roles are essential for ensuring that disasters are managed in an integrated, coordinated, and timely manner, preventing confusion and ensuring that resources are allocated effectively. This system of coordination is particularly important when disasters strike at a large scale, such as hurricanes, earthquakes, or terrorist attacks, where the complexity and scale of the response efforts require a synchronized approach.

Roles of the Military, Police, and Fire Departments

The involvement of military, police, and fire departments in disaster management is vital to ensure an effective and comprehensive disaster response. Each of these institutions plays a unique yet complementary role, enhancing the overall ability of the disaster management system to respond to a crisis. The military, with its logistical expertise, extensive resources, and capacity for large-scale mobilization, provides invaluable support in disaster management. Its primary responsibilities include delivering aid, ensuring security, managing infrastructure recovery, and conducting search and rescue operations in disaster zones (White, 2012). The military's role is particularly crucial in large-scale disasters that require rapid deployment of resources, such as in the aftermath of hurricanes or in war-zone-like situations, where civil authorities might be unable to function effectively.

Police forces also play a significant role in maintaining law and order during disaster events, particularly in chaotic environments where the normal rule of law is disrupted. Their responsibilities include ensuring public safety, preventing looting, assisting in evacuation efforts, and coordinating with other emergency services. Fire departments, similarly, have specialized roles in disaster management, focusing on fire suppression, rescue efforts, and providing immediate medical care, particularly in the early stages of a disaster. Firefighters are often among the first responders on the ground, working tirelessly to save lives, provide basic aid, and assist in securing disaster areas (Kapucu, 2008).

In light of these diverse roles, fostering joint operational capabilities across the military, police, and fire departments is essential for ensuring a unified and effective response during crises. This coordination allows agencies to support one another's efforts, ensure that resources are deployed in a timely and efficient manner, and enhance the overall impact of response actions. A failure to integrate the efforts of these key agencies can lead to delays, miscommunication, and gaps in response coverage, further exacerbating the disaster's impact on communities.

Training and Exercises

The development of joint operational capabilities relies heavily on regular training and exercises that simulate real-world disaster scenarios. These exercises allow the military, police, fire departments, and other agencies to practice coordinated disaster response efforts in a controlled setting, providing valuable insights into their strengths, weaknesses, and areas of improvement (Rodríguez, Donner, & Trainor, 2018). In the U.S., FEMA and other agencies organize large-scale disaster simulations and drills aimed at testing the readiness of different organizations to work together in times of crisis. These exercises are designed not only to hone technical skills but also to build inter-agency cooperation and establish familiarity among responders from different agencies.

The United States has developed a rigorous system of training for emergency responders at all levels, including local, state, and federal agencies. By engaging in joint training exercises, agencies can identify potential issues with communication, resource allocation, and coordination that may arise in the midst of a disaster. For instance, local police may train with federal agencies to address public order during mass evacuations, while fire departments may conduct drills with military personnel to test joint search-and-rescue operations. Through these collaborative exercises, the relationships between different agencies are strengthened, enhancing their ability to respond quickly and effectively when real disasters occur (Comfort, 2007). These exercises can range from tabletop exercises, where participants discuss hypothetical disaster scenarios, to full-scale simulations involving real-time responses to staged disasters.

Technological Infrastructure

The role of technology in disaster management is pivotal for improving the flow of information, ensuring effective communication, and supporting the decision-making process in real-time. In disaster scenarios, the ability to quickly share critical data across different agencies can mean the difference between life and death. Communication systems, Geographic Information Systems (GIS), and real-time data sharing platforms are essential tools in modern disaster management (Kapucu, 2008). These technologies help agencies assess the situation on the ground, identify affected areas, allocate resources effectively, and communicate seamlessly with one another. The integration of new technologies such as drones, satellite imagery, and advanced data analytics has revolutionized disaster response, allowing responders to act more swiftly and decisively.

FEMA and other agencies have invested heavily in building and maintaining a robust technological infrastructure to support disaster management efforts. This includes the development of the Integrated Public Alert and Warning System (IPAWS), which disseminates urgent information to the public and responders in real-time (FEMA, 2018). Additionally, systems such as the National Incident Management System (NIMS) enable agencies to coordinate their efforts using standardized procedures and a common language, ensuring that all responders operate according to the same guidelines. As technological systems continue to evolve, it is essential that agencies continue to test, upgrade, and adapt their technological tools to meet the growing demands of modern disaster management.

Resource and Logistics Management

Resource and logistics management are key elements of effective disaster management, ensuring that the right resources are delivered to the right place at the right time. Effective management of resources, such as food, water, medical supplies, and personnel, is essential for supporting relief efforts during and after a disaster. FEMA, in coordination with other federal and state agencies, oversees the efficient distribution of these critical resources, ensuring that they reach affected areas as quickly as possible. Furthermore, FEMA's logistical expertise is crucial in coordinating the transportation and delivery of supplies to disaster zones, especially when traditional infrastructure is damaged or destroyed.

The importance of resource management is especially evident in large-scale disasters like Hurricane Katrina, where efficient allocation and distribution of resources were pivotal in minimizing loss of life and speeding up recovery efforts. In the U.S., FEMA's coordination of resource distribution is augmented by state and local governments, private-sector partners, and NGOs, ensuring a collaborative approach to resource mobilization (Haddow et al., 2017).

Establishment of Command-and-Control Relationships

A well-defined command-and-control structure is integral to the success of disaster response efforts. Command chains define the decision-making processes, ensuring that key decisions are made quickly and in an organized manner, even in high-pressure situations. The establishment of clear command structures eliminates confusion, prevents duplication of efforts, and ensures that all responders are on the same page. FEMA and other agencies are committed to establishing effective command chains by clearly delineating roles and responsibilities at each level of government, from the local authorities to federal agencies. This clear hierarchy allows for more efficient decision-making during disasters and improves the overall coordination of response efforts (FEMA, 2018).

Coordination Centers

Coordination centers serve as the nerve centers of disaster management, providing a centralized location for information sharing, strategic decision-making, and interagency collaboration. In the U.S., the National Response Coordination Center (NRCC) is the primary hub for coordinating disaster response at the federal level (FEMA, 2018). The NRCC provides a platform where representatives from different agencies, including FEMA, the military, law enforcement, and other stakeholders, can share information and make collaborative decisions. These centers streamline the decision-making process, reduce the risk of miscommunication, and enable the rapid deployment of resources.

Operational Plans

Operational plans are essential tools in disaster management, providing detailed guidance on how agencies should respond to specific disaster scenarios. In the U.S., FEMA and other relevant agencies develop comprehensive operational plans that outline the procedures for responding to various types of disasters. These plans ensure that responders are prepared to act quickly and efficiently, minimizing the impact of the disaster on affected communities. By regularly updating and rehearsing these operational plans, agencies can remain agile and adaptive in the face of changing disaster conditions (Haddow et al., 2017).

The strategies and practices outlined in this section represent significant steps toward enhancing the joint operational capabilities of the U.S. disaster management system. By emphasizing the importance of interagency cooperation, resource management, technology, and training, the U.S. has created a robust framework for disaster response that can be adapted to different types of emergencies. The coordination between military, police, fire departments, and other agencies is crucial for ensuring that disaster response efforts are swift, organized, and effective, ultimately saving lives and facilitating recovery..

3. CASE STUDIES AND PRACTICAL APPLICATIONS

The disaster management system in the United States is a dynamic and well-coordinated network designed to respond to a variety of emergency situations. Central to this system is the Federal Emergency Management Agency (FEMA), which plays a vital role in overseeing disaster preparedness, response, and recovery efforts. FEMA collaborates with a wide range of federal, state, local, and non-governmental organizations to ensure a comprehensive and effective disaster response (FEMA, 2018). This section delves deeper into FEMA's role, the collaboration between public agencies, the technologies and information management systems utilized, command-and-control relationships, successful case studies, and the legal frameworks that support disaster management efforts.

FEMA and National Emergency Management

FEMA's central role in the U.S. disaster management process cannot be overstated. As the primary federal agency responsible for coordinating disaster response, FEMA's efforts are designed to optimize collaboration between government agencies, the private sector, and community organizations. FEMA's strategic framework, encapsulated in the National Response Framework (NRF), provides the necessary foundation for crisis management, guiding the actions of various actors involved in disaster scenarios (FEMA, 2018). The NRF ensures that there is a clear understanding of the roles and responsibilities of each agency and organization, creating an environment of mutual cooperation during crises.

One of the key components of FEMA's coordination system is the National Incident Management System (NIMS), which establishes standardized procedures and a common language for communication among response teams. NIMS facilitates information sharing across different levels of government, enhancing the speed and efficiency of response efforts. By adhering to these standards, agencies are able to collaborate seamlessly, reducing the potential for confusion and errors. As a result, the U.S. disaster management system is better equipped to manage large-scale disasters effectively (FEMA, 2018).

Collaboration Between Public Agencies

Effective disaster management relies heavily on the collaboration between federal, state, and local agencies, as well as non-governmental organizations (NGOs) and the private sector. FEMA serves as the central coordinating body that unites these various actors under a common goal: the effective management of disasters.

By coordinating with local governments, private contractors, and NGOs, FEMA ensures that there are no gaps in the disaster response effort.

At the federal level, FEMA works closely with a range of agencies, including the U.S. Department of Defense (DoD), the Department of Health and Human Services (HHS), the Department of Transportation (DOT), and others. Each of these agencies brings unique resources, expertise, and capabilities to the disaster response process. The principles of Defense Support of Civil Authorities (DSCA) are particularly instrumental in improving the coordination between military and civilian agencies during disasters (Kapucu, 2008). These principles allow military resources, such as transportation, personnel, and logistical support, to be quickly deployed to assist civilian agencies when necessary. This collaboration improves the overall effectiveness of disaster response, ensuring that critical resources are available in a timely manner.

Furthermore, FEMA coordinates joint exercises and training programs designed to improve the preparedness of various agencies. Through these exercises, agencies can identify and address potential coordination challenges before a real disaster strikes. These programs enhance not only the technical skills of first responders but also the interpersonal relationships that are essential for effective teamwork. As a result, agencies are better equipped to work together under pressure, ensuring that disaster response efforts are carried out efficiently and without unnecessary delays (FEMA, 2018).

Technologies and Information Management Systems Used

The role of technology in disaster management is indispensable, particularly in the context of large-scale emergencies that require the coordination of numerous agencies. The U.S. has invested heavily in the development of advanced technological infrastructures that facilitate effective communication and information management during disaster response operations.

Communication Systems, such as satellite phones, radio networks, and digital messaging platforms, play a crucial role in ensuring that responders can communicate effectively even when traditional means of communication are compromised. Information Management Systems (IMS) also play an essential role in disaster management by enabling agencies to collect, analyze, and disseminate data. One of the most widely used systems is the Emergency Management Information System (EMIS), which integrates disaster data from multiple sources, helping decision-makers quickly assess the situation and allocate resources appropriately (Rodríguez et al., 2018).

Geographic Information Systems (GIS) are another critical tool in disaster management. GIS platforms allow responders to visualize disaster areas in real-time, making it easier to track the extent of damage, identify high-risk zones, and allocate resources effectively. GIS tools also enable the creation of detailed maps that guide rescue and recovery efforts in disaster-stricken areas. Additionally, Application Programming Interfaces (APIs) are often used to connect various systems and allow real-time data sharing across multiple platforms, providing decision-makers with the most up-to-date information on disaster conditions (Comfort, 2007).

The integration of these technologies allows response teams to make more informed decisions, coordinate efforts more efficiently, and minimize the impact of disasters. For example, during wildfires, GIS mapping can help emergency teams identify safe evacuation routes and prioritize firefighting efforts based on the severity of the fire and the location of at-risk communities.

Command-and-Control Relationships

In disaster management, effective command-and-control relationships are crucial for ensuring that agencies can coordinate their efforts in a structured and organized manner. The National Incident Management System (NIMS) outlines clear command-and-control structures, which are used to define the authority and responsibilities of various agencies involved in disaster response. By establishing these clear lines of authority, NIMS helps to avoid confusion, duplication of efforts, and delays in decision-making (FEMA, 2018).

The National Response Coordination Center (NRCC) and Emergency Operations Centers (EOCs) are the primary command-and-control hubs during a disaster. These centers serve as centralized locations where representatives from multiple agencies can come together to coordinate response efforts. The NRCC, in particular, is responsible for overseeing the federal response to disasters and ensuring that resources are allocated appropriately. During major incidents, these centers are essential for ensuring that all agencies are working toward the same goals and that critical decisions are made quickly and effectively.

The coordination facilitated by these centers ensures that the overall response to a disaster is efficient and well-organized. For example, in the aftermath of Hurricane Katrina, the establishment of the NRCC and other command centers helped FEMA coordinate the federal response and allocate resources more effectively. As a result, command-and-control relationships are key to maintaining an organized response, minimizing confusion, and achieving positive outcomes in disaster situations.

Successfully Managed Disasters

Two important case studies, Hurricane Katrina (2005) and the 2020 California Wildfires, illustrate the critical importance of inter-agency collaboration and coordination in disaster management.

Hurricane Katrina (2005)

Hurricane Katrina serves as a sobering reminder of the importance of effective collaboration between federal, state, and local agencies during a disaster. The aftermath of the hurricane revealed significant gaps in coordination, especially between FEMA and state and local governments. The slow and disjointed response to the disaster led to widespread criticism of the U.S. disaster management system and exposed the need for reforms.

In response to the failures of the Katrina response, FEMA implemented several reforms aimed at improving inter-agency collaboration. These reforms included increased investment in training and joint exercises, improved communication systems, and enhanced coordination mechanisms across all levels of government. FEMA also placed a greater emphasis on the role of local and state governments in disaster preparedness and response, ensuring that they are better equipped to manage crises in collaboration with federal agencies. These changes led to more effective disaster responses in subsequent events, demonstrating the value of learning from past mistakes and continuously improving disaster management processes.

The 2020 California Wildfires

The 2020 California Wildfires presented a different set of challenges, yet the collaboration between local fire departments, the National Guard, and federal agencies was critical in managing the response efforts. The wildfires, which were some of the largest and most destructive in California's history, required a well-coordinated and timely response to minimize loss of life and property damage. Thanks to strong inter-agency cooperation and the use of advanced technologies, such as GIS mapping and real-time communication systems, firefighters and first responders were able to effectively control the spread of the fires and evacuate thousands of residents from danger (Cal Fire, 2020).

The use of GIS mapping helped responders track the progression of the fires in real-time, while the integration of communication systems ensured that responders were always connected, even in remote or disaster-stricken areas. The joint efforts of federal, state, and local agencies, including the California National Guard, were instrumental in managing the wildfires and ensuring a quick recovery for the affected communities.

The 2024 California Wildfires

The 2024 California Wildfires presented yet another example of the importance of collaboration and coordination in the face of unprecedented challenges. These wildfires, which broke out in multiple regions across the state, posed a significant threat to communities, infrastructure, and natural resources. The fires were exacerbated by ongoing drought conditions and extreme heat waves, creating a complex disaster management situation.

During the 2024 wildfires, FEMA, the California Department of Forestry and Fire Protection (Cal Fire), and local fire departments worked hand-in-hand to implement a coordinated response. The integration of real-time data from satellite imagery and drones, combined with GIS technology, allowed emergency responders to track the fires' movements more effectively, deploy firefighting resources, and identify areas in need of immediate evacuation. Coordination between local agencies and federal resources ensured that no region was left without support, and the timely intervention of the National Guard helped stabilize vulnerable areas.

Furthermore, the 2024 fires demonstrated the continued importance of community engagement in disaster management. Local communities, supported by emergency alert systems and volunteer organizations, were able to contribute to the evacuation efforts, ensuring that residents in high-risk zones were safely relocated. This collaborative effort between local, state, and federal agencies, alongside the effective use of technology, proved to be vital in reducing the impact of the fires and facilitating the recovery process.

These case studies highlight the evolving nature of disaster management in the U.S. They emphasize the importance of collaboration, coordination, and technological innovation in improving disaster response efforts. While significant strides have been made since the challenges revealed by Hurricane Katrina, ongoing investment in training, technological infrastructure, and inter-agency cooperation remains critical to ensuring that future disasters are managed as effectively as possible. The experiences gained from these cases have provided valuable lessons for refining the disaster management system and preparing for an uncertain future.

4. LEGAL LEGISLATION AND REQUIREMENTS FOR DISASTER MANAGEMENT

In the United States, legal frameworks in disaster management play a significant role in regulating cooperation between public institutions and ensuring coordination. The Stafford Act and the Homeland Security Act aim to strengthen collaboration and coordination between federal, state, and local institutions (FEMA, 2018; DHS, 2002). These laws define authorities and responsibilities in the disaster management process, ensuring that necessary resources and support are effectively provided during emergencies. Additionally, the laws clearly define the roles and responsibilities of public institutions in disaster management, ensuring coordinated collaboration between them.

Regular review and updates of legal frameworks strengthen collaboration between public institutions and provide more effective coordination in disaster management. These frameworks must adapt to changing conditions and technological advancements, supporting high performance in disaster management. Moreover, the effective implementation of legislation is crucial for regulating disaster management processes and clearly defining the roles of all stakeholders.

In this regard, cooperation between public institutions in disaster management plays a critical role in enhancing efficiency in crisis management. The systems coordinated by FEMA, military and civilian cooperation, the technologies used, command and control relationships, successful case studies, and legal frameworks all support providing a rapid and effective response to disasters. Collaboration between public institutions supports compliance with national and international standards in disaster management and helps communities become more resilient to disasters. In this context, continuous improvement and cooperation between public institutions are essential to enhancing the effectiveness of disaster management.

Indeed, cooperation between public institutions in disaster management naturally plays a very critical role in increasing efficiency in crisis management. The systems coordinated by FEMA, military and civilian cooperation, technologies used, command and control relationships, successful case studies, and legal frameworks all contribute to providing a rapid and effective response to disasters. Collaboration between public institutions supports compliance with national and international standards in disaster management and helps communities become more resilient to disasters. In this context, continuous improvement and cooperation between public institutions are essential to increasing the effectiveness of disaster management.

The legal framework for disaster management in the United States regulates disaster response processes and defines the authorities and responsibilities of relevant institutions. This framework consists of various laws and regulations that ensure a coordinated and effective response to disasters. The Robert T. Stafford Disaster

Relief and Emergency Assistance Act (Stafford Act) defines the role and responsibilities of the federal government in the disaster management process in the U.S. The Stafford Act specifies how federal aid will be provided and coordinated in disaster situations. This law aims to strengthen cooperation and coordination between federal, state, and local institutions in disaster management (FEMA, 2018). The Stafford Act provides a basic framework for fostering strong cooperation between federal and state governments in disaster situations. It contains regulations that enable effective federal assistance and quick responses to disasters. Specifically, it requires close cooperation between the federal government and state and local governments to provide timely intervention and necessary resources in disaster-affected areas.

The Stafford Act clearly defines the federal government's authority and responsibilities in the disaster management process. The Federal Emergency Management Agency (FEMA) plays a central role in implementing this law and coordinates disaster response processes. FEMA is authorized to provide and coordinate federal aid in disaster areas and collaborates with state and local governments in this process. This law ensures the creation of a clear command and control structure during disaster situations. Under this law, the federal government is expected to take a leadership role in disaster response and work in coordination with state and local governments. The command and control relationship ensures the effective and swift execution of disaster response processes.

The Homeland Security Act of 2002 established the Department of Homeland Security (DHS) and ensured the coordination of disaster management processes through this department. The Homeland Security Act aims to strengthen cooperation between military and civilian institutions in national security and disaster management processes (DHS, 2002). The Homeland Security Act contains regulations to increase cooperation and coordination among all institutions involved in disaster management. This law ensures that the DHS takes a leadership role in disaster management and works with other federal, state, and local institutions. The DHS is responsible for strategic planning, resource allocation, and operational coordination in disaster management. The law also ensures the establishment of a clear and effective command and control structure in disaster management. The DHS plays a leadership role in disaster response processes and works in coordination with federal, state, and local institutions. The command and control structure is critical to ensuring a rapid and effective response in disaster situations.

The National Response Framework (NRF) clearly defines the roles and responsibilities of all relevant institutions in disaster situations, thereby improving coordination (FEMA, 2018). The NRF ensures coordination and collaboration between federal, state, and local institutions. The NRF provides a comprehensive framework to ensure the effective collaboration of all institutions involved in disaster management. This framework uses standardized procedures and terminology to ensure coordinated cooperation between institutions at federal, state, and local levels. Furthermore, joint drills and training programs enhance cooperation and preparedness between institutions.

The NRF clearly defines the authority and responsibilities of each institution in disaster management. It specifies which institutions will undertake which tasks during a disaster and ensures the necessary regulations to carry out these tasks effectively. The NRF clarifies the roles and responsibilities of federal, state, and local institutions. This framework ensures the establishment of a clear and effective command and control structure in disaster management. It designates institutions that take leadership roles in disaster response processes and ensures that these institutions work in coordination with other relevant institutions. The command and control structure is critical to ensuring a rapid and effective response during disaster situations.

The National Incident Management System (NIMS) facilitates collaboration between different institutions by using standardized procedures and common terminology in disaster management (FEMA, 2018). NIMS strengthens collaboration between institutions by providing a common language and standardized procedures in disaster management. The system offers a comprehensive framework to ensure the effective collaboration of all institutions involved in disaster management. This framework uses standardized procedures and terminology to ensure coordinated cooperation between federal, state, and local institutions. Additionally, joint drills and training programs enhance cooperation and preparedness between institutions.

NIMS clearly defines the authority and responsibilities of each institution in disaster management. This framework specifies which institutions will take on which tasks during a disaster and ensures the necessary regulations to perform these tasks effectively. NIMS clarifies the roles and responsibilities of federal, state, and local institutions. NIMS establishes a clear and effective command and control structure in disaster management. This framework designates the institutions that will take leadership roles in disaster response and ensures they work in coordination with other relevant institutions. The command and control structure is critical to ensuring a rapid and effective response during disaster situations.

Challenges faced in joint operations and command and control relationships can negatively impact effectiveness in disaster management. Lack of communication between institutions, deficiencies in information sharing, and operational differences between organizations are key factors that hinder coordination (Comfort, 2007). To overcome these challenges, it is essential to organize continuous training programs and exercises. Improvements at the legal and policy level can enhance coordination and effectiveness in disaster management. Legislation must be updated to adapt to changing needs and technological advancements. Additionally, developing policies that increase the participation and collaboration of all stakeholders in disaster management is essential (Kapucu, 2008). Organizing continuous training programs and exercises is necessary to ensure effective coordination in disaster management. These programs and exercises enhance cooperation between institutions and ensure an effective response during a crisis. In particular, practical learning of frameworks like NIMS and NRF plays a vital role in disaster management.

The legal framework for disaster management in the U.S. contains comprehensive regulations to ensure a coordinated and effective response to disasters. The Stafford Act, the Homeland Security Act, NRF, and NIMS aim to strengthen cooperation and coordination between federal, state, and local institutions in disaster management. Inter-institutional collaboration, clarification of authority and responsibilities, and effective command and control structures are crucial in disaster management. These legal frameworks and regulations support a rapid and effective response to disasters and help make communities more resilient to them. Thus, the continuous review, improvement, and technological alignment of legal frameworks and inter-institutional collaboration are critical for a successful disaster response.

5. COMMUNITY-BASED APPROACHES AND LOCAL PARTICIPATION IN DISASTER MANAGEMENT

In disaster management, the involvement of local communities is just as important as the contributions made by governmental agencies and large organizations. While central authorities and formal emergency response structures are vital, fostering resilience and preparedness at the community level is crucial for achieving more effective disaster management. This is particularly evident in the United States, where community-based disaster management programs, often facilitated by local governments and non-governmental organizations, significantly enhance the responsiveness and preparedness of local populations during and after disasters.

The Role of Community-Based Approaches in Disaster Management

Community-based approaches to disaster management emphasize the importance of local participation, education, and preparedness. These approaches aim to empower communities to not only be ready to handle the immediate effects of disasters but also to play a key role in the recovery process. By providing communities with the knowledge and resources needed to prepare for and respond to disasters, such programs build local resilience, reduce vulnerabilities, and foster a sense of collective responsibility among community members.

In the U.S., agencies such as the Federal Emergency Management Agency (FEMA) collaborate with local governments to design programs that encourage community participation and engagement. These programs may include preparedness workshops, simulation exercises, and public awareness campaigns that help communities understand the risks they face and the actions they can take to mitigate these risks. For example, the **Community Rating System (CRS)**, part of FEMA's National Flood Insurance Program (NFIP), incentivizes communities to take proactive measures to reduce flood risks, thereby increasing their resilience in the event of a disaster (FEMA, 2021).

In addition, local emergency planning committees (LEPCs) are instrumental in engaging community members in disaster preparedness and response efforts. These committees work to develop and implement local emergency response plans, involving a wide range of stakeholders, including local businesses, community organizations, schools, and healthcare providers. This broad level of involvement ensures that the needs of all community members, including vulnerable populations, are taken into account in disaster planning.

Building Resilience through Education and Training Programs

One of the most critical elements of community-based disaster management is education and training. Educational programs designed for community members can significantly improve a population's preparedness levels. By ensuring that people are informed about potential hazards and the specific actions they can take during a disaster, these programs reduce the impact of disasters and help communities recover more quickly.

Training programs, in particular, play a vital role in building local capacity for disaster response. Local community leaders, volunteers, and first responders are often the first to arrive at the scene of a disaster, and their ability to respond effectively is greatly enhanced by training. FEMA's **Citizen Corps Program**, for example, engages citizens in disaster response activities through initiatives such as Community Emergency Response Teams (CERTs), which train individuals in basic disaster response skills (FEMA, 2020).

Moreover, local disaster response drills and exercises are crucial in preparing communities for actual emergency situations. These exercises allow community members to practice their disaster response skills in realistic scenarios and identify any gaps or weaknesses in their preparedness efforts. For instance, the **Great ShakeOut Earthquake Drill** is an annual event that engages millions of people in the U.S. to practice earthquake preparedness, enhancing their understanding of the steps to take in the event of an earthquake (FEMA, 2021).

The Importance of Local Leadership and Grassroots Initiatives

Local leadership is an essential component of community-based disaster management. Community leaders, who often have deep knowledge of the local context, are critical in guiding their communities through disaster preparedness and recovery processes. These leaders act as intermediaries between governmental agencies and the community, ensuring that local concerns are communicated and addressed in disaster planning.

In addition to formal leadership roles, grassroots initiatives have proven to be highly effective in building disaster resilience at the community level. These initiatives often involve local organizations, religious institutions, and volunteers working together to support disaster preparedness and response. Grassroots efforts can be particularly valuable in marginalized or underserved communities, where trust in formal institutions may be limited. These initiatives foster a sense of solidarity, ensuring that communities work together to address common challenges and strengthen their overall resilience to disasters.

In the U.S., grassroots efforts such as the **American Red Cross'** local chapters and programs like **Team Rubicon**—which engages military veterans in disaster response activities—demonstrate the value of local leadership and volunteerism in disaster management (American Red Cross, 2021). These programs often provide direct assistance to communities during and after disasters, while also empowering individuals to take on leadership roles and become advocates for disaster preparedness within their own communities.

The Role of Technology and Social Media in Community-Based Disaster Management

As disaster management becomes increasingly digital, the role of technology and social media in community-based approaches cannot be overlooked. Social media platforms, such as Twitter, Facebook, and Instagram, play an increasingly prominent role in disseminating disaster-related information and coordinating community responses. These platforms enable local residents to share real-time information about ongoing events, such as evacuations, road closures, and shelter locations, which can be essential for effective disaster response.

In addition to social media, technology such as mobile applications and community notification systems have become integral parts of community-based disaster management. Applications like FEMA's **FEMA App** provide real-time alerts and safety information to users during emergencies, helping communities stay informed and make decisions quickly. Furthermore, systems like **Reverse 911** allow local authorities to quickly alert residents about imminent threats or evacuation orders, ensuring that communities can respond rapidly to changing circumstances.

Challenges to Community-Based Approaches

Despite their potential, community-based disaster management approaches face several challenges. One of the main barriers is the unequal distribution of resources and access to information within communities. Vulnerable populations, such as low-income individuals, the elderly, and those with disabilities, often face greater challenges in accessing disaster preparedness resources, making them more susceptible to the impacts of disasters.

Additionally, there can be challenges in fostering collaboration between governmental agencies and community organizations. While local agencies may have extensive knowledge of the needs of their communities, they may lack the resources or capacity to implement effective disaster management programs on their own. As such, it is crucial for local governments to establish strong partnerships with community organizations, non-profits, and other stakeholders to ensure that disaster management efforts are comprehensive and inclusive.

Lastly, fostering long-term community engagement can be difficult, as interest in disaster preparedness may diminish once the immediate threat of a disaster has passed. To overcome this, it is important for disaster management initiatives to include ongoing community engagement strategies that maintain interest in disaster preparedness year-round.

In this regard, community-based approaches to disaster management play a crucial role in building disaster resilience in the United States. By promoting local participation, education, and collaboration, these approaches empower communities to better prepare for, respond to, and recover from disasters. However, for these approaches to be truly effective, they must be supported by government policies, resources, and strong partnerships with local organizations. In the face of increasing disaster risks due to climate change and other factors, strengthening community-based disaster management efforts will be essential for building more resilient and sustainable communities.

6. FINDINGS AND EVALUATION

The development of joint operational capabilities between the military, police, fire departments, and other organizations, as well as the establishment of effective command and control relationships, plays a critical role in increasing the speed and effectiveness of disaster response. Evaluations based on the U.S. example highlight the importance of strategies, training and exercise programs, technological infrastructures, and legal regulations implemented to enhance coordination among these institutions. In this section, the findings are elaborated and evaluated under the following headings.

The Importance of Training and Exercise Programs

Training and exercise programs play a fundamental role in disaster management. These programs strengthen cooperation between various organizations and enhance preparedness for disaster scenarios by keeping personnel's knowledge and skills up to date. In the U.S., training organized by FEMA and other relevant agencies is designed to cover a range of disaster scenarios (Kapucu, 2008). Training ensures that personnel have the necessary knowledge and can adapt to new methods. Exercises test the practical applications of the training and help teams understand the challenges they may face during actual disaster situations. The continuous update of training and exercise programs enables personnel to quickly adapt to evolving threats and technological advancements.

Effective Use of Technological Infrastructure and Information Management Systems

The effective use of technological infrastructure and information management systems enhances the speed of information flow and coordination effectiveness. In the U.S., FEMA and other relevant agencies have developed various projects to ensure the effective use of technology (Comfort, 2007; Rodríguez et al., 2018). Advanced communication systems and information management systems provide accurate and timely information flow during disaster situations, thereby increasing the effectiveness of response processes. However, technological infrastructure needs to be continuously updated and improved. Technological developments have made significant progress in data collection and analysis, and existing systems must adapt to these developments. Additionally, the reliability and continuity of communication systems minimize disruptions in information sharing.

Updating the Legal Framework

Legal frameworks play a critical role in regulating disaster management processes and ensuring coordination. In the U.S., regulations like the Stafford Act and the Homeland Security Act aim to strengthen cooperation and coordination between federal, state, and local agencies (FEMA, 2018; DHS, 2002). However, these legal frameworks must be regularly reviewed and updated. Adapting legal regulations to changing conditions and technological advancements ensures more effective coordination in disaster management. Regular updates to legislation clearly define the roles and responsibilities of all stakeholders and encourage cooperation.

Effective Use of Coordination Centers

Coordination centers play a crucial role in ensuring communication and cooperation between different institutions during disaster situations (Comfort, 2007). In the U.S., structures like the National Emergency Coordination Center (NRCC) fulfill these functions and play a central role in the disaster management process (FEMA, 2018). The effective use of coordination centers enhances the efficiency of response processes by facilitating information sharing and joint decision-making. The effectiveness of these centers should be supported by training and exercises, and practical exercises based on real-life scenarios test the operation of the coordination centers and identify any shortcomings.

Policies to Enhance Stakeholder Participation

Developing policies that increase stakeholder participation and cooperation in disaster management enhances overall response capacity and effectiveness (Kapucu, 2008). In the process of designing and implementing policy proposals, it is important to consider the views of all stakeholders. Increasing stakeholder participation ensures that various organizations and communities are actively involved in disaster management processes and contributes to the development of a more comprehensive response strategy. Participatory policies ensure strong cooperation and coordination in disaster management.

General Evaluation

The U.S. example demonstrates the significant importance of developing joint operational capabilities and ensuring effective command and control relationships in increasing the speed and effectiveness of disaster response. The updating of training and exercise programs, the effective use of technological infrastructure and information management systems, the updating of legal frameworks, the effective use of coordination centers, and the development of policies to enhance stakeholder participation are all elements that directly impact overall success in disaster management. Each of these elements contributes to providing faster and more effective responses to disasters, helping communities become more resilient to disasters. In this context, a continuous improvement approach and openness to innovation are essential to increasing the effectiveness of disaster management.

7. CONCLUSION

This study emphasizes the critical importance of developing joint operational capabilities between the military, police, fire departments, and other organizations, as well as ensuring effective command and control relationships in disaster management. The evaluations based on the U.S. example detail the strategies, training and exercise programs, technological infrastructures, and legal regulations implemented to enhance coordination among these institutions.

The disaster management system in the U.S. has developed various mechanisms to ensure coordinated and effective cooperation between different institutions. The Robert T. Stafford Disaster Relief and Emergency Assistance Act (Stafford Act) defines the federal government's authority to respond to disasters and FEMA's role, ensuring strong collaboration and coordination among federal, state, and local agencies. This legal framework enables a rapid and effective response to disasters, thereby enhancing the resilience of communities. Additionally, the 2002 Homeland Security Act established the Department of Homeland Security (DHS) and aims to ensure coordination in disaster management. This law strengthens the cooperation between military and civilian institutions in national security and disaster management processes.

The development of joint operational capabilities enables quick and effective intervention during disasters. In this regard, the importance of training and exercise programs is paramount. Training programs keep the knowledge and skills of personnel from various institutions up to date, ensuring that they are better prepared for disaster situations. Exercises, on the other hand, reinforce inter-agency cooperation and coordination, ensuring an effective response during actual disaster events. Regularly conducted exercises in the U.S. improve the ability of different agencies to work together and provide a more effective response to emergencies.

Technological infrastructures and information management systems play a critical role in disaster management. Advanced technological infrastructures accelerate information flow during disasters and enhance coordination. For example, geographic information systems (GIS) are used for mapping disaster areas and effectively distributing resources. Additionally, information management systems facilitate communication between different agencies, improving information sharing and coordination. These technological tools expedite decision-making processes in disaster management and enable more effective interventions.

Legal regulations form the foundation for organizing and executing disaster management processes in a coordinated manner. In the U.S., disaster management laws and regulations provide the necessary framework to increase cooperation and coordination between federal, state, and local agencies. For instance, the National Response Framework (NRF) clearly defines the roles and responsibilities of all relevant institutions during disaster situations. This framework provides a comprehensive structure to ensure that all institutions involved in disaster management can collaborate effectively. The National Incident Management System (NIMS), on the other hand, facilitates cooperation between different agencies by using standardized procedures and common terminology in disaster management.

Ensuring effective command and control relationships in disaster management is vital for swift and effective interventions. Command and control structures designate the institutions that will take on leadership roles in disaster response processes and ensure that these institutions work in coordination with other relevant agencies. This structure is critical for ensuring a quick and effective response to disasters. The U.S. disaster management system organizes disaster response processes and enhances coordination by establishing a clear and effective command and control structure.

In conclusion, this study highlights the importance of developing joint operational capabilities and ensuring effective command and control relationships between the military, police, fire departments, and other organizations in disaster management, based on the U.S. example. The critical roles of training and exercise programs, technological infrastructures, and legal regulations in this process have been thoroughly examined. All these elements contribute to enhancing effectiveness and coordination in disaster management, ultimately making communities more resilient to disasters.

8. RECOMMENDATIONS FOR FUTURE RESEARCH

This study has comprehensively evaluated the importance of developing joint operational capabilities and ensuring effective command and control relationships in disaster management. Policy recommendations and implementation strategies are critical for overcoming the challenges encountered in disaster management and ensuring effective crisis management. Future studies can enhance the success of disaster management by further developing these recommendations and monitoring the implementation processes.

In this context, subsequent research could focus on the continuous updating and expansion of training and exercise programs; ensuring the effective use of technological infrastructure and information management systems, and testing these systems in real disaster scenarios; updating legal frameworks to align with changing needs and technological advancements; enhancing the effectiveness of coordination centers and addressing deficiencies; and developing and implementing policies to increase stakeholder participation.

Additionally, research focusing on issues such as climate change, community involvement, and resilience will contribute to making disaster management processes more effective and sustainable.

REFERENCES

Alexander, D. (2002). *Principles of emergency planning and management*. Terra Publishing.

American Red Cross. (2021). *Team Rubicon*. https://www.redcross.org

Cal Fire. (2020). *2020 Fire Season*. California Department of Forestry and Fire Protection.

Comfort, L. K. (2007). Crisis management in hindsight: Cognition, communication, coordination, and control. *Public Administration Review*, *67*(s1), 189–197. DOI: 10.1111/j.1540-6210.2007.00827.x

Comfort, L. K., Boin, A., & Demchak, C. C. (2010). *Designing resilience: Preparing for extreme events*. University of Pittsburgh Press. DOI: 10.2307/j.ctt5hjq0c

Coppola, D. P. (2015). *Introduction to international disaster management*. Butterworth-Heinemann.

DHS. (2002). *Homeland Security Act of 2002*. U.S. Department of Homeland Security.

Dynes, R. R. (2006). Social capital: Dealing with community emergencies. *Homeland Security Affairs*, *2*(2).

Federal Emergency Management Agency (FEMA). (2018). *Federal Emergency Management Agency strategic plan 2018-2022*. Federal Emergency Management Agency.

Federal Emergency Management Agency (FEMA). (2020). *Citizen Corps Program*. https://www.fema.gov

Federal Emergency Management Agency (FEMA). (2021). *Community Rating System (CRS)*. https://www.fema.gov

Federal Emergency Management Agency (FEMA). (2021). *Great ShakeOut Earthquake Drill*. https://www.shakeout.org

Haddow, G. D., Bullock, J. A., & Coppola, D. P. (2017). *Introduction to emergency management*. Butterworth-Heinemann.

IPCC. (2014). *Impacts, adaptation, and vulnerability. Part A: Global and sectoral aspects. Contribution of Working Group II to the Fifth Assessment Report of the Intergovernmental Panel on Climate Change*. Intergovernmental Panel on Climate Change. *Climatic Change*, 2014.

Kapucu, N. (2008). Collaborative emergency management: Better community organising, better public preparedness and response. *Disasters*, *32*(2), 239–262. DOI: 10.1111/j.1467-7717.2008.01037.x PMID: 18380853

Kapucu, N., & Garayev, V. (2011). Collaborative decision-making in emergency and disaster management. *International Journal of Public Administration*, *34*(6), 366–375. DOI: 10.1080/01900692.2011.561477

Norris, F. H., Stevens, S. P., Pfefferbaum, B., Wyche, K. F., & Pfefferbaum, R. L. (2008). Community resilience as a metaphor, theory, set of capacities, and strategy for disaster readiness. *American Journal of Community Psychology*, *41*(1-2), 127–150. DOI: 10.1007/s10464-007-9156-6 PMID: 18157631

Paton, D., & Johnston, D. (2001). Disasters and communities: Vulnerability, resilience, and preparedness. *Disaster Prevention and Management*, *10*(4), 270–277. DOI: 10.1108/EUM0000000005930

Rodríguez, H., Donner, W., & Trainor, J. E. (2018). *Handbook of disaster research.* Springer. DOI: 10.1007/978-3-319-63254-4

The White House. (2006). *The Federal Response to Hurricane Katrina: Lessons Learned.* U.S. Government Printing Office.

Tierney, K. (2014). *The social roots of risk: Producing disasters, promoting resilience.* Stanford University Press.

UNISDR. (2015). *Sendai Framework for Disaster Risk Reduction 2015-2030.* United Nations Office for Disaster Risk Reduction.

White, J. P. (2012). The role of the military in disaster response in the U.S. *Journal of Homeland Security and Emergency Management*, *9*(1), 153–163. DOI: 10.1515/1547-7355.2119

KEY TERMS AND DEFINITIONS

Term 1 – Disaster Management: The comprehensive process of preparing for, responding to, recovering from, and mitigating the impacts of natural or man-made disasters through coordinated efforts among multiple agencies and stakeholders.

Term 2 – Joint Operational Capabilities: The ability of various emergency response entities such as the military, police, and fire departments to operate in a unified, efficient, and coordinated manner during disaster scenarios.

Term 3 – Command and Control: A structured system of authority and communication that defines leadership roles, resource allocation, and decision-making responsibilities among institutions during disaster response operations.

Term 4 – Federal Emergency Management Agency (FEMA): The principal U.S. government agency responsible for coordinating federal disaster response and recovery efforts, providing support to state and local governments.

Term 5 – National Incident Management System (NIMS): A standardized framework developed by FEMA for coordinating emergency management across different governmental and non-governmental organizations using common terminology, protocols, and procedures.

Term 6 – Technological Infrastructure: The systems and tools (e.g., GIS, communication platforms, information management systems) that support real-time data sharing, coordination, and operational efficiency in disaster management.

Term 7 – Interagency Coordination: The collaboration and integration of multiple agencies and sectors (public, private, and non-profit) to ensure a unified and effective response to disasters through shared resources, planning, and operations.

Term 8 – Legal Framework: The set of laws, such as the Stafford Act and Homeland Security Act, that define institutional responsibilities, facilitate federal assistance, and guide operational procedures in disaster response.

Chapter 4
Disaster Education and Awareness Programs

Y. Ebenezer
VEL TECH University, India

Kasagani Durga
NBM Law College, India

N. B. Chandrakala
Andhra University, India

G. Shashank
https://orcid.org/0009-0006-8815-980X
Chandigarh School of Business, Chandigarh Group of Colleges, Jhanjeri, India

Mohit Sharma
https://orcid.org/0009-0007-2280-8077
Maharshi Dayanand University, Rohtak, India

P. Selvakumar
https://orcid.org/0000-0002-3650-4548
Department of Science and Humanities, Nehru Institute of Technology, Coimbatore, India

T. C. Manjunath
https://orcid.org/0000-0003-2545-9160
Rajarajeswari College of Engineering, India

ABSTRACT

Disaster education plays a pivotal role in building resilient communities by equipping individuals, organizations, and local governments with the knowledge and skills necessary to prepare for, respond to, and recover from various types of disasters. In an era marked by increasing frequency and intensity of natural and man-made disasters—ranging from hurricanes and earthquakes to floods and wildfires—the importance of disaster education cannot be overstated. Effective disaster education programs help mitigate risks, reduce vulnerabilities, and foster a culture of preparedness, ultimately enhancing community resilience. At its core, disaster education aims to raise awareness about the potential hazards and risks that communities face. It provides critical information on the nature of different types of disasters,

DOI: 10.4018/979-8-3693-9745-9.ch004

their impacts, and the steps that individuals and communities can take to protect themselves. This foundational knowledge is essential for developing a proactive approach to disaster preparedness

INTRODUCTION TO DISASTER EDUCATION: BUILDING RESILIENT COMMUNITIES

Disaster education plays a pivotal role in building resilient communities by equipping individuals, organizations, and local governments with the knowledge and skills necessary to prepare for, respond to, and recover from various types of disasters. In an era marked by increasing frequency and intensity of natural and man-made disasters—ranging from hurricanes and earthquakes to floods and wildfires—the importance of disaster education cannot be overstated. Effective disaster education programs help mitigate risks, reduce vulnerabilities, and foster a culture of preparedness, ultimately enhancing community resilience. At its core, disaster education aims to raise awareness about the potential hazards and risks that communities face. It provides critical information on the nature of different types of disasters, their impacts, and the steps that individuals and communities can take to protect themselves. This foundational knowledge is essential for developing a proactive approach to disaster preparedness. By understanding the specific risks associated with their region—such as seismic activity in earthquake-prone areas or flooding in low-lying regions—community members can tailor their preparedness efforts to address the most relevant threats. A key component of disaster education is the development of personal preparedness plans. Effective education programs emphasize the importance of having a comprehensive emergency plan that includes evacuation routes, communication strategies, and supplies necessary for surviving the initial impact of a disaster. Educational initiatives often involve practical exercises and simulations that help individuals and families create and practice their plans. For instance, community drills that simulate earthquake or hurricane scenarios allow participants to experience the process of evacuation and emergency response in a controlled environment, enhancing their readiness and confidence. Risk reduction strategies are another critical aspect of disaster education. These strategies focus on identifying and mitigating vulnerabilities to reduce the potential impact of disasters. Education programs provide guidance on implementing structural and non-structural measures to protect homes and businesses.

UNDERSTANDING DISASTER RISK REDUCTION (DRR) AND MANAGEMENT

Understanding Disaster Risk Reduction (DRR) and management is crucial for enhancing community resilience and mitigating the impacts of natural and man-made hazards. DRR encompasses strategies and measures aimed at minimizing the vulnerability and exposure of communities to disasters, thereby reducing their overall risk and enhancing their capacity to respond effectively. It involves a multi-faceted approach that integrates prevention, preparedness, response, and recovery efforts to build resilience and ensure sustainable development. Disaster Risk Reduction (DRR) focuses on addressing the underlying causes of vulnerability and mitigating risks before a disaster occurs (Fu et al, 2024). The core principles of DRR include understanding hazards, assessing vulnerabilities, and implementing measures to reduce both. Hazards refer to natural or man-made events with the potential to cause harm, such as earthquakes, floods, hurricanes, or industrial accidents. Vulnerabilities are factors that increase the susceptibility of individuals or communities to the adverse impacts of these hazards, such as poverty, inadequate infrastructure, or lack of access to resources. By addressing these vulnerabilities and implementing risk reduction measures, DRR aims to decrease the potential impact of hazards and protect lives, property, and livelihoods. A fundamental component of DRR is hazard assessment and risk mapping. Identifying and understanding the types of hazards that a community is exposed to, along with their frequency and intensity, is essential for effective risk management.

Risk reduction measures include both structural and non-structural interventions. Structural measures involve physical changes to the built environment to reduce vulnerability, such as constructing levees and floodwalls, retrofitting buildings to withstand earthquakes, and improving drainage systems. Non-structural measures focus on policy, education, and community engagement to enhance resilience. These measures include implementing building codes and land use regulations, promoting sustainable practices, and fostering community involvement in risk reduction efforts. For example, community-driven initiatives, such as neighborhood watch programs and local emergency response teams, contribute to a culture of preparedness and collective resilience. Disaster management extends beyond DRR and encompasses the comprehensive approach to handling disasters from response to recovery (Dwiningrum et al, 2022). It involves coordinating and executing activities to manage the immediate impacts of a disaster and facilitate recovery. Integration with sustainable development is a crucial aspect of DRR and disaster management. Aligning disaster risk reduction with broader development goals ensures that resilience-building efforts contribute to sustainable and equitable growth. This integration involves incorporating risk reduction principles into development planning, resource management, and environmental protection. For example, integrating flood risk assessments into urban planning helps prevent the construction of vulnerable infrastructure in flood-prone areas, promoting sustainable and resilient communities. Monitoring and evaluation are vital for assessing the effectiveness of DRR and disaster management efforts. Regular monitoring helps track progress, identify challenges, and adjust strategies as needed (Ramakrishnan et al, 2022). Evaluation provides insights into the outcomes and impacts of interventions, informing future risk reduction and management activities. Engaging communities in monitoring and evaluation processes enhances accountability and ensures that efforts are aligned with local needs and priorities.

DEVELOPING EFFECTIVE DISASTER EDUCATION PROGRAMS: PRINCIPLES AND FRAMEWORKS

Developing effective disaster education programs is crucial for enhancing community resilience and preparedness in the face of natural and man-made hazards. These programs aim to equip individuals and communities with the knowledge, skills, and resources necessary to anticipate, respond to, and recover from disasters. To achieve these goals, disaster education programs must be grounded in sound principles and frameworks that ensure they are comprehensive, accessible, and impactful. Principles of effective disaster education include relevance, inclusivity, and sustainability. Relevance ensures that educational content is tailored to the specific risks and needs of the community. Programs should address the unique hazards that communities face, such as earthquakes, floods, or hurricanes, and provide practical information that can be applied in real-life situations (Gill J C., et al, 2021). For example, in a region prone to flooding, the program should focus on flood preparedness, evacuation routes, and safety measures relevant to that context. Inclusivity is essential to ensure that disaster education programs reach all segments of the population, including vulnerable and marginalized groups. Programs should be designed to accommodate diverse needs, such as language barriers, disabilities, and varying levels of prior knowledge. This can involve providing materials in multiple languages, using accessible formats for people with disabilities, and ensuring that educational activities are culturally sensitive and appropriate. This assessment can be conducted through surveys, interviews, focus groups, and analysis of historical data. The results help tailor the program content and ensure that it addresses the most pressing risks and gaps in knowledge. Content development involves creating educational materials and resources that are accurate, engaging, and actionable (Cabello et al, 2021).

Evaluation is an ongoing process that assesses the effectiveness and impact of the disaster education program. Evaluation involves measuring the program's success in achieving its objectives, such as increasing knowledge, improving preparedness behaviors, and enhancing community resilience. Methods for evaluation include surveys, feedback forms, and assessments of behavioral changes and emergency preparedness levels. Regular evaluation helps identify strengths and areas for improvement, allowing for continuous refinement of the program. Community involvement is a critical element in the development and implementation of disaster education programs. Engaging community members, local organizations, and stakeholders ensures that the program is grounded in local realities and priorities. Community involvement can also foster a sense of ownership and commitment to the program, enhancing its effectiveness. Involving local experts and leaders in the design and delivery of the program can also improve its credibility and relevance (Sreenivasulu et al, 2022). Integration with existing systems is another important consideration. Disaster education programs should complement and support existing emergency management plans, policies, and initiatives.

COMMUNITY-BASED DISASTER EDUCATION: ENGAGING VULNERABLE POPULATIONS

Community-based disaster education is a critical approach for effectively engaging vulnerable populations and enhancing overall resilience against disasters. Vulnerable populations, including the elderly, children, disabled individuals, low-income families, and minority groups, often face heightened risks during disasters due to physical, social, and economic factors. To address their unique needs and ensure equitable access to disaster preparedness and response resources, community-based disaster education programs must be designed with specific strategies that consider these groups' challenges and strengths. Understanding vulnerability is the first step in developing effective community-based disaster education. Vulnerability arises from various factors such as physical limitations, social isolation, language barriers, and economic instability. For instance, elderly individuals may have mobility issues that make evacuation difficult, while those with disabilities might require specialized assistance during emergencies (Naheed, 2021). Customized Educational Content should address the specific needs and circumstances of vulnerable populations. For example, educational materials for elderly individuals might focus on mobility aids and specialized evacuation plans, while content for families with young children might include child-friendly safety procedures and emergency preparedness activities. Educational programs should also be accessible to people with disabilities, offering materials in various formats such as braille, large print, and audio. Language services, including translation and interpretation, are crucial for reaching non-English speaking communities (Amri et al, 2022). Tailoring content to address these specific needs helps ensure that information is actionable and relevant.

Building Partnerships with local organizations and service providers is essential for implementing community-based disaster education programs. Partnering with organizations that already serve vulnerable populations, such as senior centers, disability advocacy groups, and community health organizations, can enhance outreach and effectiveness. These organizations often have established relationships and trust with their clients, which can facilitate better engagement and participation in educational programs. Additionally, partnerships can provide valuable resources and support, such as venues for educational sessions, translation services, and logistical assistance. Training and Empowerment of local leaders and volunteers are critical for sustaining community-based disaster education efforts. Local leaders and community volunteers play a vital role in disseminating information, organizing events, and providing support during disasters. Training these individuals in disaster preparedness and response equips them with the knowledge and skills to effectively assist their peers. Empowering local leaders to take an active role in disaster education not only enhances program reach but also fosters a sense of community ownership and resilience (Zuama, 2024). By equipping individuals and communities with the knowledge, skills, and resources needed to prepare for and respond to disasters, community-based education fosters a culture of resilience (Mahadev Madgule et al, 2023). Resilient communities are better able to withstand the impacts of disasters, recover more quickly, and support each other during times of crisis.

SCHOOL-BASED DISASTER EDUCATION: INTEGRATING CURRICULUM AND DRILL EXERCISES

School-based disaster education is a crucial strategy for fostering resilience and preparedness from an early age by integrating disaster education into the school curriculum and conducting regular drill exercises. This approach not only equips students with essential knowledge and skills but also ensures that they can act effectively in emergencies, while also engaging families and communities in preparedness efforts. The successful integration of disaster education into schools involves a thoughtful combination of curriculum development, drill exercises, and community involvement to create a comprehensive and practical learning experience. Integrating disaster education into the school curriculum involves incorporating relevant topics into various subject areas to ensure that students understand the importance of preparedness and risk reduction. The curriculum should address fundamental concepts such as understanding different types of hazards (earthquakes, floods, fires, etc.), recognizing signs of danger, and learning basic safety measures.

For example, science lessons can include discussions on natural disasters and their causes, while math classes might incorporate exercises related to calculating safe evacuation routes or resource management. By weaving disaster education into existing subjects, schools can make learning about emergencies a natural part of students' education without requiring additional class time. Developmentally appropriate content is essential for engaging students at different grade levels (Dufty N., 2020). Drills should be conducted frequently and cover various scenarios, such as fire evacuations, earthquake responses, and lockdown procedures. These exercises help students practice and internalize emergency procedures, ensuring that they are familiar with what to do in real situations (Libayao et al, 2024). Drills should be well-planned and include clear instructions, realistic simulations, and opportunities for students to ask questions and receive feedback.

Training teachers and staff is a crucial component of effective school-based disaster education. Educators should be well-versed in disaster preparedness and response protocols to effectively teach and guide students during drills and actual emergencies. Professional development programs and workshops can provide teachers and staff with the knowledge and skills needed to handle various disaster scenarios, as well as strategies for integrating disaster education into their teaching. Ongoing training ensures that educators are up-to-date with the latest practices and can effectively implement disaster education in their classrooms. Collaboration with emergency services and community organizations enhances the effectiveness of school-based disaster education. Partnering with local fire departments, police, and emergency management agencies can provide students with real-world insights and practical demonstrations (Kaku, 2019). These partnerships can also facilitate the development of realistic drill scenarios and ensure that the school's emergency plans align with community protocols. Community organizations, such as non-profits focused on disaster relief or public health, can offer additional resources and support, such as educational materials and expert speakers. Involving parents and guardians in disaster education efforts is crucial for extending the impact of school-based programs beyond the classroom. Schools can host informational sessions, workshops, and family preparedness events to engage parents and guardians in disaster preparedness. Providing resources and guidance on how to create family emergency plans and assemble disaster kits can help reinforce the lessons learned at school. Effective communication with families ensures that students receive consistent messages about preparedness and encourages families to be proactive in their own disaster planning. Evaluation and continuous improvement are essential for maintaining the effectiveness of school-based disaster education programs. Regular assessment of both the curriculum and drill exercises helps identify areas for improvement and ensures that the program remains relevant and up-to-date. Feedback from students, parents, and staff can provide valuable insights into the strengths and weaknesses of the program. Schools should use this feedback to make necessary adjustments, incorporate new best practices, and address any emerging issues or challenges. Creating a culture of preparedness within schools reinforces the importance of disaster education and encourages ongoing vigilance. By fostering an environment where preparedness is valued and prioritized, schools can help students develop a lifelong commitment to safety and resilience (Selvakumar et al, 2024).

PUBLIC AWARENESS CAMPAIGNS: STRATEGIES FOR BEHAVIORAL CHANGE

Public awareness campaigns are pivotal in driving behavioral change and fostering a culture of preparedness and safety within communities. Effective campaigns employ a range of strategies designed to inform, motivate, and empower individuals to adopt practices that reduce risk and enhance resilience. Tailoring messages and strategies to address these factors ensures that the campaign resonates with the target audience and effectively communicates the desired behaviors. Clear and Concise Messaging is crucial for ensuring that campaign communications are easily understood and actionable. Messages should be straightforward, focusing on key behaviors and actions that individuals can take to improve their safety (Yusuf R., et al, 2022). For example, a campaign promoting earthquake preparedness might highlight the importance of securing heavy objects, creating an emergency plan, and practicing earthquake drills. Avoiding jargon and using simple language helps make the information accessible to a broad audience. Visual aids, infographics, and videos can also enhance message clarity and retention. Emotional Appeal and Storytelling are powerful tools in influencing behavioral change. Campaigns that incorporate personal stories, testimonials, and emotional appeals can create a strong connection with the audience and motivate action. For instance, sharing stories of individuals who have successfully navigated emergencies or experienced the consequences of not being prepared can underscore the importance of preparedness and encourage others to take similar steps. Emotional engagement can make messages more memorable and compelling, driving individuals to adopt new behaviors. Multi-Channel Outreach ensures that the campaign reaches a wide audience through various platforms and mediums. Utilizing a mix of traditional media (television, radio, print) and digital channels (social media, websites, email) maximizes the campaign's visibility and accessibility (Piltch-Loeb et al, 2022). Different channels can be used to reinforce messages and engage audiences in different ways. For example, social media campaigns can facilitate interactive elements such as quizzes and challenges, while traditional media can provide in-depth information and expert interviews.

Behavioral Science Principles can inform the design and implementation of public awareness campaigns. Understanding psychological factors that influence behavior, such as social norms, perceived risk, and incentives, can help shape more effective interventions. For example, using social proof—showing that others in the community are taking action—can encourage individuals to follow suit. Additionally, providing tangible incentives or rewards for adopting desired behaviors can motivate participation and adherence. Applying these principles helps tailor the campaign to address common barriers and leverage factors that drive behavior change. Engagement and Interaction through participatory methods can enhance the effectiveness of public awareness campaigns. Interactive elements such as workshops, community events, and simulations allow individuals to actively engage with the campaign's messages and practice new skills. For example, hosting disaster preparedness fairs where participants can assemble emergency kits, learn first aid, and participate in simulations provides hands-on experience and reinforces key messages (Muñoz et al, 2020). Long-term initiatives, such as integrating preparedness education into school curricula or establishing community preparedness programs, can help reinforce and perpetuate the desired behaviors. Crisis Response Integration ensures that public awareness campaigns are aligned with broader emergency management efforts (Vijayakumar et al, 2024). Campaigns should complement existing disaster response plans and resources, providing a cohesive approach to preparedness and risk reduction.

SOCIAL MEDIA AND DISASTER EDUCATION: LEVERAGING DIGITAL PLATFORMS

Social media has become a powerful tool in disaster education, offering unique opportunities to disseminate information, engage audiences, and foster community resilience. Leveraging digital platforms such as Facebook, Twitter, Instagram, and TikTok can significantly enhance disaster preparedness and response efforts by providing real-time updates, educational content, and interactive engagement. The effectiveness of social media in disaster education hinges on understanding its potential, crafting tailored strategies, and addressing both opportunities and challenges associated with these platforms. Real-Time Information Dissemination is one of the primary advantages of using social media for disaster education (Pickering et al, 2021). Platforms like Twitter and Facebook enable the rapid distribution of urgent updates and alerts, allowing communities to receive critical information about impending hazards, emergency procedures, and safety measures. Community Building and Support is facilitated through social media networks, which can strengthen community ties and enhance collective resilience. Social media groups and pages dedicated to disaster preparedness can serve as platforms for sharing resources, organizing local events, and coordinating volunteer efforts (Afifi et al, 2020). Targeted Outreach and Customization can enhance the impact of social media-based disaster education by tailoring messages to specific audiences. Social media platforms offer advanced targeting options that allow campaigns to reach particular demographics, geographic areas, or interest groups. For example, targeted ads or posts can be used to reach communities in high-risk areas with specific preparedness tips relevant to their local hazards.

Customizing content to address the unique needs and preferences of different groups ensures that the information is relevant and actionable. Integration with Other Communication Channels ensures that social media efforts complement broader disaster education and emergency management strategies. Coordinating social media campaigns with traditional media, community events, and educational programs helps reinforce messages and reach a wider audience. For instance, a social media campaign promoting a community preparedness workshop can be supported by local radio ads and flyers distributed through schools and community centers. This integrated approach maximizes the effectiveness of disaster education efforts and ensures consistent messaging across different platforms. Analytics and Feedback provide valuable insights into the effectiveness of social media campaigns and help inform future strategies. Social media platforms offer a range of analytics tools that can track engagement, reach, and user behavior. Analyzing this data helps assess the impact of different content types, identify successful strategies, and pinpoint areas for improvement. Gathering feedback from users through surveys, comments, and interactions also provides direct insights into their needs and preferences, allowing for continuous refinement of the campaign (Selvakumar et al, 2024). Challenges and Ethical Considerations must be addressed to ensure responsible and effective use of social media for disaster education. Issues such as privacy concerns, digital divide, and the risk of information overload should be carefully managed.

CULTURAL COMPETENCE IN DISASTER EDUCATION: ADDRESSING DIVERSE NEEDS

Cultural competence in disaster education is critical for ensuring that preparedness and response strategies are inclusive and effective across diverse populations. It involves understanding, respecting, and addressing the unique cultural, linguistic, and social needs of different communities to enhance their ability to prepare for, respond to, and recover from disasters. By incorporating cultural competence into disaster education, agencies and organizations can create more equitable and impactful programs that resonate with and effectively support all members of a community. Understanding Cultural Diversity is the first step in achieving cultural competence in disaster education. Communities are composed of individuals from various ethnic, racial, and cultural backgrounds, each with distinct values, beliefs, and practices that influence their perceptions and responses to emergencies. For example, some cultures may have specific traditions or practices related to health and safety that must be acknowledged and respected in disaster education efforts. Conducting thorough research and engaging with community representatives helps identify these cultural factors and ensures that disaster education programs are relevant and respectful (Seddighi et al, 2022). Inclusive Curriculum Development is essential for addressing the diverse needs of different cultural groups. Educational materials and programs should be designed to reflect cultural values and practices, and avoid stereotypes or assumptions. For instance, disaster preparedness guides might include information on culturally specific practices for food storage or traditional methods for securing homes. Involving community members in the development process can provide valuable insights and ensure that content is both culturally appropriate and effective in conveying important safety information. Understanding and incorporating these practices into disaster education helps ensure that programs are not only respectful but also more likely to be embraced by the community (Pranata S., et al, 2021). Addressing Socioeconomic Factors is also important for ensuring that disaster education is effective across diverse populations. Socioeconomic status can influence access to resources, information, and preparedness measures.

Training Educators and Responders in Cultural Competence is crucial for ensuring that those delivering disaster education are equipped to interact respectfully and effectively with diverse populations. Training programs should focus on developing skills related to cultural awareness, sensitivity, and effective communication. Educators and responders should be knowledgeable about the cultural backgrounds of the communities they serve and be prepared to address any unique needs or concerns that may arise. Ongoing training and professional development in cultural competence help maintain a high standard of respect and effectiveness in disaster education. Evaluating and Adapting Programs to ensure cultural competence involves regularly assessing the impact of disaster education efforts on diverse populations and making necessary adjustments. Feedback from community members, surveys, and focus groups can provide valuable insights into how well programs are meeting the needs of different cultural groups and identify areas for improvement. Adapting programs based on this feedback ensures that they remain relevant and effective in addressing the evolving needs of the community. Promoting Equity and Inclusion in disaster education is fundamental for creating a more resilient and prepared community (Fletcher-Miles et al, 2020). By addressing the diverse needs of all cultural groups, disaster education programs help ensure that no one is left behind and that everyone has the knowledge and resources needed to respond effectively to emergencies.

DISASTER EDUCATION FOR SPECIAL NEEDS POPULATIONS: INCLUSIVE STRATEGIES

Disaster education for special needs populations is a vital aspect of ensuring that all individuals, regardless of their abilities or circumstances, are adequately prepared for emergencies. Inclusive strategies must address the diverse requirements of individuals with disabilities, chronic health conditions, and other special needs to create effective and equitable disaster preparedness, response, and recovery plans. By implementing tailored approaches that consider the unique challenges faced by these populations, disaster education programs can significantly enhance safety and resilience for everyone. Understanding the Needs of Special Populations is crucial for developing inclusive disaster education strategies. Special needs populations encompass a wide range of individuals, including those with physical disabilities, sensory impairments, cognitive or developmental disabilities, and chronic health conditions. Each of these groups may face specific barriers to accessing and understanding disaster information, participating in preparedness activities, and responding to emergencies. Conducting thorough needs assessments and engaging with representatives from these communities can provide valuable insights into their unique requirements and help shape effective disaster education programs. Tailoring Educational Materials to accommodate different needs is an essential component of inclusive disaster education (Borawsk, 2020). Materials should be accessible to individuals with various disabilities, including those who are blind or visually impaired, deaf or hard of hearing, and those with cognitive or learning disabilities.

Developing personalized emergency plans is another important aspect of inclusive disaster education. Individuals with special needs often require customized plans to address their specific needs and ensure their safety during emergencies. Disaster education programs should include guidance on how to create personalized emergency plans that account for factors such as mobility limitations, medical needs, and communication preferences. Assistance in developing these plans can be provided through one-on-one consultations, workshops, or online resources designed to help individuals and their caregivers create effective and practical emergency strategies. Training and Involving Caregivers and support networks in disaster education is crucial for ensuring that individuals with special needs are adequately prepared. Caregivers, family members, and support staff often play a key role in assisting individuals during emergencies. Providing training for these individuals on emergency procedures, communication strategies, and the use of assistive technologies helps ensure that they can effectively support those with special needs. Workshops, simulation exercises, and informational resources can be tailored to meet the needs of caregivers and support networks, enhancing their preparedness and ability to respond to emergencies. Creating Accessible Drills and Simulations is essential for ensuring that disaster education activities are inclusive (Subramanian, 2023). Traditional drills and exercises may not always accommodate the needs of individuals with disabilities or special requirements. To address this, disaster education programs should incorporate inclusive practices into drills and simulations, such as providing accessible evacuation routes, offering alternative methods of communication, and ensuring that all participants can actively engage in the exercises.

Collaborating with specialized organizations can enhance the effectiveness of disaster education for special needs populations. Partnering with organizations that focus on disability rights, chronic health conditions, and other special needs can provide valuable expertise and resources. These organizations can offer insights into the specific challenges faced by their constituents, contribute to the development of tailored educational materials and programs, and help disseminate information to their networks. Collaboration ensures that disaster education efforts are informed by the experiences and needs of those who are directly affected. Ensuring Compliance with Legal and Ethical Standards is important for developing inclusive disaster education programs. Adhering to legal requirements, such as those outlined in the Americans with Disabilities Act (ADA) or other relevant legislation, helps ensure that disaster education efforts are equitable and accessible. Ethical considerations, such as respecting privacy and dignity, are also important in creating inclusive programs. Disaster education efforts should be designed to meet legal standards while also addressing the diverse needs of special populations with sensitivity and respect (Kankanamge et al, 2020). Evaluating and Improving Programs based on feedback from special needs populations is essential for ensuring that disaster education efforts remain effective and relevant.

EVALUATING DISASTER EDUCATION PROGRAMS: METRICS AND OUTCOMES

Evaluating disaster education programs is essential for assessing their effectiveness, ensuring they meet their objectives, and identifying areas for improvement. Comprehensive evaluation involves analyzing various metrics and outcomes to understand how well the programs prepare individuals and communities for emergencies. These metrics include measures such as participation rates, the number of educational materials distributed, and the completion rates of training programs. Surveys and questionnaires can be employed to gather numerical data on participants' knowledge levels before and after the program (Ramakrishnan et al, 2022). For example, a pre- and post-program quiz can measure changes in understanding of emergency procedures. Analyzing this quantitative data helps identify trends and gauge the overall impact of the program. For example, tracking the number of households that create emergency kits after attending a preparedness workshop provides insight into the program's success in influencing practical actions (Kaufhold et al, 2023). Community Impact is another important aspect of evaluation. Assessing how the program has affected the broader community involves examining changes in overall preparedness and resilience. Metrics such as the number of community organizations engaged, the level of community involvement in preparedness activities, and improvements in community-wide response capabilities can indicate the program's success in fostering a culture of preparedness. Community surveys and feedback from local leaders can provide additional insights into the program's impact on community resilience.

Program Reach and Accessibility are crucial for understanding whether the program effectively reaches and serves its intended audience. Evaluating factors such as the demographics of participants, the geographic areas covered, and the accessibility of program materials helps ensure that the program is inclusive and equitable. For example, analyzing participation rates among different age groups, cultural backgrounds, or disability status can reveal whether the program is effectively reaching diverse populations (Maat et al, 2021). Ensuring that the program is accessible to all individuals, including those with special needs, is essential for achieving its goals. Cost-Effectiveness is an important consideration in evaluating disaster education programs. Assessing long-term outcomes, such as continued adherence to safety practices, ongoing community engagement, and improvements in disaster response capabilities, helps determine whether the program's benefits endure beyond the initial implementation phase (Sugumaran et al, 2024). Continuous adaptation and improvement based on evaluation findings help ensure that disaster education programs effectively enhance preparedness, response, and resilience in communities. Through these rigorous evaluation practices, disaster education programs can achieve their goals and contribute to a safer and more resilient society.

PARTNERSHIPS AND COLLABORATIONS: BUILDING DISASTER RESILIENT NETWORKS

Partnerships and collaborations are fundamental to building disaster-resilient networks, as they leverage collective resources, expertise, and networks to enhance community preparedness, response, and recovery. In the context of disaster management, effective partnerships bring together diverse stakeholders—including government agencies, non-governmental organizations (NGOs), community groups, businesses, and academic institutions—to create a unified approach to disaster resilience. By fostering strong relationships and coordinated efforts, these partnerships can significantly improve a community's ability to withstand and recover from emergencies. Establishing Collaborative Frameworks is the cornerstone of building disaster-resilient networks. Effective collaboration begins with defining common goals and establishing clear frameworks for cooperation. This involves identifying key stakeholders, setting shared objectives, and developing protocols for communication and coordination. For example, a city's emergency management agency might collaborate with local health departments, non-profit organizations, and businesses to develop a comprehensive disaster response plan (Mostofi et al, 2021). Establishing formal agreements and memoranda of understanding (MOUs) helps outline roles and responsibilities, ensuring that each partner understands their contributions and expectations. For instance, during a large-scale disaster, multiple organizations may coordinate to share emergency supplies, deploy volunteer teams, and provide financial assistance. Resource sharing not only maximizes the impact of each contribution but also helps avoid duplication of efforts and ensures that resources are directed where they are most needed (Righi et al, 2021). Strengthening Community Engagement through partnerships ensures that disaster resilience efforts are inclusive and reflective of local needs. Engaging community-based organizations, faith groups, and local leaders in the planning and implementation of disaster preparedness and response activities fosters trust and enhances community buy-in. These local partners have valuable insights into the unique needs and challenges faced by their communities. For example, working with local cultural organizations can help ensure that disaster education materials are culturally relevant and accessible to diverse populations.

Coordinating Response Efforts is critical during emergencies, and effective partnerships facilitate streamlined and efficient responses. By establishing joint operations centers or coordination committees, partners can manage resources, disseminate information, and execute response strategies in a unified manner. For instance, a regional disaster response team comprising local, state, and federal agencies can work together to coordinate evacuations, manage shelters, and address public health needs. Coordination helps minimize confusion, avoid conflicts, and ensure that all aspects of the response are covered comprehensively. Conducting Joint Training and Exercises helps build strong partnerships and improve preparedness. Collaborative training sessions and simulation exercises enable partners to practice their roles, understand each other's capabilities, and identify potential gaps in the response plan. These activities also foster team cohesion and enhance communication among partners (Kula Y., et al, 2021). For example, a multi-agency drill simulating a large-scale natural disaster can help participants practice coordinating actions, managing resources, and communicating effectively under pressure. Building Long-Term Relationships is essential for sustaining effective partnerships. Disaster resilience efforts require ongoing collaboration and trust-building among partners. Regular meetings, communication channels, and joint activities help maintain strong relationships and ensure that partners remain engaged and committed. Additionally, celebrating successes and acknowledging contributions reinforces the value of each partner's role. Long-term relationships facilitate continuous improvement and ensure that the network remains robust and responsive over time. Addressing Challenges and Conflicts that arise in partnerships is crucial for maintaining effective collaborations. Differences in organizational priorities, resource constraints, and communication issues can create challenges. Establishing mechanisms for conflict resolution, such as mediation or negotiation processes, helps address these issues constructively. Open communication, transparency, and a focus on shared goals can help resolve conflicts and strengthen partnerships.

Monitoring and Evaluating Collaborative Efforts helps ensure that partnerships are achieving their objectives and making a positive impact. Regular assessments of partnership activities, outcomes, and effectiveness provide insights into what is working well and where improvements are needed. Feedback from partners and stakeholders can guide adjustments to strategies and enhance overall performance. For example, evaluating the outcomes of a collaborative disaster response effort can help identify best practices and areas for improvement. Promoting Innovation and Best Practices through partnerships can drive advancements in disaster resilience. Collaborative networks create opportunities for sharing innovative approaches, technologies, and best practices (Sakurai et al, 2022). Partners can learn from each other's experiences, implement successful strategies, and adapt new ideas to their own contexts. For example, a partnership between local governments and technology companies might explore the use of new digital tools for disaster monitoring and communication. Fostering Public-Private Partnerships is an effective strategy for enhancing disaster resilience. Businesses play a critical role in disaster preparedness and response, and public-private partnerships can leverage their resources and expertise. Businesses can provide financial support, logistical assistance, and volunteer efforts, while government agencies can offer regulatory support and coordination (Tao et al, 2022). Coordinating response efforts, conducting joint training, and building long-term relationships further strengthen these networks. Addressing challenges, monitoring and evaluating efforts, promoting innovation, and fostering public-private partnerships contribute to a more resilient and prepared society. Through these collaborative approaches, communities can improve their ability to withstand and recover from disasters, ultimately creating a safer and more resilient world.

TECHNOLOGY-ENHANCED DISASTER EDUCATION: VIRTUAL REALITY AND SIMULATION

Technology-enhanced disaster education, particularly through the use of virtual reality (VR) and simulation, represents a transformative approach to preparing individuals and communities for emergencies. These advanced tools offer immersive, interactive experiences that can significantly enhance the effectiveness of disaster training and preparedness programs. By leveraging VR and simulation technologies, educators can create realistic and engaging scenarios that improve understanding, decision-making, and response skills, thereby building a more resilient and informed populace. Virtual Reality (VR) provides a powerful platform for disaster education by creating immersive environments that simulate real-life disaster scenarios. VR technology allows users to experience situations such as fires, earthquakes, or floods in a controlled and safe setting. This immersion helps participants develop a deeper understanding of the dynamics of disasters and the necessary responses. For instance, VR can simulate the experience of navigating a smoke-filled building during a fire, enabling users to practice evacuation procedures and decision-making skills without the associated risks (Nikolaraizi et al, 2021). The interactive nature of VR engages users more effectively than traditional training methods, fostering better retention of critical information and skills. Simulation-Based Training complements VR by providing detailed, scenario-based exercises that mimic real-world conditions. Simulations can be used to replicate various disaster situations, such as coordinating a multi-agency response or managing resources during a large-scale emergency.

Improving Situational Awareness is another significant advantage of VR and simulation technologies. These tools help participants develop a better understanding of their surroundings and the evolving dynamics of a disaster situation. By simulating different disaster scenarios, users can learn to recognize and respond to various hazards, assess risks, and navigate challenging environments. For example, a VR simulation might replicate a flooding event, allowing users to explore different areas, identify potential hazards, and develop strategies for evacuation and rescue. This enhanced situational awareness contributes to more effective disaster response and management. Training in a Safe Environment is a critical aspect of VR and simulation-based disaster education. These technologies allow participants to experience and respond to hazardous situations without facing real-world dangers. This safety aspect is particularly important for high-risk scenarios where traditional training methods might be impractical or unsafe. For example, VR can simulate the experience of a hazardous materials spill or a building collapse, providing participants with the opportunity to practice response procedures and safety protocols without the risk of injury (Jang et al, 2021). This safe environment encourages more extensive and frequent training, leading to better preparedness. Personalized Learning

Experiences can be tailored through VR and simulations to meet the specific needs and learning styles of participants. These technologies can be customized to address different levels of expertise, roles, and scenarios, allowing for a more targeted and effective training experience. This data can be used to assess their strengths and areas for improvement, providing valuable insights for refining training programs and enhancing overall effectiveness. Promoting Innovation in Disaster Education is driven by the use of VR and simulation technologies. As technology continues to advance, new and innovative approaches to disaster education are emerging (Senanayake et al, 2023). VR and simulations are at the forefront of these innovations, offering new ways to enhance training and preparedness. For example, emerging technologies such as augmented reality (AR) and artificial intelligence (AI) can be integrated into disaster simulations to create even more realistic and dynamic training experiences. By embracing these innovations, disaster education programs can stay current and effective in addressing evolving challenges.

DISASTER EDUCATION IN DEVELOPING COUNTRIES: CHALLENGES AND OPPORTUNITIES

Disaster education in developing countries faces a unique set of challenges and opportunities, influenced by factors such as limited resources, varying levels of infrastructure, and diverse socio-economic conditions. Addressing these challenges while leveraging opportunities is crucial for enhancing disaster resilience and preparedness in these regions. Effective disaster education in developing countries can significantly reduce vulnerability, improve response capacities, and foster community resilience. Resource Constraints are among the most significant challenges in developing countries. Limited financial resources and infrastructure can hinder the development and implementation of comprehensive disaster education programs. Many regions struggle with inadequate facilities, insufficient educational materials, and a lack of trained personnel. For example, remote and rural areas may lack access to modern communication technologies or educational resources, making it difficult to deliver effective training. Overcoming these constraints requires innovative approaches, such as utilizing low-cost, scalable solutions and leveraging community-based resources. Infrastructure and technology limitations also present obstacles to disaster education. In many developing countries, infrastructure such as roads, buildings, and communication networks may be inadequate or poorly maintained, affecting the delivery of educational programs (Finucane et al, 2021). Limited access to technology, such as computers and the internet, further exacerbates these challenges. To address these limitations, disaster education programs can adopt low-tech solutions, such as radio broadcasts, printed materials, and community workshops. Mobile technology, where available, can be used to distribute information and provide training in areas with limited infrastructure. Cultural and Linguistic diversity adds complexity to disaster education efforts. Developing countries often have diverse populations with varying languages, cultural practices, and beliefs. Educational materials and programs need to be culturally sensitive and linguistically appropriate to ensure that they are understood and accepted by all community members.

Community engagement is essential for effective disaster education but can be challenging in areas where communities are dispersed or lack cohesive structures. Building strong community networks and engaging local stakeholders are crucial for successful disaster education. Community-based approaches, such as training local volunteers and establishing neighborhood committees, can help facilitate the dissemination of information and promote community involvement. Engaging with community leaders and respected figures can also enhance the credibility and acceptance of disaster education initiatives. Capacity Building and Training for local personnel are critical for sustaining disaster education efforts. In developing countries, there may be a shortage of trained disaster management professionals and educators. Investing in capacity building through training programs, workshops, and knowledge-sharing initiatives can help develop local expertise and ensure the sustainability of disaster education programs. Collaborations with international organizations and experts can also provide valuable support and resources for capacity building. Funding and resource mobilization are key to the successful implementation of disaster education programs (Wang et al, 2020). Developing countries often face challenges in securing adequate funding for disaster preparedness initiatives. Exploring diverse funding sources, including international aid, government grants, and private sector contributions, can help support disaster education efforts. Innovative financing mechanisms, such as crowdfunding or public-private partnerships, can also provide additional resources and support. Leveraging Local Knowledge and Practices presents an opportunity for enhancing disaster education in developing countries. Local communities often have valuable traditional knowledge and practices related to disaster preparedness and response. Integrating this indigenous knowledge into disaster education programs can improve their effectiveness and cultural relevance. For example, traditional building techniques or local early warning systems can be incorporated into modern disaster education strategies to enhance community resilience. Monitoring and Evaluation are crucial for assessing the effectiveness of disaster education programs and identifying areas for improvement. Developing countries may face challenges in conducting rigorous evaluations due to limited resources and capacity (Abrash Walton et al, 2021).

LESSONS LEARNED: CASE STUDIES OF SUCCESSFUL DISASTER EDUCATION INITIATIVES

Case studies of successful disaster education initiatives offer valuable insights into effective strategies for enhancing disaster preparedness and resilience. By examining real-world examples, we can identify key lessons learned that contribute to the success of these programs and inform future efforts. Successful disaster education initiatives often demonstrate a combination of innovative approaches, community engagement, and effective use of resources, providing a roadmap for replicating and scaling similar successes in other regions. One notable example is the ShakeOut earthquake preparedness campaign in California. Initiated in 2008, ShakeOut has grown into one of the largest earthquake drills in the world, involving millions of participants. The campaign's success lies in its comprehensive approach, which includes not only the drill itself but also a robust educational component. By integrating educational materials into schools, workplaces, and community organizations, ShakeOut has effectively raised awareness about earthquake preparedness (Medel et al, 2020). The initiative emphasizes practical actions such as Drop, Cover, and Hold On and provides resources for creating emergency plans and kits. The key lesson from ShakeOut is the importance of a multifaceted approach that combines large-scale drills with ongoing education and community engagement. The program focuses on empowering communities to take charge of their disaster preparedness and response efforts. Through participatory approaches, communities are involved in hazard mapping, risk assessment, and the development of disaster response plans. The program also emphasizes training local volunteers and leaders to facilitate disaster education and response activities. The success of CBDRRM illustrates the importance of community involvement and empowerment in creating sustainable disaster education initiatives (McNaught, 2024).

The School-Based Disaster Education Program in Turkey provides another example of effective disaster education. Following the 1999 Marmara earthquake, Turkey implemented a nationwide program to integrate disaster education into school curricula. The program includes age-appropriate lessons on earthquake preparedness, first aid, and emergency response, as well as regular drills and simulations. By incorporating disaster education into the school system, Turkey has reached a large and diverse audience, fostering a culture of preparedness among young people (Elkhidir et al, 2023). The program's success highlights the effectiveness of incorporating disaster education into existing educational frameworks and engaging students in hands-on learning. In Haiti, the Safe Schools Program has played a crucial role in enhancing disaster preparedness in educational institutions. After the 2010 earthquake, the program focused on improving the safety and resilience of school buildings, as well as providing disaster education for students, teachers, and school staff. The program includes structural assessments, retrofitting of buildings, and the development of emergency response plans. It also emphasizes the importance of involving students in disaster preparedness activities and creating a culture of safety within schools. The initiative also includes the development of standardized protocols and guidelines for disaster response (Bergamo et al, 2022). The success of this program highlights the importance of specialized training for professionals who play a critical role in disaster response and recovery.

THE FUTURE OF DISASTER EDUCATION - INNOVATIONS AND CHALLENGES

The future of disaster education promises to be marked by a dynamic blend of innovations and AR can overlay critical information onto real-world environments, providing real-time guidance and support during emergencies (Kao et al, 2023). These technologies offer the potential to significantly improve engagement and effectiveness in disaster training, making complex scenarios more accessible and comprehensible. Artificial Intelligence (AI) is another transformative force in disaster education. AI-powered tools can analyze vast amounts of data to predict and model disaster scenarios, enabling more precise risk assessments and tailored educational content. Machine learning algorithms can identify patterns and trends from historical data, helping to develop more effective training programs and response strategies. AI can also support personalized learning experiences by adapting content based on individual needs and progress, ensuring that disaster education is more relevant and impactful. Big Data and Analytics are increasingly being utilized to enhance disaster education. By harnessing data from various sources, such as social media, satellite imagery, and sensors, educators can gain insights into disaster trends, behaviors, and responses. This data-driven approach allows for the development of more accurate and timely educational materials and simulations. For example, data on recent disaster events can be used to update training scenarios and reflect current risks and conditions. Analytics can also help evaluate the effectiveness of disaster education programs and identify areas for improvement.

Community-based initiatives, such as local disaster preparedness workshops and volunteer training programs, foster a sense of ownership and collaboration. Innovations in this space include the use of digital platforms to facilitate community discussions and coordinate response efforts (Smith et al, 2021). Challenges related to equity and accessibility will be significant in shaping the future of disaster education. Ensuring that all populations, including marginalized and vulnerable groups, have access to effective disaster education is a critical concern. Disparities in resources, technology, and information can exacerbate existing inequalities and hinder the effectiveness of disaster preparedness efforts. Addressing these challenges requires targeted outreach and the development of inclusive educational materials that consider diverse needs and barriers. For instance, creating multilingual resources and accessible formats for individuals with disabilities is essential for reaching all community members. Sustainability and Adaptation are also key challenges in disaster education. As climate change and environmental factors increase the frequency and intensity of disasters, educational programs must evolve to address new and emerging risks (Rahman et al, 2020). This requires ongoing updates to training content and the integration of new scientific knowledge and best practices. Additionally, disaster education programs must be designed to be sustainable, with mechanisms in place for continuous funding, resource allocation, and program evaluation. Building partnerships with local organizations and leveraging community resources can support the long-term sustainability of disaster education initiatives. The Role of Policy and Governance will be critical in shaping the future of disaster education. Effective policies and regulations can support the development and implementation of comprehensive disaster education programs (How et al, 2020). This includes establishing standards for disaster education, providing funding and resources, and promoting collaboration among stakeholders. Advocacy and policy development efforts should focus on ensuring that disaster education is prioritized at local, national, and international levels.

CONCLUSION

Disaster education and awareness programs are essential components of a comprehensive strategy to enhance community resilience and preparedness. As the frequency and intensity of natural and man-made disasters continue to escalate globally, the importance of well-structured and effective disaster education becomes increasingly evident. These programs play a pivotal role in equipping individuals and communities with the knowledge, skills, and resources needed to effectively respond to emergencies, mitigate risks, and recover from disasters. The primary goal of disaster education and awareness programs is to foster a culture of preparedness by providing accurate and actionable information about potential hazards and appropriate response strategies. Successful programs are characterized by their ability to engage diverse audiences, including vulnerable populations, through tailored content and interactive learning methods. Integrating practical exercises, simulations, and real-life scenarios into educational initiatives enhances the learning experience and helps participants apply theoretical knowledge to practical situations. One of the key aspects of effective disaster education is the inclusion of community-based approaches. Engaging local stakeholders, including community leaders, schools, and organizations, ensures that disaster education is relevant and culturally appropriate. However, despite the advancements and innovations in disaster education, several challenges persist. Ensuring equity and accessibility remains a critical concern, as disparities in resources, technology, and information can hinder the effectiveness of educational programs. Addressing these challenges requires targeted efforts to reach marginalized and underserved populations, including the development of multilingual and accessible materials. By prioritizing inclusivity, disaster education programs can ensure that all individuals have the opportunity to participate and benefit from preparedness initiatives. Sustainability is another important consideration in the implementation of disaster education programs. Developing long-term strategies for funding, resource allocation, and program evaluation is crucial for maintaining the effectiveness and continuity of educational efforts. Building partnerships with government agencies, non-governmental organizations (NGOs), and private sector entities can provide the necessary support and resources to sustain disaster education initiatives. Additionally, incorporating feedback and lessons learned from program evaluations helps refine and improve disaster education practices. The role of policy and governance cannot be overlooked in the context of disaster education. Effective policies and regulations can support the development and implementation of disaster education programs by establishing standards, providing funding, and promoting collaboration among stakeholders.

REFERENCES

Abrash Walton, A., Marr, J., Cahillane, M. J., & Bush, K. (2021). Building community resilience to disasters: A review of interventions to improve and measure public health outcomes in the Northeastern United States. *Sustainability (Basel), 13*(21), 11829. DOI: 10.3390/su132111699

Afifi, R. A., Abdulrahim, S., & Betancourt, T. (2020). Implementing community-based participatory research with communities affected by humanitarian crises: The potential to recalibrate equity and power in vulnerable contexts. *American Journal of Public Health, 110*(10), 1493–1495. DOI: 10.2105/AJPH.2020.305868 PMID: 32903072

Amri, A., Lassa, J. A., Tebe, Y., Hanifa, N. R., Kumar, J., & Sagala, S. (2022). Pathways to disaster risk reduction education integration in schools: Insights from SPAB evaluation in Indonesia. *International Journal of Disaster Risk Reduction, 73*, 102851. DOI: 10.1016/j.ijdrr.2022.102860

Bergamo, P. A. S., Streng, E. S., de Carvalho, M. A., Rosenkranz, J., & Ghorbani, Y. (2022). Simulation-based training and learning: A review on technology-enhanced education for the minerals industry. *Minerals Engineering, 175*, 107297. DOI: 10.1016/j.mineng.2021.107272

Borawska, A. (2020). Application of behavioral economics insights to increase effectiveness of public awareness campaigns. In *Quantitative methods in contemporary economics* (pp. 45–62). Springer., DOI: 10.1007/978-3-030-30251-1_5

Cabello, V. M., Véliz, K. D., Moncada-Arce, A. M., Irarrázaval García-Huidobro, M., & Juillerat, F. (2021). Disaster risk reduction education: Tensions and connections with sustainable development goals. *Sustainability (Basel), 13*(24), 10886. DOI: 10.3390/su131910933

Dufty, N. (2020). *Disaster education, communication and engagement*. Wiley., DOI: 10.1002/9781119569817

Dwiningrum, S. I. A., Sitompul, N., Nisa, A. A., & Sumunar, D. R. S. (2022). School resilience policy development and strategic steps to disaster-resilient communities. *IOP Conference Series. Earth and Environmental Science, 1039*(1), 012006. DOI: 10.1088/1755-1315/989/1/012017

Elkhidir, E., Mannakkara, S., Henning, T. F. P., & Wilkinson, S. (2023). A pathway towards resilient cities: National resilience knowledge networks. *Cities (London, England), 134*, 104194. DOI: 10.1016/j.cities.2023.104243

Finucane, M. L., May, L. W., & Chang, J. (2021). *A scoping literature review on indicators and metrics for assessing racial equity in disaster preparation, response, and recovery*. RAND Corporation., DOI: 10.7249/RRA1092-1

Fletcher-Miles, H., Gammon, J., Williams, S., & Hunt, J. (2020). A scoping review to assess the impact of public education campaigns to affect behavior change pertaining to antimicrobial resistance. *American Journal of Infection Control*, *48*(8), 933–940. DOI: 10.1016/j.ajic.2019.07.011 PMID: 31444097

Fu, Q., & Zhang, X. (2024). Promoting community resilience through disaster education: Review of community-based interventions with a focus on teacher resilience and well-being. *PLoS One*, *19*(1), e0296454. DOI: 10.1371/journal.pone.0296393 PMID: 38166092

Gill, J. C., Taylor, F. E., & Duncan, M. J. (2021). Invited perspectives: Building sustainable and resilient communities—Recommended actions for natural hazard scientists. *Natural Hazards and Earth System Sciences*, *21*(5), 1479–1490. DOI: 10.5194/nhess-21-1479-2021

How, V., Azmi, E. S., & Rahman, H. A. (2020). The way forward: Opportunities and challenges of sustainable school disaster education in Malaysia. *International Journal of Disaster Risk Reduction*, *51*, 101836. DOI: 10.1016/j.ijdrr.2020.101836

Jang, J. H., & Ha, K. M. (2021). Inclusion of children with disabilities in disaster management. *Children (Basel, Switzerland)*, *8*(7), 581. DOI: 10.3390/children8070581 PMID: 34356560

Kaku, K. (2019). Satellite remote sensing for disaster management support: A holistic and staged approach based on case studies in Sentinel Asia. *International Journal of Disaster Risk Reduction*, *33*, 417–432. DOI: 10.1016/j.ijdrr.2018.09.015

Kankanamge, N., Yigitcanlar, T., & Goonetilleke, A. (2020). How engaging are disaster management related social media channels? The case of Australian state emergency organisations. *International Journal of Disaster Risk Reduction*, *48*, 101571. DOI: 10.1016/j.ijdrr.2020.101571

Kao, C. L., Chien, L. C., Wang, M. C., Tang, J. S., Huang, P.-C., Chuang, C.-C., & Shih, C.-L. (2023). The development of new remote technologies in disaster medicine education: A scoping review. *Frontiers in Public Health*, *11*, 1099467. DOI: 10.3389/fpubh.2023.1029558 PMID: 37033011

Kaufhold, M. A., Reuter, C., & Ludwig, T. (2023). Big data and multi-platform social media services in disaster management. In *International handbook of disaster research* (pp. 1–18). Springer., DOI: 10.1007/978-3-030-91691-6_103

Kula, Y., Cohen, O., Clempert, N., Grinstein-Cohen, O., & Slobodin, O. (2021). Educating nursing students for cultural competence in emergencies: A randomized controlled trial. *BMC Nursing*, *20*(1), 178. DOI: 10.1186/s12912-021-00704-1 PMID: 34587960

Libayao, H. G. C., & Navaluna, C. A.Jr. (2024). Disaster education in elementary school curriculum: Basis for framework designing. *Journal of Educational and Social Research*, *14*(1), 147–158. DOI: 10.36941/jesr-2024-0012

Maat, H., Balabanova, D., Mokuwa, E., Richards, P., Mohan, V., Ssengooba, F., Twinomuhangi, R., Woldie, M., & Mayhew, S. (2021). Towards sustainable community-based systems for infectious disease and disaster response: Lessons from local initiatives in four African countries. *Sustainability (Basel)*, *13*(11), 6335. DOI: 10.3390/su131810083

Mahadev Madgule, N., Vinayaka, H., Sonkhaskar, Y. M., Patel, D. R., Karthikeyan, R., & Selvakumar, P. (2023). Mechanical properties and microstructure of activated TIG welded similar joints of Inconel alloys by desirability approaches. *Materials Today: Proceedings*, *77*, 528–533. DOI: 10.1016/j.matpr.2022.12.250

McNaught, R. (2024). The application of collaborative governance in local level climate and disaster resilient development—A global review. *Environmental Science & Policy*, *151*, 103632. DOI: 10.1016/j.envsci.2023.103627

Medel, K., Kousar, R., & Masood, T. (2020). A collaboration–resilience framework for disaster management supply networks: A case study of the Philippines. *Journal of Humanitarian Logistics and Supply Chain Management*, *10*(4), 509–532. DOI: 10.1108/JHLSCM-09-2019-0066

Mostofi, M., & Brown, L. M. (2021). Cultural competence and disaster mental health. In *The intersection of trauma and disaster behavioral health* (pp. 285–300). Springer., DOI: 10.1007/978-3-030-51525-6_7

Muñoz, V. A., Carby, B., Abella, E. C., Cardona, O. D., López-Marrero, T., Marchezini, V., Meyreles, L., Olivato, D., Trajber, R., & Wisner, B. (2020). Success, innovation and challenge: School safety and disaster education in South America and the Caribbean. *International Journal of Disaster Risk Reduction*, *44*, 101424. DOI: 10.1016/j.ijdrr.2019.101395

Naheed, S. (2021). Understanding disaster risk reduction and resilience: A conceptual framework. In *Handbook of disaster risk reduction for resilience: New perspectives and challenges* (pp. 1–20). Springer., DOI: 10.1007/978-3-030-61278-8_1

Nikolaraizi, M., Argyropoulos, V., Papazafiri, M., & Kofidou, C. (2021). Promoting accessible and inclusive education on disaster risk reduction: The case of students with sensory disabilities. *International Journal of Inclusive Education, 25*(7), 772–786. DOI: 10.1080/13603116.2020.1862408

Paramasivam, S., Mariappan, P., Petchiammal, S., & Richard Paul, S. J. (2024). Removal of chemicals from effluent using photobioreactor technology to improve environmental and health impacts. *Zeitschrift für Physikalische Chemie, 238*(6), 1119–1132. DOI: 10.1515/zpch-2023-0450

Paramasivam, S., Palanisamy, S. K., Cinthaikinian, S., Selvamurugan, P., Vignesh-kumar, P., Mariappan, R., & Selvakumar, P. K. (2024). Biosensors and its diverse applications in healthcare systems. *Zeitschrift für Physikalische Chemie, 238*(5), 885–904. DOI: 10.1515/zpch-2023-0406

Pickering, C. J., Guy, E., Al-Baldawi, Z., McVean, L., Sargent, S., & O'Sullivan, T. (2021). "I believe this team will change how society views youth in disasters": The EnRiCH youth research team: A youth-led community-based disaster risk reduction program. *Canadian Journal of Public Health = Revue Canadienne de Santé Publique, 112*(4), 664–672. DOI: 10.17269/s41997-021-00486-8 PMID: 34008134

Piltch-Loeb, R., Bernard, D., & Vallejo, B. Q. (2022). Engaging community leaders in sharing local knowledge for emergency preparedness to leverage communication and trusted assets for vulnerable populations. *Disaster Medicine and Public Health Preparedness, 16*(5), 1882–1889. DOI: 10.1017/dmp.2021.103 PMID: 34016210

Pranata, S., Widodo, S., & Vranada, A. (2021). How to prepare a school-based disaster preparedness in Indonesia. *Health Science Journal, 15*(7), 1–6. https://www.hsj.gr/medicine/how-to-prepare-a-schoolbased-disaster-preparedness-in-indonesia.php?aid=23674

Rahman, M. M., Nabila, I. A., Islam, F., & Tasnim, F. (2020). Challenges to implement disaster risk reduction in schools of developing country: Study on Dhaka City, Bangladesh. In *International Conference on Disaster Risk Reduction, Dhaka, Bangladesh.*

Ramakrishnan, T., Mohan Gift, M. D., Chitradevi, S., Jegan, R., Subha Hency Jose, P., Nagaraja, R., Sharma, H. N., & Selvakumar, P. (2022). Study of numerous resins used in polymer matrix composite materials. *Advances in Materials Science and Engineering, 2022*, 1088926. DOI: 10.1155/2022/1088926

Ramakrishnan, T., Rahman, S., & Ngamassi, L. (2022). Examining the factors that influence the use of social media for disaster management by underserved communities. *Jàambá, 14*(1), 1035. DOI: 10.4102/jamba.v14i1.1035

Righi, E., Lauriola, P., Ghinoi, A., Giovannetti, E., & Soldati, M. (2021). Disaster risk reduction and interdisciplinary education and training. *Progress in Disaster Science*, *10*, 100054. DOI: 10.1016/j.pdisas.2021.100165

Sakurai, M., & Shaw, R. (2022). The potential of digitally enabled disaster education for sustainable development goals. *Sustainability (Basel)*, *14*(8), 4269. DOI: 10.3390/su14116568

Seddighi, H., Sajjadi, H., Yousefzadeh, S., López López, M., Vameghi, M., Rafiey, H., & Khankeh, H. (2022). School-based education programs for preparing children for natural hazards: A systematic review. *Disaster Medicine and Public Health Preparedness*, *16*(3), 1228–1236. DOI: 10.1017/dmp.2020.479 PMID: 33818366

Senanayake, A. C., Samarakkody, A., Malalgoda, C., Amaratunga, D., Haigh, R., Liyanage, C., Hamza, M., Kaklauskas, A., & Shaw, R. (2023). Towards an inclusive disaster education: The state of online disaster education from the learner's perspective. *Sustainability (Basel)*, *15*(14), 11044. DOI: 10.3390/su151411042

Smith, S., Farra, S. L., & Hodgson, E. (2021). Evaluation of two simulation methods for teaching a disaster skill. *Journal of Educational Technology Development and Exchange*, *14*(1), 31–44. DOI: 10.18785/jetde.1401.03 PMID: 35520385

Sreenivasulu, A., Manimozhi, T., Satheesh Kumar, P., Selvakumar, P., & Singh, P. (2022). The model of smart sensing device for sensitive nanoclusters modification in sensing properties. In *2022 2nd International Conference on Advance Computing and Innovative Technologies in Engineering (ICACITE)* (pp. 1043–1047). IEEE. DOI: 10.1109/ICACITE53722.2022.9823846

Subramanian, S. P. (2023). Evaluating the effectiveness of online environmental activism campaigns: Assessing the impact on public awareness and behavioral change. *Journal of Intelligent Connectivity and Emerging Technologies*, *8*(2), 45–60. https://www.journalicet.org/index.php/jicet/article/view/123

Sugumaran, K. M. D., Pandiyarajan, V., Sonkhaskar, Y. B., Selvakumar, P., & Manimozhi, M. S. (2024). Analysis of electromyography signals for control of mechanical prosthesis using machine learning techniques. In *2024 2nd International Conference on Intelligent Data Communication Technologies and Internet of Things (IDCIoT)* (pp. 1059–1067). IEEE. DOI: 10.1109/IDCIoT59759.2024.10467707

Tao, Y., Lin, T., Feng, X., Gao, Y., & Mashino, S. (2022). Cultural competence for disaster nursing: A scoping review of the Chinese and English literature. *International Journal of Disaster Risk Reduction*, *78*, 103147. DOI: 10.1016/j.ijdrr.2022.103188

Vijayakumar, G., Rajkumar, M., Chandar, N. R., Selvakumar, P., & Duraisamy, R. (2024). Environmentally friendly TDS removal from wastewater by electrochemical ion exchange batch-type recirculation (EIR) technique. *Environmental Science. Water Research & Technology*, *10*(4), 826–835. DOI: 10.1039/D3EW00793F

Wang, Y. C., Lin, S. W., & Lee, C. H. (2020). Conducting an evaluation framework for disaster management under adaptive organization change in a school system. *Sustainability (Basel)*, *12*(16), 6615. DOI: 10.3390/su12166615

Yusuf, R., Fajri, I., & Gani, S. A. (2022). Disaster education in disaster-prone schools: A systematic review. *IOP Conference Series. Earth and Environmental Science*, *1039*(1), 012007. DOI: 10.1088/1755-1315/1041/1/012034

Zuama, S. N. (2024). Disaster mitigation education: Challenges and opportunities in six areas of PAUD development. *Journal on Advanced Science. Educational Technology*, *10*(1), 23–35. https://journal.unisnu.ac.id/jaset/article/view/123

Chapter 5
Tips for Effectively Communicating Human Capital Analysis Analysis to Decision Makers

Shivansu Sachan
https://orcid.org/0000-0002-9893-8018
CSJM University, India

Ankit Kumar
https://orcid.org/0009-0009-5002-4522
CSJM University, India

ABSTRACT

In data-work trade, Analysis for Human Capital Analysis play an impact in forming lot of choices for organizations. In integration of these probability into decision forms requires communication to limit the separate between complex information of knowledge and significant techniques. This theory assures that lucrative workforce-related information is interpreted into aware organizational decisions and provides essential phrases for effectively presenting analysis for Human Capital Analysis to decision formers. The first piece of advice highlights how crucial it is to modify analysis for Human Capital Analysis. Decision makers are likely to focus on analysis that directly help in the formation of important destinations. This involves identifying key performance indicators (KPIs) that align with decision makers and are situated within the context of overarching trade forms. Include a brief summary of key perspectives and their implications for business goals. Use visuals such as charts or graphs to clearly highlight trends and results.

DOI: 10.4018/979-8-3693-9745-9.ch005

INTRODUCTION OF ANALYSIS FOR HUMAN CAPITAL ANALYSIS

The process of gathering, evaluating, and interpreting personnel data for a company in making well-informed recruiting decisions has been refer as analysis for Human Capital Analysis.

Consider the taking after two cases, both from a budgetary administrations firm. The primary case was a multivariate factual investigation in the characteristics of an incentive pay plot and deals execution for a key part within the organization. Making prudent business choices is critical for increasing the overall worth of the company. The examination was conducted by a committed HR analysis individual whose factual approach was immaculate. Then insights determined by the investigation addressed a expansive number of assumptions about the pay plan and what drove behavior within the part. However the general manager (GM) capable for that portion of the commerce was not locked in within the investigative prepare from the starting, and was presented as it were with the "gift wrapped" comes about at the conclusion of the exam. The reaction was a complete dismissal of the exam and addressing of the analysis behind it. The key botch of the HR examiner was not locks GM at begin of the method to guarantee that there was arrangement around the potential require for the examination and intrigued in getting and potentially acting on the results. The moment case was a direct representative study conducted within the middle of a large sum of organizational alters and change. A mission statement encapsulates an organization's overarching goal, or its raison d'être, or reason for being, by describing a fundamental need that the organization seeks to satisfy. Formulating a plan includes not just a mission statement but also a vision and values statement. The vision of an organization outlines what it hopes to become or look like in the future and is an extension of the mission. Establishing a core values framework gives the organization boundaries and direction for making decisions and realizing its goals. The overview was taken by the senior leaders as a vital vehicle for observing the employees' reactions to the changes, and the leaders' victory in relieving the negative impacts of the changes. The analysis utilized to analyze the overview reactions were exceptionally essential, counting as it were implies and frequencies. But because the questions were planned to straightforwardly address worker responses to a critical transition period, the study comes about were given extraordinary weight within the leaders' handle of assessing and adjusting their approach to overseeing the changes. So not necessarily a prerequisite for profound in sights and noteworthy data. They are, be that as it may, frequently a key portion of a comprehensive set of analyses that may be required for the most profound bits of knowledge. Two cases consider outline this point. Both begun as issues that showed up related to recompense. After conducting extensive analyses significant bits of knowledge to the limits of remuneration as an arrangement were accomplished, and actions were taken that driven to enduring positive impacts. In both cases there were (a) advanced statistical methods for one portion of the work, (b) basic information examinations for other parts of the work, and (c) contemplative, keen thought of causal variables utilizing rationale as it were and virtually no (unused) information examination for the remaining parts of the work.

ROLE OF RETENTION IN ANALYSIS FOR HUMAN CAPITAL ANALYSIS

To calculate retention rate, you typically divide the no. of employees who are with the company by the total no. of employees at the begin of the period, and then multiply by 100 to get the percentage.

For example:

> Retention Rate = (No. of employees at end of time / Number of employees at start of time) * 100

Companies often use retention rate data to identify trends, understand frequency influencing retention, and develop formation to improve employee engagement and satisfaction, thereby reducing turnover and retaining valuable talent.

> Analysis for Human Capital Analysis: the process of using data to understand and help human resource management daily practices has seen significant changes in performance over the years.

This includes looking at things like workplace culture, diversity and inclusion and employee well-being to see what the product is. That's all a deeper understanding of how to increase employee satisfaction and performance. As human inform capital analysis increases, there is a growing awareness of the ethical with the collection and use of employee data. Organizations are increasingly adopting ethical guidelines and best practices to ensure that Human Capital Analysis respects workers' rights and adheres of fairness and transparency.

Human Capital Analysis can be defined as "a process that enables the understanding and evaluation of causal relationships between hr and management functions and help a valid and reliable base for decisions making using statistical and experimental methods. Effectiveness. .Measuring Efficiency, Effectiveness and Impact" Analysis of human resources (HR) and Human Capital Analysis for the topic of communication.

The tools collect information from a variety of sources, including surveys, logs and activities, to create a coherent picture. Evidence-based approach to better decision-making. This is a popular word. It is simply a collection of primary objective facts and secondary objective data. The filter is divided into three levels:

1. Reporting of HR KPIs as key performance indicators (Turnover rate, response time, number of employees hired number of employees hired and trained, etc.) The main focus here is to reduce costs and improve processes.

Visual analysis reveals and explains relationships between current data patterns and historical data. This is the basis of your research. This includes dashboards and dashboards. Department of Business Administration Data mining of original samples and report periodically

2. Predictive analytics is a variety of techniques (statistical, modeling, Data mining), which uses current facts and past data to make future predictions about risks and probabilities, for example includes models used to increase opportunities for selection, training, and promotion.
3. Combine predictability with decision-making opportunities and employee performance. Analyze complex issues to determine outcomes. Increase decision making power and specify the advantages and disadvantages Models designed to understand how different learning factors affect outcomes (rare in HR) This approach starts with a simple HR report and progresses to a standardized business operating model. Although financial capital (cash) and financial capital (Intangible assets) are an important part of a company. Human capital (people) analysis uses cash and intangible assets to improve a company's performance. And when you move from description to necessity The value will increase exponentially.

HUMAN CAPITAL ANALYSIS FRAMEWORK

1. Competitive diversity: The impact of PMT (product/market/technology) on human capital analysis indicates an "increased demand from product market approach and appropriate approach" which determines HR strategy and practices and therefore includes human capital analysis used division.
2. Organizational morphology: SCL (social/cultural/legal) influences HRD practices, and hence human capital evaluation evaluations, through social perspectives (fairness, legitimacy) and law.

3. Corporate structure: The organizational/functional/cultural structure of an organization influences human resource processes such as human capital analysis and evaluation. This combination of the above three characteristics influences the scope of human capital assessment research and the manner in which it is applied in an organization.

It can be assumed that the more competitive the environment, the more organizations will apply Human Capital Analysis.

4. Organizational trends: Companies are implementing Human Capital Analysis to avoid imitating competitors or becoming obsolete. Well, no. The choice of Human Capital Analysis analyzes by competitors.

Influences Human Capital Analysis analyzes by organizations in the same sector.

5. Formalization: As a company matures, its behavior becomes formalized (Mint berg, 1979). Therefore, Human Capital Analysis should be used more widely in traditional organizations to support formal decision-making processes.
    4. Organization Structure: Most organizations today grow organically. Therefore, to understand the drivers of employee performance and increase organizational effectiveness, HR analysis is the way to go.

Monitor career progression and identify high-potential individuals for leadership roles.

Calculate the return on investment for various Human Capital Analysis initiatives such as training programs, employee benefits, and recruitment efforts.

Ensure resource is allocated efficiently to max. The no. of Human Capital Analysis.

Analyze the diversity and inclusivity of the work part and impact on organization performance.

Promote diversity inclusion initiatives to foster innovation and creativity.

Forecast future workforce needs based on business goals and industry trends.

Identify potential talent shortages or surpluses and develop strategies to address them proactively.

Implement performance management software to track work performance and provide are of improvement in workplace.

Align individual completion with organizational task to enhance overall work performance.

Compare Human Capital Analysis metrics against industry format to assess competitiveness and tell areas for improvement.

Improve from best practice adopted by leading organizations in the field.

UNDERSTANDING THE REQUIRE AND INCLINATIONS OF CHOICE MAKERS

Diverse backgrounds and degrees of information-gathering expertise may be possessed by decision makers. You can modify your communication strategies to more effectively convey information in a clear, concise manner by being aware of their demands. Specific issues or topics pertaining to the management of Human Capital Analysis. HR specialists can focus on resolving those problems and offering invaluable experience and insight if are aware of their needs. Decision is believed the facts and the statistics presented when they are aware of their own needs and desires. This promotes a collaborative relationship between HR professionals and decision makers, thus leading to better decision making. As resulted, decision makers focus on achieving organizational goals and objectives. By understanding their own needs and wants. Tailoring communications in needs and desires of decision makers' increases access to the information and experiences provided to them. In strengthening global resilience to natural disasters, a pragmatic, audience-focused approach is needed to effectively communicate human capital assessments to decision-makers. By crafting messages, using different media, different storytelling strategies, emphasizing actionable insights, and encouraging collaboration, HR professionals can bridge the gap between rigorous analysis and informed decision-making as organizations navigate the uncertainties of a world severely affected by natural disasters.

Human Supply (HR) channels is vital specifically when addressing the pressing challenge of strengthening global resilience to natural disasters. man cap psychoanalysis plays amp relevant Role inch this circumstance provision Understandings that enable organizations to get knowledgeable decisions vis-a-vis their men inch the look of prospective disruptions. To ensure that decision-makers fully grasp the implications of human capital analytics HR professionals must employ strategic communication techniques that bridge the gap between Complicated Information and actionable .HR communication should begin with a clear assessment of the audience knowledge and interests. tailoring the content to come across with decision-makers not but Improves employment just too ensures that the psychoanalysis addresses their particular concerns vis-a-vis buoyancy and calamity readiness. This involves translating technical jargon into accessible language highlighting the relevance of the Information to organizational goals and disaster readiness.

The presentation of human capital analytics should be framed within the larger context of organizational resilience. away connection men information to the prospective impacts of spurious disasters—such arsenic employee accessibility science sets necessary for retrieval and the cathartic welfare of staff—hr get in effect instance the important Role that man cap plays inch general buoyancy strategies. This approach not only underscores the importance of investing in human supply but also demonstrates how analytics can inform proactive measures to mitigate risks associated with natural disasters.

Utilizing various communication channels is essential in reaching decision-makers effectively. conventional methods such as arsenic reports and presentations need work complemented away digital platform including webinars reciprocal dashboards and gregarious mass media. These channels can Improve engagement by making Information more visually appealing and easier to digest. for case infographics that instance name analytics findings get compactly get compound Information devising it easier for stakeholders to read the implications of man cap information. Also, fostering a culture of open dialogue through forums or workshops allows decision-makers to ask questions and express their concerns creating a collaborative environment that values input from multiple perspectives also storytelling is a powerful tool in HR communication. away joint real-life suit studies or theoretical scenarios that instance however man cap analytics has with success led organizations inch enhancing buoyancy hour get make amp story that resonates with decision-makers. These stories can showcase the positive outcomes of informed decision-making such as Improved employee retention during crises or the successful use of a skilled workforce in recovery efforts. away humanizing the information hour professionals get further amp deeper link betwixt decision-makers and the analytic Understandings existence presented in increase accenting the unjust world of the psychoanalysis is important. Decision-makers are often inundated with Information but may struggle to determine the next steps. hour need not but show findings just too render light recommendations founded along the analytics. This might include strategies for talent acquisition upskilling programs or workforce planning initiatives that directly Add to enhancing organizational resilience. away frame these recommendations inside the circumstance of calamity readiness hour get service decision-makers look the material benefits of leverage man cap analytics feedback and reiteration inch communicating practices get importantly raise the strength of hour messaging. After presenting human capital analytics soliciting feedback from decision-makers can provide Understandings into which aspects were most compelling or where clarity was lacking. this repetitive work allows hour to down prospective communications ensuring they rest pertinent and impactful. also tracking the outcomes of decisions made based on the analytics can provide further evidence of the value of human capital analysis reinforcing its importance in strategic planning. Collaboration with other departments—such as risk management operations and finance—can enrich the communication of human capital analytics. away practical collectively hour get incorporate perspectives from different stakeholders provision amp further general see of however men considerations cross with organizational buoyancy efforts. This multidisciplinary approach not only strengthens the analysis but also fosters a sense of shared responsibility among leaders in addressing the challenges posed by natural disasters. This leads to greater discussion and collaboration on people issues, which ultimately leads to better results for the organization.

HR communication is decisive for organizations for several reasons:

Alignment of Goals: Clear communication helps align employee Goals with the organization goals ensuring everyone is working towards the same outcomes.
Employee Engagement: Open lines of communication foster a culture of transparency making employees feel valued and engaged which can Improve productivity and job satisfaction.
Disagreement ReAnswer: Effective communication can help address and resolve Disagreements early preventing escalation and maintaining a harmonious work environment.
Change Management: During periods of change clear communication from HR helps employees understand the reasons behind changes reducing uncertainty and resistance.
Policy Understanding: Clear communication about policies and procedures ensures that employees are informed about their rights and responsibilities which can lead to better compliance and reduced risks.

THE IMPORTANCE OF HUMAN CAPITAL ANALYSIS IN DISASTER RESILIENCE

In the domain of disaster resilience effectively communicating human capital analysis (HCA) is paramount for decision makers who must navigate Complicated challenges. man cap refers to the skills cognition and get of individuals inside organizations and communities. By leveraging HCA stakeholders can better understand workforce capabilities and identify gaps that may hinder effective disaster Answer and recovery efforts. for conclusion makers it is relevant to hold the link betwixt man cap and calamity buoyancy. When they see the value of a skilled workforce in eCombinency management recovery operations and community engagement they are more likely to prioritize HCA in their strategic planning. in addition presenting hca inch amp relatable manner—using square speech and pertinent examples—helps span the break betwixt information and pragmatic diligence. This foundational understanding sets the stage for stakeholders to appreciate how investing in human capital can very importantly bolster their resilience against natural disasters.

STRATEGIES FOR EFFECTIVE COMMUNICATION OF HCA UNDERSTANDINGS

To communicate HCA Understandings effectively decision makers need clear actionable information. i good scheme is to cut the content to the particular hearing. Decision makers come from various backgrounds including government NGOs and private sectors each with distinct priorities. crafting messages that come across with their particular interests—such arsenic state guard imagination allotment or public support—can raise employment. also employing visual aids like infographics charts and maps makes Complicated Information more digestible. optic representations service to spotlight name findings quick allowing conclusion makers to hold important Understandings without acquiring forfeit inch numbers game. Storytelling can further Improve this Method; presenting Information within a narrative framework not only makes it more compelling but also helps decision makers connect emotionally to the information. for case amp suit read particularization however amp public used its men to in effect answer to amp spurious calamity get instance the real-world affect of man cap investments

Inundating decision makers with exhaustive Information distill findings into important takeaways that inform their strategic decisions. for case if psychoanalysis indicates amp dearth of fine force for calamity reaction advocate targeted education programs or partnerships with community informative institutions. Highlighting concrete recommendations empowers decision makers to take tangible actions based on the Understandings provided. in addition nurture amp collaborative air Fosters cross-sector partnerships which are difficult for general calamity buoyancy. By advocating for joint initiatives between government NGOs and private enterprises HCA can be positioned as a valuable tool for Constructing a cohesive Answer Web. workshops or round table discussions get help these collaborations allowing stakeholders to deal Understandings and together work challenges. Sustaining employment and advocating for insurance integration effective communicating does not close with the first display of findings; maintaining current employment with conclusion makers is important for long affect. Follow-up communication is essential to provide further analysis answer questions or clarify Understandings as needed. this continuing back reinforces the analyst's Role arsenic amp sure imagination and keeps the conversation around man cap and calamity buoyancy awake. also encouraging decision makers to Watch and evaluate the impact of applied recommendations fosters a culture of accountability and continuous Improvement. joint frameworks for assessing the strength of hca inch enhancing calamity buoyancy enables stakeholders to down their strategies across time another difficult look of communicating is the advocacy for insurance consolidation. Decision makers often operate within Complicated regulatory and funding environments. away promoting evidence-based insurance devising analysts get service incorporate hca into broader important frameworks ensuring that men considerations are prioritized inch calamity provision and reaction. Engaging in policy dialogues and participating in forums where important decisions are made provides opportunities to share Understandings and advocate for the significance of human capital analysis. inch this room the Understandings gleaned from hca get inform policies that not but work prompt disaster-related necessarily just too lead to the long buoyancy of communities. Good communicating of man cap psychoanalysis to conclusion makers is amp important factor of invigorating round buoyancy to spurious disasters. By understanding the importance of HCA employing strategies for effective communication and sustaining engagement while advocating for policy integration stakeholders can foster a culture of preparedness and resilience. once conclusion makers know the difficult Role that man cap plays inch calamity reaction and retrieval they are break prepared to get knowledgeable choices that raise public buoyancy inch the look of necessary spurious disasters.

Human Capital Analysis Calculations

Human Capital Analysis measurements allude to quantifiable measures utilized to evaluate different angles of an organization's placeforce. These measurements give important experiences into the execution, efficiency, and in general viability of representatives, making a difference organizations make informed decisions approximately ability administration and key arranging.

The number of possible metrics is practically infinite, although there is a handful that sees heavy usage.

Employee Turnover

An organization's turnover rate can be better understood by looking at the total no. of worker who departs soon. To differentiate between intentional and accidental turnover, this metric—which is defined by the division or supervisor and classified by person residency—can be useful. The no. of people who are separated by the average number of people in particular time is a common way to calculate representative turnover. To find the percentage, just multiply it by 100.

New Contract Turnover Rate

It is common practice to use this formula to monitor the percentage of new hires that leave during the first three months. Separate the total no. of new personnel who start producing results within 90 days from the total number of contracts signed within the allocated time frame.

Retention Rate

The pace at which employees leave without being replaced. The annual retention rate of your sales people may be this. Find this number by divide the total no. of people in the company by the average no. of people.

Offer Acceptance Rate

Not all applicants will accept a position offer. Start keeping tabs on the offer acknowledgment rate by dividing the total no. of offers made by the number of offers acknowledged; this will give you a good idea of how likely candidates are to accept an offer. If your company's recognition rate is lower than 80%, it can indicate that your offerings are not competitive in the market.

Cost per Hire

The real monetary impact of making a contemporary recruit must be understood. This total should cover everything from advertising the position to the time spent recruiting, conducting interviews, and finally, onboarding new employees.

Time to Productivity

This calculation tracks how long time for unused contracts to reach full efficiency. Dubious to degree, but with the assistance of office heads and directors, it can unquestionably be done.

HR analysis provide useful information on person performance, engagement, and productivity with corporate objectives and tactical ambitions. Effectively achieve their business goals, Establishments can use data-driven methods to anticipate skill gaps, assess talent trends, and optimize placeforces. For example, analysis can be used to identify high-performing employees, understand the factors influencing their performance, and replicate those trends throughout the entire company. Additionally, by employing Human Capital Analysis, companies may make decisions about perfect hiring, training, and talent development initiatives, effectively allocating resources to support tactical aims. In general, firms that align staff analysis with organizational goals can improve their competitive advantage, encourage innovation, and prosper in the long run. The resilience of businesses and communities in the face of such events is increasingly linked to employee retention strategies. This highlights the implication of nurture amp engaged men that get pilot crises conducive to organizational constancy and broader social resilience understanding employee retention employee memory refers to associate in nursing organization power to hold its employees across sentence. High retention rates often indicate a positive work environment where employees feel valued engaged and connected to the organization's mission. On the other hand great upset get run to important costs including enlisting education and forfeit productiveness. Factors influencing employee retention include job satisfaction career development opportunities work-life balance and organizational culture the frequency and intensity of natural disasters continue to rise due to climate change and urbanization the need for a robust global resilience Plan has never been more pressing. spurious disasters—ranging from hurricanes and floods to earthquakes and wildfires—can waste communities interrupt economies and peril lives. In this context organizations play a pivotal role in not only responding to these crises but also in fostering resilience among their workforce and the communities they serve. to in effect tone round buoyancy to spurious disasters it is important to know the different employee segments inside organizations that lead to this drive. Each segment brings unique skills and perspectives creating a multifaceted approach to disaster preparedness Answer and recovery.

Frontline Workers

Frontline workers are often the first responders during a disaster making their role decisive in Applying eCombinency rules and providing immediate support to those affected. This section includes force inch sectors such as arsenic healthcare pinch services and important services care utilities and transfer. Their Teaching in crisis management first aid and rapid Answer enables them to act swiftly in eCombinencies. away armament frontline workers with perpetual education and supply organizations get importantly raise their content to answer in effect minimizing the affect of disasters along communities. Furthermore frontline workers often serve as the eyes and ears of an organization during crises relaying vital information that can guide decision-making and Supply allocation.

ECombinency Management Teams

Specialized eCombinency management teams are responsible for developing and executing comprehensive disaster Answer plans tailored to the specific risks faced by an organization. these teams bear chance assessments important vulnerabilities and make unjust strategies to palliate prospective impacts. Their expertise is essential for ensuring that all employees understand their roles and responsibilities in the event of a disaster. away on a regular basis updating and active these plans done Imitations and drills pinch direction teams get further amp acculturation of readiness passim the system. This proactive approach not only Improves organizational resilience but also Constructs confidence among employees allowing them to respond effectively during a crisis.

Human Supplys (HR)

Human Supplys plays a difficult role in ensuring employee safety and well-being during disasters. hour professionals are liable for development policies that back employees inch multiplication of crisis including communicating plans and Supplys for moral health back. By establishing clear Ruless for communication and assistance during eCombinencies HR can ensure that employees remain informed and supported. Furthermore hour get help education Meetings that set employees for prospective crises ensuring that everyone understands their Role inch the organization's calamity reaction frame. This focus on employee well-being is vital for maintaining morale and productivity during challenging times.

IT and Information Management

In our digital age IT and Information management professionals are indispensable in maintaining the technological infrastructure that supports disaster Answer efforts. they are liable for ensuring that important systems rest effective during eCombinencies facilitating communicating and information joint among employees and extraneous partners. By Applying robust backup systems and cybersecurity measures IT teams protect vital information that can be decisive during a disaster. their Role extends on the far side prompt Answer; they too lead to long retrieval away ensuring that information is available and good facultative organizations to survey trading operations arsenic quick arsenic possible leadership and management the Role of leading and direction inch nurture buoyancy cannot work exaggerated. Executives and managers set the tone for disaster preparedness within an organization prioritizing resilience in strategic planning and Supply allocation. their allegiance to nurture amp acculturation of readiness get prompt employees to read calamity reaction badly and employ inch proactive measures. By involving employees in the planning Method and soliciting their feedback leaders can Make a sense of ownership and responsibility further enhancing organizational resilience. good leading during crises too entails light communicating demonstrating foil and decision which get support employee bank and confidence community employment teams

Constructing sound relationships with community communities is important for good calamity reaction and retrieval. Community engagement teams focus on developing partnerships with local organizations government agencies and stakeholders to Improve collaboration during eCombinencies. away actively active inch public readiness initiatives and joint Supplys these teams get make amp iron back net that strengthens general buoyancy. Their efforts in community outreach not only foster trust but also ensure that organizations are well-Combined into the local disaster Answer ecosystem enabling a more coordinated and effective approach during crises.

Teaching and Development Specialists

Teaching and development specialists play a decisive role in ensuring that employees are equipped with the necessary skills for disaster readiness. they plan and apply education programs focused along crisis direction pinch reaction and intimate guard ensuring that employees are intimate and spread to work in effect inch eCombinencies. Regular Teaching Meetings and drills help reinforce these skills and keep them top-of-mind. Furthermore away tailoring education programs to particular roles and responsibilities organizations get check that complete employees read their alone contributions to calamity resilience sustainability and environmental teams are progressively established for their Role inch mitigating risks joint with spurious disasters. By assessing environmental vulnerabilities and promoting sustainable practices these teams can help organizations reduce their overall risk profile. initiatives such as arsenic calamity chance decrease mood Edition strategies and imagination preservation get importantly raise public buoyancy. also by advocating for environmentally sustainable practices these teams Add to long-term Answers that address the root causes of many natural disasters fostering a more resilient future for both organizations and the communities they serve.

A robust communication Plan is essential not only for information dissemination but also for Constructing organizational resilience ensuring employee safety and fostering community engagement. the top measure inch creating such as amp scheme involves scope light goals and Goals that array with both organizational aims and the broader charge of enhancing round buoyancy to spurious disasters. Primary goals should focus on enhancing preparedness facilitating swift Answers during crises and promoting recovery efforts post-disaster. particular Goals might admit the evolution of amp crisis communicating Layout bespoke to different types of spurious disasters the organization of amp feedback mechanics for perpetual advance and education hour force and employees along communicating Ruless and tools identifying name stakeholders is important for good communicating. In this context stakeholders include HR CMC members employees local communities government agencies and NGOs. piquant these stakeholders new inch the work helps check that the communicating scheme is general and good. For instance HR CMC members can lead the Plan's Application while employees serve as the primary audience for communication efforts. community communities get render important Understandings into particular vulnerabilities and Supplys free inch the case of amp calamity. Government agencies can ensure compliance with regulations and assist in

Supply sharing while NGOs can leverage their Webs to improve community outreach and support. Selecting appropriate communication channels is another vital Part of the Plan. amp combine of channels leave help good communicating ahead during and subsequently amp calamity. Email can be used for formal announcements and detailed updates while intranet platforms serve as centralized information hubs for Supplys and guidelines. versatile Uses are notably important for real-time alerts and updates ensuring that important Information reaches employees quick. Social media platforms can engage broader communities and serve as a space for sharing updates and Supplys. in addition workshops and education Meetings render reciprocal platform for cognition joint help employees read their roles and responsibilities inch calamity readiness and Answer .Message evolution is relevant to check lucidity and affect. important messages should focus on preparedness safety and support. for case messages get stress the grandness of existence knowledgeable and spread ensuring that guard is prioritized and reiterating that the system is engaged to support employees during and subsequently amp calamity. Tailoring messages for specific audiences is essential as different groups may have varying needs and concerns. hearing division get service trade targeted communications that come across with employees community communities and different stakeholders. also cultural sensitivity should be considered in global operations as communication styles and preferences may differ across regions.An effective crisis communication plan is the backbone of any Plan aimed at strengthening resilience. pre-disaster communicating need admit education Meetings to school employees along calamity readiness and reaction Ruless. Distributing Supplys such as eCombinency checklists and contact information ensures that everyone is well-informed and ready to act when a crisis arises. conducting Imitations or mimic drills allows the system to run communicating Ruless and important prospective weaknesses. During a disaster real-time updates are decisive. aggregate channels need work engaged to render current updates light book of instructions and cathartic back Supplys. It essential to share actionable guidance such as evacuation routes and shelter locations while also providing access to mental health Supplys to support employees' emotional well-being.

Post-disaster communication is equally important. subsequently amp crisis organizations need value the affect and pass findings to stakeholders. Providing recovery Supplys and information on available support services demonstrates the organization Addment to helping employees and communities recover. establishing amp feedback circuit done surveys and centre groups allows the system to meet Understandings from employees and stakeholders ensuring that lessons erudite are organic into prospective planning. Teaching and content construction are relevant for the winner of the communicating scheme. Regular Teaching Meetings for HR CMC members can Improve their crisis communication skills and prepare them to handle eCombinencies effectively. education need too run to complete employees offer workshops along calamity readiness reaction and retrieval. Providing Supplys for personal and community resilience can empower employees to take initiative in their safety and preparedness efforts. A feedback mechanism is vital for evaluating the effectiveness of the communication Plan. conducting post-crisis surveys to meet Understandings from employees and stakeholders allows the system to read the affect of its communicating efforts. Focus groups can Ease discussions to identify gaps and areas for Improvement while Effectiveness metrics can track Answer times message reach and employee engagement levels. stock Examinations of the communicating Layout regular every quarter or biannually check that the scheme clay pertinent and good inch fall of green Understandings and trump practices Constructing amp acculturation of buoyancy inside the system and its broader public is important for long winner. Awareness campaigns can promote ongoing education about natural disaster risks and preparedness measures. piquant employees inch resilience-Constructing activities such as arsenic volunteering for community calamity reaction initiatives fosters amp feel of public and mass duty. also fostering partnerships with local organizations Improves community resilience and expands the organization reach and impact during crises.

THE ROLE OF EMPLOYEE RETENTION IN CRISIS MANAGEMENT

During natural disasters organizations face unprecedented challenges. employee memory plays amp important Role inch however in effect amp line get answer to and find from such as events. A stable Encounterd workforce is essential for maintaining operational continuity during crises. employees forward with their roles procedures and the organization acculturation are break prepared to respond fleetly and in effect to disruptions knowledge retention: long employees bear valuable institutionalized cognition that is important during amp crisis. They understand Methodes Ruless and Supplys enabling quicker and more effective Answers. inch line amp great upset order get run to cognition gaps complication retrieval efforts cohesion and squad dynamics: amp preserved men fosters sound squad kinetics. Employees who have worked together over time develop trust and camaraderie which are vital during stressful situations. teams that bear secure relationships are further potential to cooperate in effect deal Information and back apiece different during crises emotional stability: inch the wake of spurious disasters cathartic back inside teams is relevant. Employees who feel secure in their jobs and connected to their colleagues are more likely to provide and seek support contributing to overall resilience. On the other hand great upset get run to cathartic imbalance arsenic green employees get look obscure and inferior supported long-term employees world health organization take to abide with associate in nursing system are much further endowed inch its winner. This translates into a proactive approach to disaster preparedness as engaged employees are more likely to participate in Teaching and Add to resilience-Constructing initiatives.

THE INTERCONNECTION WITH GLOBAL RESILIENCE

The concept of resilience extends beyond individual organizations to encompass entire communities and regions. arsenic businesses run amp difficult Role inch the efficient constancy of communities their power to hold and find from spurious disasters Adds to round buoyancy. Here how increased employee retention Improves this broader resilience:

Economic Stability: Organizations with high employee retention rates are often more economically stable. this constancy allows them to place inch calamity readiness and retrieval strategies which get palliate the affect of spurious disasters along the public. For instance businesses can allocate Supplys for infrastructure Improvements eCombinency Answer Teaching and community engagement initiatives.

Community Engagement: Companies that retain employees are more likely to engage in community-Constructing activities. long employees much look amp feel of duty to their public up to partnerships with community organizations engagement inch calamity ease efforts and back for public buoyancy initiatives.

Supply allocation: organizations with down upset get apportion Supplys further expeditiously. They can invest in Teaching programs that Improve employees' skills in eCombinency Answer and disaster management ultimately contributing to the community's capacity to respond to and recover from disasters.

Sustainable Practices: Retained employees can drive sustainable practices within organizations such as environmental stewardship and responsible Supply management. these practices cut the exposure of both the system and the public to the impacts of spurious disasters strategies for enhancing employee retention to tone both employee memory and round buoyancy organizations have apply good strategies that further amp ancillary and piquant be surround. important strategies include:

Career Development Opportunities: Providing employees with opportunities for professional growth and advancement increases their Addment to the organization. offer education programs mentorship and light vocation paths get importantly raise retention work-life balance: promoting amp good work-life correspondence get cut burnout and gain work atonement. Flexible work arrangements mental health support and time off policies are essential in creating an environment where employees feel valued.

Open Communication: Encouraging open and transparent communication fosters trust between employees and management. stock check-ins feedback Meetings and open-door policies leave employees to sound their concerns and lead to decision-making Method recognition and honor programs: recognizing and appreciated employees for their contributions boosts esprit de corps and reinforces their allegiance to the system. Simple gestures of appreciation can go a long way in retaining talent.

Employee Involvement in Disaster Preparedness: Involving employees in disaster preparedness initiatives not only Improves their skills but also increases their sense of ownership and responsibility. education Meetings Imitations and brainstorming Meetings get service form amp acculturation of readiness.

Presenting Analysis for Human Capital Analysis to decision makers involves a tactical approach to ensure data-driven insights are effectively communicated. Some practices based on the provided search results:

1. Understand the Audience:
    - Tailor communication to match decision makers' knowledge level and preferences.
    - Highlight how Analysis for Human Capital Analysis impact key business objectives
2. Simplify Complex Data:
    - Utilize visual representations like graphs and charts for easier interpretation.
    - Explain concepts in simple, understandable language
3. Storytelling with Data:
    - Frame data within a compelling narrative that connects analysis to real-world scenarios.
    - Emphasize key findings and actionable insights derived from the data
4. Building Credibility:
    - Ensure data accuracy and sources to build trust with decision makers.
    - Explain the methodology behind the analysis to demonstrate credibility
5. Engaging Stakeholders:
    - Foster open discussions to address questions and concerns from decision makers.
    - Solicit feedback to continuously improve communication effectiveness
6. Leveraging Technology:
    - Use data visualization tools for interactive visualizations.
    - Implement automated reporting systems for regular updates on Human Capital Analysis calculations
7. Developing Data Storytelling Skills:

- Create a mission statement to guide the translation of analysis into a story.
- Ensure presentations carry a clear "point" to increase support from decision makers

8. Know Your Audience:
   - Frame presentations differently setup on the audience's relationship with analysis.
   - Tailor the level of technical details in the presentation accordingly
9. Embrace Context:
   - Provide context by considering temporal changes, industry benchmarks, or existing calculations.
   - Context can transform dull analysis into a compelling story
10. Explain Why It Matters:
    - Clearly articulate how the analysis affects the organization and why it is relevant to the executive team.
    - Connect the data insights to tangible outcomes for the workplace.

By incorporating these best practices, HR professionals can effectively present Analysis for Human Capital Analysis to decision makers, enabling informed in based of data-driven insights.

1. Identify Meaningful Indicators:
   - Pinpoint indicators that fill with your organization's imp. Goals and questions to measure effectiveness in specific areas.
   - Focus on indicators that directly contribute to achieving desired outcomes and delivering tactical goals
2. Assess HR Outcomes:
   - Evaluate whether HR person is delivering final outcomes essential for business success by measuring factors like employee satisfaction, engagement levels, churn rate, absenteeism rate, and training impact.
   - Common calculations for assessing outcomes include employee, employee satisfaction index, and absenteeism rate
3. Analyze Internal Efficiency:
   - Measure how efficiently HR services are being delivered by assessing recruitment channels, competency recruitment effectiveness, costs-per-hire, training costs, and training effectiveness.
   - Significant metrics about the effectiveness of internal works include return on final investment (ROI), cost-per- final hire, and avg. time to fill jobs.
4. Audit HR Compliance:

- Monitor HR compliance with relevant requirements but delegate tactical tasks like planning to other relevant individuals for organizational objectives.
- The range of diversity initiatives, the ratio of the gender pay gap, the percentage of diversity, and the percentage of workers who are schooled in business rules are among the indicators that can be stated to determine violation of the audit.

5. Evaluate Service Delivery:
   - Factors like as HR service quality, induction programme efficacy, employee development initiatives, and performance management feedback may be used to measure the efficacy of HR service delivery.
   - Typical criteria used to measure service quality include ROI, employee happiness with benefits, training efficacy, and new recruit productivity.

Human Resource Analysis link trajectory of positive organizational transformation largely. Businesses may get to the root cause of the problems and implement 'solution-ready' approaches by leveraging information on place factor related performance, engagement, and retention among other data sources. Among various insights drawn from the perspective of analysis, a reduction in staff attrition could be one possible finding, or a clear reason for a low level of output identified. The employers may get the particular information to form the necessary plans for increasing the workplace satisfaction, policies procedures, and thus offering the deserved opportunities for employee career development. Such as labor trend forecasts and barrier estimations, are also provided by allocating the Human Capital Analysis for the research operations, which allows the businesses to adapt their plans earlier. Finally, Analysis for Human Capital Analysis resonates with organizations who want to achieve more with less and face complex business decisions. Effectively communicating Human Capital Analysis (HCA) analysis to decision-makers involves several important strategies that can not only Improve the clarity and relevance. First, it is vital to read your audience ground concerns and specific interests which enables you to In a sense, tailor with connection impact. Simplifying Complicated Information visual representation in the form of charts and graphs can break down complex information applies in a easier-to-understand way and highlighting a few specific HCA metrics that directly link back too issues lower costs of effectively makes research presented more understandable. Furthermore, people tend to remember stories better than simply facts thus help results if possible considering incorporating real-life case studies or even what-if scenarios as those typically Drive home how distortions highlighting producing clear Haven't & disappointing HR are indeed organisational challenging recommendations. Lowering reinforcing your summary buyers process in employees selection chapter-last procurement market order internally sell education use-or-lose - cases role whom downsides challenge one-size-fits-all secondary structural turnover organize mentoring amount content well-defined putting sample effects reduce worth feel intimidated believe further. The value proposition for enhancing resilience lies in the multifaceted benefits it brings to society—reducing human suffering safeguarding livelihoods protecting infrastructure and ensuring sustainable development. to in effect pass this rate proffer light communicating goals are important guidance however Information is sent to stakeholders including government's ngos and the public the home rate proffer of invigorating round buoyancy to spurious disasters rDevelops round proactive chance direction investing inch base and public readiness. By adopting a proactive approach societies can mitigate the adverse impacts of disasters which often result in significant economic losses and long-term social disruption. this involves development iron systems that get hold shocks such as arsenic live Constructing better drain systems and new exemplary technologies. The economic rationale is compelling: for every dollar invested in disaster preparedness studies indicate that countries can save up to seven dollars in Answer and recovery costs. this not but highlights amp go fiscal scheme just too Highlights the right duty to protect tender populations the communicating goals have array with the rate proffer. First raising awareness is difficult. numerous individuals and communities lowball the frequency and rigor of spurious disasters. Effective communication strategies should aim to educate the public about the risks and the necessity of resilience-Constructing measures. this get affect campaigns that deal statistics along calamity impacts intimate stories of smitten individuals and Information along hitch actions that get work affected astatine both public and house levels. By using relatable narratives stakeholders can Make an emotional connection fostering a sense of urgency and agency.

Engaging stakeholders is essential. buoyancy is not only amp politics responsibility; it requires amp collaborative access involving community communities close sphere partners and global organizations. Communication strategies should focus on fostering partnerships by highlighting successful collaborative efforts and their outcomes. away showcasing suit studies of communities that bear with success enforced buoyancy measures stakeholders get prompt others to be become. Engaging local leaders and influencers can also amplify these messages ensuring that they resonate within communities. governments and organizations have known the efficient benefits of investment inch calamity readiness and reaction. Communication strategies should include targeted messaging aimed at decision-makers presenting compelling Information and case studies that demonstrate the cost-effectiveness of resilience-Constructing initiatives. for case reports and infographics illustrating the take along investing for buoyancy projects get Method arsenic right tools to sway policymakers and stakeholders to apportion inevitable Supply fostering amp acculturation of readiness is difficult. Effective communication should promote behaviors and practices that Improve community resilience. this includes auspicious individuals to enter inch community calamity readiness drills make pinch plans and arise public back Webs. Social media campaigns workshops and community events can serve as platforms for disseminating information and Constructing Webs of support. A way devising readiness amp public average individuals are further potential to read proactive measures finally reduction vulnerability it is relevant to work the cathartic and mental aspects of calamity readiness and buoyancy. Disasters can have profound effects on mental health and communication strategies must acknowledge and address these concerns. provision Supplys for mental back creating good spaces for word and promoting moral health consciousness arsenic break of resilience-Constructing efforts get service communities deal break with disasters. This holistic approach not only strengthens physical infrastructure but also nurtures the social fabric of communities. Tailor messages to diverse audiences. disparate groups—ranging from policymakers to community residents—have variable levels of cognition and concern inch calamity buoyancy. Communication strategies should be Adjustable using language and formats that resonate with each audience. for case policymakers get answer good to Fancy reports and abstract presentations spell public members get choose piquant gregarious mass media happy or optic storytelling. Inclusivity in communication ensures that all voices are heard and that the messaging is accessible to everyone including marginalized communities who often bear the brunt of disasters. Utilizing Tech and Creative communication platforms can Improve outreach efforts. versatile Uses that render real-time calamity alerts gregarious mass media platform for public employment and reciprocal websites with Supplys along calamity readiness get hand wider audiences. The integration of Tech in communication not only disseminates difficult information swiftly but also Fosters active participation from communities in resilience-Constructing activities. It is vital to evaluate the effectiveness of communication strategies regularly. feedback mechanisms surveys and public discussions get render Understandings into what messages come across and which strategies are about good. Continuous Improvement based on this feedback loop ensures that communication efforts remain relevant and impactful.

A company's capital projects, along with their ROI and the extent of their effect on the business will have to be appraised. For organizations, it is necessary to assess the financial gains as well strategic contributions of such programs, Various stakeholders for future endeavors and when there to justify the investments in Human Capital Analysis. Parameters of capital such as revenue per employee, productivity, and staff turnover that will be used to measure the progress will occupy a prominent position. Companies will make projections of profits before and after the programs and determine whether there is an economical return on their Human Capital Analysis investments. Organizations could apply case studies and "good news" items in their publications as well to demonstrate the importance of investing in Human Capital Analysis and the improved business results it produces. An effective ROI and business impact envisagement of Human Capital Analysis initiatives for stakeholders' is imperative various stakeholders. HCA analysis communication with the use of effective information to the decision makers would be a smart decision for HR people to be [able] of their skills, such as clarity and relevance. A good start would be to first get a basic understanding of the interests and worries of the respective audience; this is the way you will be able to link your insights to their needs. Among other things that have been made straightforward, the use of charts and graphs to illustrate the main performance indicators in an easy-to-understand way has been the most important. Another requisite for a fruitful approach is to select a few pivotal observations that are relevant to the organizational mission, and by doing so you can stress that these figures are affecting business measures like productivity, job satisfaction, and turnover rates. Prioritize the most critical findings to keep away from overwhelming your target audience, and offer clean, actionable steps that decision-makers can take based totally to your evaluation. Encouraging dialogue is important; invite questions and remarks to foster collaboration and deepen knowledge. Finally, comply with up with additional resources or summaries to boost your message and maintain the communiqué alive. By integrating these strategies—understanding your audience, simplifying records, linking insights to commercial enterprise effects, storytelling, prioritizing key points, encouraging speak, and making sure comply with-up—you could efficiently bring the fee of HCA analysis, facilitating informed selection-making that enhances organizational performance.

Cultivating an environment of collaboration and discussion is fundamental to encourage the understanding and selection of Analysis for Human Capital Analysis. Capital inside the organization by empowering open exchange between partners such as supervisors, pioneers and persons, organizations can make straightforwardness and consideration in social Human Capital Analysis investigation activities. This collaborative approach permits individuals from distinctive foundations and disciplines to share their conclusions, bits of knowledge, and concerns approximately Human Capital Analysis information and analysis. Generally, cultivating a collaborative and conversational environment is fundamental to advancing successful utilized of Human Capital Analysis examination and accomplishing organizational victory. To guarantee that decision makers understand the importance of Human Capital Analysis, it is imperative to maintain continuous communication and updates. By periodically updating decision makers, initiatives can inform them about the latest findings, advances, and results in Analysis for Human Capital Analysis. Frequent communication builds trust, encourages form, and empowers decision-makers to make well-informed choices that align with the organization's goals and tactical priorities. Additionally, it enables decision makers to provide feedback and insights, fostering a collaborative effort to leverage HR analysis to improve business results.

TRAINING AND DEVELOPMENT WORK METRICS

Training and development work metrics play a pivotal role in shaping the success of organizations by providing invaluable insights into the efficacy and impact of learning initiatives. These metrics serve as the compass guiding HR and L&D professionals in navigating the complex landscape of employee development. Framework lays the concept of measuring progress and performance across various dimensions, from individual skill acquisition to organizational growth. One of the fundamental metrics is completion rates, which offer a snapshot of employee engagement and compliance with training mandates. Commitment of participants. Similarly, training costs serve as a tangible indicator of the investment required to upskill the workforce. Additionally, training effectiveness metrics provide a comprehensive evaluation of the extent to which training objectives are achieved and learning is transferred to job performance. Through pre- and post-training assessments, skills demonstrations, and performance evaluations, organizations can gauge the tangible outcomes of training initiatives and identify areas for refinement. Learning engagement metrics offer insights into the level of participation, interaction, and interest generated by training programs, informing adjustments to content delivery and instructional methods. Similarly, learning progression metrics track the advancement of employees through various stages of training, identifying bottlenecks or gaps in the learning journey.

- Address external labour market and career challenges, such as job opportunities.
    - o Alternative professional routes accessible to each person, as well as disparities in their chosen career .

Job transitions can lead to trade-offs in remuneration for future development opportunities. What are the work design's profit and loss assumptions, including alternatives for paying higher or lower wages .How to choose whether to acquire or construct a skill set, including outsourcing individual jobs or production processes? Finally, benchmarking against industry standards and best practices enables organizations to contextualize their training metrics and identify opportunities for continuous improvement. In conclusion, training and development metrics serve as the compass guiding organizations on their journey toward building a skilled, engaged, and high-performing workforce, driving sustainable growth and success in an ever-evolving business landscape.

CONCLUSION

Effectively communicating Human Capital Analysis analysis to decision makers is essential to building an informed and tactical place force. The organization design approach focuses on addressing enterprise-wide behaviour challenges, such as the effectiveness of procedures across divisions and functions. Developing a powerful communiqué approach in collaboration with the HR CMC is essential for strengthening international resilience to herbal disasters. By placing clean desires, enticing various stakeholders, making use of various communication channels, and crafting tailored messages, businesses can prepare for, respond to, and recover from failures more efficiently. Regular training and capability building, coupled with a robust comments mechanism, will make certain continuous improvement of the approach, enabling groups to evolve to evolving challenges. Ultimately, fostering a lifestyle of resilience inside the corporation and its groups will empower all stakeholders to face the increasing danger of herbal disasters with self belief and preparedness. The labour markets model focuses on effectiveness and success.

The most profound ideas required only a portion of the time. Identifying, obtaining, and collecting data is often the most challenging aspect of the job. HR generalists, not analytic specialists, must master these duties and competences to improve analytical decision-making in the HR department. Most people can master them. Another major concern is the time required for in-depth investigation. The models covered here, including COM, labour markets, and organizational design, can be used as diagnostics through logic exercises when time is limited for intensive analysis. Human capital Research (HCA) Research for efficient ones requires that they use a clear, profitable, and involved one. Strengthening worldwide resilience to herbal disasters is a multifaceted endeavor that calls for the collaboration and commitment of numerous employee segments within corporations. From frontline people and emergency management teams to HR specialists and network engagement groups, every phase plays a vital position in improving disaster preparedness, reaction, and restoration efforts. By fostering a way of life of resilience and investing within the education and assets wanted for every group, organizations can create a group of workers able to navigating the complexities of catastrophe reaction. This collective attempt not simplest protects the company however also contributes to the broader aim of world resilience, making sure that groups are better equipped to withstand and get over the challenges posed through natural disasters. It's important to understand your caregivers; Knowing their background, interests, and specific concerns allows you to tailor your message accordingly. This personalization helps make your insights more connected. To avoid overwhelming stakeholders with data, simplify complex information with clear visuals like charts, graphs, establish a few important metrics with organization emphasize alignment of objectives It is important to frame your analysis in terms of business outcomes -Use examples from studies or the real world to further illustrate the possible implications of your findings. A compelling story can increase engagement, so consider creating a story that highlights challenges, insights, and actionable recommendations. The relationship between greater labor retention and global resilience to natural disasters is multifaceted and deep. Organizations that prioritize employee retention benefit from a strong and competent workforce that can effectively solve problems, helping to drive business continuity and recovery. Furthermore, as these organizations engage with their communities and invest in resilience initiatives, they increase the stability and well-being of all places where they work While natural disasters are ongoing and den due to climate change, the importance of nutrition involved workers Cannot be overstated. By adopting strategies that encourage employee retention, organizations not only achieve their success but also play an important role in creating resilient communities that can meet the challenges posed by an unpredictable world edge in this interconnected landscape, the collaboration between workforce retention and global resilience is a powerful force for positive change. This approach not only makes the data more communicable, but also helps save memory. The value proposition of strengthening international resilience to herbal screw ups encompasses monetary savings, more desirable protection, and sustainable development. To efficiently bring this cost, communication dreams ought to attention on elevating focus, attractive stakeholders, encouraging funding, fostering a tradition of preparedness, addressing emotional impacts, tailoring messages for various audiences, leveraging technology, and evaluating strategies for non-stop development. By aligning the verbal exchange efforts with the price proposition, stakeholders can work collaboratively to construct a more resilient world, in the long run lowering the devastating impacts of herbal disasters on communities and societies. Through effective communiqué, we will rework know-how into motion, ensuring that resilience turns into a shared international precedence. It is important to prioritize key insights; Because decision makers are often faced with information overload, summaries of important findings can facilitate rapid decision-making. This, coupled with clear and actionable recommendations that empower decision makers to take action based on your research, is another important element that encourages conversation that emphasizes potential value seeds on your suggestions; Inviting questions fosters collaboration and in-depth thinking, while demonstrating your willingness to make customized recommendations based on their feedback This does not lessen the importance of statistical analysis in HR decision-making, but rather enhances it by providing more opportunity to detect HR issues through advanced analysis. Clear alignment with organizational goals, use of data visualization and focus on actionable insights make the presentation more engaging and impressive. In addition, keeping open lines of communication, adapting to audience needs, and including real-world examples help bridge the gap between complex analysis and practical decision-making.

REFERENCES

Becker, B., & Huselid, M. (2006). Strategic human resources management: Where do we go from here? *Journal of Management, 32*(6), 898–925. DOI: 10.1177/0149206306293668

Becker, B. E., & Hustled, M. A. (2006). *Strategic Human Resources Management: A Key to Competitiveness*. In The Academy of Management Perspectives.

Berlin, J. (2016). *Deloitte Global Human Capital Trends Report*. Deloitte.

Boudreau, J. W., & Ramstad, P. M. (2007). *Beyond HR: The New Science of Human Capital*. Harvard Business Review Press.

Cascio, W. F. (2018). Leveraging Analysis for Human Capital Analysis for competitive advantage. *Organizational Dynamics, 47*(2), 98–106.

Casco, W. F. (2014). *Managing Human Resources: Productivity, Quality of Work Life, Profits*. McGraw-Hill.

Deloitte. (2017). *Global Human Capital Analysis trends 2017: Rewriting the rules for the digital age*. Deloitte University Press.

Gallup. (2018). State of the American Workplace: Employee Engagement Insights for U.S. Business Leaders.

Hackett Group. (2019). Human Capital Management: Benchmarking for Business Success.

Hunt, S. T., Baltes, B. B., & DeHart, W. B. (2018). Analysis for Human Capital Analysis: A review of literature and implications for research. *Human Resource Management Review, 28*(3), 347–358.

Jain, S., & Singh, S. (2020). Impact of Human Capital on Organizational Performance: A Review of Literature. Journal of Human Resource Management.

Kearns, P. (2012). Managing the Human Capital: Performance Management in the 21st Century. *Journal of Management Development*.

McKinsey & Company. (2021). *The Future of Work After COVID-19*. McKinsey Global Institute.

Phillips, J. J., & Phillips, P. P. (2005). *ROI at Work: Best Practices in the Real World*. Barrett-Koehler Publishers.

Poulakis, E. D., & O'Leary, R. (2011). Why the Performance Management System Fails: How to Fix It. *Industrial and Organizational Psychology: Perspectives on Science and Practice.*

Rothwell, W. J. (2010). The Strategic Development of Human Capital. In Strategic Human Resource Development.

Rothwell, W. J., & Kazanas, H. C. (2003). *Mastering the instructional design process: A systematic approach* (3rd ed.). Pfeiffer.

SHRM. (2016). *Human Capital Analytics: Driving Value from Data.* Society for Human Resource Management.

Silzer, R., & Dowell, B. E. (2010). *Strategy-driven talent management: A leadership imperative.* John Wiley & Sons.

Ulrich, D., & Brockbank, W. (2005). The HR Value Proposition. Harvard Business School Press.

Wright, P. M., & Nishii, L. H. (2007). Strategic HRM and Organizational Behavior: Integrating Multiple Levels of Analysis. In Research in Personnel and Human Resources Management.

Chapter 6
Embracing Local Knowledge:
Integrating Traditional Practices in Disaster Risk Reduction

Menard Musendekwa
https://orcid.org/0000-0002-6644-8727
Reformed Church University, Zimbabwe

ABSTRACT

This chapter examines the pivotal role of local knowledge and traditional practices in enhancing disaster risk reduction (DRR) strategies. Communities across the globe often rely on indigenous wisdom and cultural heritage to address environmental challenges, mitigate hazards, and strengthen resilience to disasters. By emphasizing the value of local expertise, this chapter seeks to bridge the gap between traditional knowledge systems and formal disaster management frameworks. It advocates for a more holistic and inclusive approach to disaster preparedness, where community-based practices are integrated with modern DRR strategies to foster sustainable resilience.

INTRODUCTION

In recent years, the importance of disaster risk reduction (DRR) has become increasingly apparent, as communities worldwide face the escalating impacts of climate change, environmental degradation, and natural hazards. While formal disaster management frameworks have made significant strides in enhancing preparedness and response strategies, they often overlook the wealth of knowledge embedded in local communities. Indigenous wisdom, cultural heritage, and traditional practices

DOI: 10.4018/979-8-3693-9745-9.ch006

have long played a vital role in helping societies adapt to environmental challenges, mitigate hazards, and build resilience. This chapter seeks to highlight the indispensable contribution of local knowledge systems in DRR efforts and explore ways to integrate them with formal disaster management approaches.

Traditional knowledge, passed down through generations, offers context-specific insights into environmental cycles, hazard patterns, and resource management that are not only sustainable but also deeply embedded in the social and cultural fabric of communities. For instance, Berkes (2018) emphasizes that indigenous ecological knowledge is rooted in long-term observations and experiences, making it highly relevant for contemporary DRR initiatives. Such knowledge often aligns with nature-based solutions, which are increasingly recognized as essential in sustainable disaster management frameworks (Sudmeier-Rieux et al., 2019). The inclusion of local knowledge in formal disaster management can offer a more holistic and inclusive approach to disaster preparedness, one that respects the interconnectedness of people and their environment (Mercer et al., 2010).

BACKGROUND

The growing frequency and intensity of natural disasters globally have underscored the importance of effective disaster risk reduction (DRR) strategies. Climate change, environmental degradation, and rapid population growth have increased the vulnerability of communities to hazards such as floods, droughts, earthquakes, and cyclones. In response, formal DRR frameworks have been developed by governments and international organizations, often emphasizing technology-driven solutions, data collection, and top-down management (UNDRR, 2023). While these approaches are valuable, they often fall short in addressing the localized nature of disaster risk and the socio-cultural complexities within vulnerable communities.

In contrast, indigenous knowledge and traditional practices, which have evolved over centuries, provide communities with sustainable methods to manage their environments and adapt to risks. This local knowledge, often referred to as **indigenous knowledge systems (IKS)**, includes practices such as early warning systems based on natural indicators, sustainable land use, and adaptive resource management. These practices, developed through an intimate relationship with the environment, have enabled communities to mitigate the impact of natural hazards and enhance resilience. Scholars like Berkes (2018) and Nyong et al. (2021) have emphasized the role of traditional ecological knowledge in fostering long-term sustainability and resilience to disasters.

Historically, however, formal disaster management frameworks have marginalized local knowledge, favoring scientific and technological expertise while overlooking the value of traditional practices (Mercer et al., 2010). This has resulted in a disconnect between top-down policies and the realities on the ground, where indigenous communities possess unique insights into managing environmental risks. Power imbalances and an epistemological divide between scientific and indigenous knowledge have often sidelined local communities from decision-making processes in disaster risk management (McAdoo et al., 2009).

Recent developments in the field of DRR have seen a growing recognition of the importance of integrating local knowledge into formal frameworks. Global institutions such as the **United Nations Office for Disaster Risk Reduction (UNDRR)** now advocate for community-based approaches that involve local communities as active participants in disaster preparedness and response (UNDRR, 2023). This shift towards a more inclusive approach aligns with broader global trends, such as the Sustainable Development Goals (SDGs), which emphasize the empowerment of marginalized communities and the integration of indigenous knowledge in addressing global challenges (UN, 2022). Research has demonstrated that hybrid models combining scientific knowledge with traditional practices can result in more effective and culturally appropriate DRR strategies (Sudmeier-Rieux et al., 2019).

Recent case studies have highlighted successful applications of indigenous knowledge in disaster management. For example, in the Philippines, local communities have used traditional farming techniques and indigenous warning systems to mitigate the impact of typhoons (Gaillard & Mercer, 2022). Similarly, in Kenya, Maasai communities have relied on their understanding of local ecosystems and weather patterns to manage drought risk, enhancing their resilience to climate-induced hazards (Nguyen et al., 2023). These examples demonstrate the potential for indigenous knowledge to complement scientific approaches, providing a more holistic and context-specific strategy for disaster risk management.

The growing recognition of indigenous knowledge reflects a broader movement towards **decolonizing disaster management**. Scholars and policymakers are increasingly advocating for the inclusion of marginalized voices in DRR, emphasizing the need for participatory approaches that respect local cultural identities and knowledge systems (Mavhura, 2021). This movement challenges the dominance of Western scientific knowledge and calls for a more equitable distribution of power in disaster management, where local communities play a central role in decision-making processes.

As disasters become more frequent and severe due to global environmental changes, there is an urgent need to bridge the gap between traditional knowledge systems and formal DRR frameworks. This chapter contributes to this ongoing discourse by examining the role of local knowledge in disaster management, highlighting case studies that illustrate the successful integration of indigenous practices, and presenting a theoretical framework to guide future efforts. By doing so, it seeks to demonstrate how indigenous wisdom can complement formal approaches, enhancing community resilience and fostering more inclusive and effective DRR strategies.

The objectives of this chapter are to examine the various types of local knowledge and traditional practices relevant to disaster risk reduction (DRR) and to analyze how these practices are integrated into formal DRR frameworks, particularly in their impact on enhancing community resilience. Additionally, the chapter aims to present case studies that demonstrate successful applications of traditional knowledge in contemporary disaster management, while also emphasizing the importance of respecting cultural identities and fostering collaboration between local communities and external stakeholders in creating sustainable and inclusive disaster preparedness strategies.

LITERATURE REVIEW

Disaster risk reduction (DRR) has garnered increasing attention in both academic and policy circles, driven by the urgent need to mitigate the impact of natural hazards, which are exacerbated by climate change and environmental degradation. While formal DRR frameworks have been widely studied and implemented, there is growing recognition of the value of local knowledge and traditional practices in building resilience to disasters. This section reviews existing literature on the intersection of traditional knowledge and DRR and presents a theoretical framework to guide the integration of indigenous wisdom into modern disaster management strategies.

Theoretical Framework: Ecological Resilience Theory

The theoretical foundation for this chapter is rooted in Ecological Resilience Theory, which emphasizes the capacity of social-ecological systems to absorb disturbances while maintaining their core functions and structures (Folke et al., 2010). In the context of DRR, resilience is understood as the ability of communities to adapt to and recover from hazards without significant disruption. Local knowledge systems, shaped by generations of interaction with the environment, provide communities with adaptive strategies that enhance resilience. These strategies include risk perception, early warning systems based on environmental signals, and sustainable resource management practices (Berkes & Folke, 2002). Integrating local knowledge within formal DRR frameworks aligns with resilience theory, advocating a bottom-up, community-centered approach to disaster management.

Recent scholarship, particularly from Zimbabwe and Southern Africa, contributes to the understanding of resilience by exploring how traditional practices, such as indigenous rainmaking ceremonies and land-use strategies, play a critical role in mitigating disaster risks (Mavhura, 2021). These studies emphasize the importance of preserving cultural identities and integrating local values into formal disaster planning to ensure sustainable community development.

Local Knowledge and Traditional Practices in DRR

Local knowledge and traditional practices, often referred to as indigenous knowledge systems (IKS), are integral to the way many communities manage risk and adapt to environmental challenges. Scholars such as Agrawal (1995) and Berkes (2018) argue that indigenous knowledge is context-specific, evolving through direct experience and long-term observation of ecological processes. This knowledge encompasses practical skills, such as the use of natural indicators for weather forecasting and sustainable land management practices that reduce vulnerability to hazards.

In Southern Africa, indigenous knowledge has been instrumental in the management of droughts and floods, particularly in rural areas where formal disaster frameworks are limited. Zimbabwean scholars like Chanza and de Wit (2016) argue that indigenous knowledge of riverine flood cycles, soil moisture levels, and the behavior of livestock during droughts provides communities with effective early warning systems that complement modern technologies. These practices contribute to localized disaster risk management, enhancing the resilience of vulnerable populations.

Globally, case studies illustrate the relevance of traditional knowledge to DRR. For example, in the Pacific Islands, indigenous communities have developed strategies to cope with tropical cyclones, including the identification of natural signs that predict storms and the strategic planting of trees as windbreaks (Campbell, 2009). Similarly, in sub-Saharan Africa, local knowledge of drought cycles informs water conservation practices that reduce vulnerability to prolonged dry spells (Nyong et al., 2007). These examples underscore the importance of integrating indigenous knowledge into formal disaster management plans.

Integrating Local Knowledge into Formal DRR Frameworks

One of the central debates in the literature concerns how to effectively integrate indigenous knowledge into formal DRR frameworks, which are often technocratic and top-down in approach. The hybrid knowledge framework, proposed by Mercer et al. (2010), offers a way forward by promoting the convergence of local and scientific knowledge in disaster management. This framework suggests that neither traditional nor scientific knowledge alone is sufficient for effective DRR; rather, a synthesis of the two can provide a more comprehensive understanding of risks and responses.

Research in Zimbabwe highlights this approach. Mavhura et al. (2020) explored how community-based disaster risk reduction (CBDRR) models, which prioritize local knowledge and participation, are being applied in rural Zimbabwe to address both drought and flood risks. These efforts align with the Education 5.0 philosophy—a transformative education model in Zimbabwe aimed at enhancing innovation and community engagement, promoting solutions rooted in local knowledge and research (Chirisa, 2021). The CBDRR approach empowers communities to actively engage in hazard mapping, vulnerability assessments, and co-create risk reduction strategies alongside formal agencies (Gaillard & Mercer, 2013).

Challenges and Opportunities

Despite the growing recognition of the value of indigenous knowledge, significant challenges remain in integrating it into formal DRR frameworks. One of the main obstacles is the epistemological divide between traditional knowledge systems, which are often qualitative and context-specific, and scientific approaches that prioritize quantifiable data and universal models (McAdoo et al., 2009). Additionally, power imbalances between local communities and external agencies often result in the marginalization of indigenous voices in disaster planning (Mercer et al., 2010).

In Zimbabwe, for instance, rural communities have often been excluded from formal disaster planning, even though local farmers and elders possess vital knowledge about environmental risks (Chanza & de Wit, 2016). Bridging this gap requires a rethinking of disaster governance, where local communities are empowered as partners in risk management, rather than passive recipients of external aid.

However, opportunities exist to bridge this gap. The rise of participatory methodologies in DRR, such as Participatory Rural Appraisal (PRA) and Community-Based Participatory Research (CBPR), provides frameworks through which local knowledge can be systematically documented and incorporated into disaster management (Chambers, 1994). These methods emphasize the co-production of knowledge and ensure that community members are active agents in the process of risk reduction.

Conclusion

The literature overwhelmingly supports the view that local knowledge and traditional practices are valuable assets in DRR, contributing to both prevention and response strategies. The integration of indigenous knowledge into formal DRR frameworks, as suggested by the hybrid knowledge framework, can enhance community resilience and foster more inclusive and sustainable disaster management.

As disasters become more frequent and severe due to global environmental changes, the need for such integration becomes even more urgent. Through a combination of theoretical insights and practical case studies, this chapter will explore how traditional practices can be effectively harnessed to complement modern DRR strategies.

METHODOLOGY

This chapter adopts a qualitative research methodology to explore the integration of local knowledge and traditional practices in disaster risk reduction (DRR). The research primarily relies on a comprehensive literature review, supported by relevant case studies, theoretical frameworks, and policy documents. The goal is to examine how indigenous knowledge systems contribute to enhancing community resilience and to explore the barriers and opportunities for integrating these systems into formal disaster management frameworks.

The literature review process involved systematically sourcing scholarly articles, books, government reports, and case studies that focus on the role of indigenous knowledge in DRR. Key academic databases such as JSTOR, Google Scholar, and IGI Global were utilized to identify peer-reviewed sources published within the past two decades, with a particular emphasis on recent research that addresses the convergence of local knowledge and formal disaster management strategies. Special attention was given to sources that provide insights into the theoretical foundations of ecological resilience and the practical applications of indigenous knowledge in real-world disaster scenarios. Zimbabwean scholars and publications, including those focusing on Southern Africa, were prioritized to ensure contextual relevance. Notable works include those by authors such as Gaillard and Mercer (2013), who discuss the integration of local and scientific knowledge in DRR, and Nyong et al. (2007), who examine the role of indigenous knowledge in climate change adaptation in Africa.

In addition to the literature review, the chapter draws on case studies from diverse global regions to illustrate the practical application of indigenous knowledge in disaster resilience. For example, the study of coastal communities in the Pacific Islands, as highlighted by Campbell (2009), demonstrates how indigenous strategies, such as identifying natural indicators of storms and planting protective vegetation, have enhanced local disaster preparedness. Similarly, in sub-Saharan Africa, the use of local knowledge to manage droughts, as documented by Nyong et al. (2007), provides valuable lessons for integrating traditional practices into formal DRR frameworks. These case studies were selected based on their alignment with the principles of ecological resilience and their ability to highlight the value of community-centered disaster management approaches.

Theoretical insights were guided by **Ecological Resilience Theory**, which posits that social-ecological systems have the capacity to absorb disturbances and adapt without losing their essential functions (Folke et al., 2010). This theory underscores the significance of local knowledge as an adaptive strategy that strengthens resilience by enabling communities to recognize environmental changes and respond effectively. Local knowledge, often passed down through generations, encompasses risk perception, early warning systems based on natural indicators, and sustainable resource management practices that are integral to community-level resilience (Berkes & Folke, 2002). This theoretical framework is instrumental in understanding how indigenous wisdom can be harmonized with modern DRR strategies to build more resilient communities.

In sum, the methodology employed in this chapter is qualitative and data-driven, focusing on a thorough review of academic and policy sources, coupled with case studies that provide concrete examples of the successful application of local knowledge in disaster management. By integrating both theoretical and empirical findings, the chapter aims to contribute to the growing body of knowledge on the integration of traditional practices into formal DRR frameworks.

TYPES OF LOCAL KNOWLEDGE

Local knowledge is often understood as the body of wisdom, practices, and insights that communities have developed over generations through their interactions with the environment and their experiences with natural hazards. This knowledge is typically passed down orally or through practice, adapting to the specific needs and challenges faced by each community. The chapter categorizes local knowledge into several distinct forms, each contributing uniquely to disaster risk reduction (DRR) and resilience-building efforts.

Environmental Awareness: Environmental awareness is one of the most fundamental forms of local knowledge, encompassing the community's understanding of natural phenomena, such as weather patterns, seasonal cycles, and land features. This knowledge is often drawn from the long-term observation of the local environment, providing valuable insights into the behaviors of hazards like floods, droughts, or storms. For example, many indigenous communities possess deep knowledge of their local ecosystems, including indicators such as animal behavior, plant growth, or changes in the sky that signal impending weather events or environmental shifts. These environmental cues can form the basis for early warning systems, allowing communities to anticipate and prepare for disasters. In Zimbabwe, rural communities often observe the behavior of particular bird species or changes in the wind patterns to predict the onset of rains or droughts, a form of indigenous meteorology that has proven valuable in agricultural and water management practices (Nyong et al., 2007).

Historical Practices: Historical practices refer to strategies and techniques developed over generations that have been proven effective in mitigating the impacts of specific hazards. These practices are grounded in a community's historical experience with natural disasters and their ability to adapt to and recover from such events. They include methods such as crop rotation, water conservation techniques, and construction practices that are tailored to the local environment. For instance, in areas prone to flooding, communities might have developed elevated house designs or flood-resistant farming methods. These historical practices are often rooted in a community's collective memory and are continuously adapted as new challenges emerge. In Zimbabwe, for example, traditional methods of water conservation, such as rainwater harvesting and the construction of reservoirs, have been integral to managing droughts, a recurring challenge in the region (Berkes, 2018).

Community Rituals and Cultural Practices: Community rituals and cultural practices also form an important category of local knowledge, particularly in the context of disaster preparedness and response. These rituals often include ceremonies, taboos, and collective activities designed to maintain harmony with the natural environment and protect the community from harm. For example, many communities have rituals that honor the spirits of the land, ancestors, or deities believed to control weather patterns, offering symbolic acts of respect to ensure favorable conditions for crops, water, and general well-being. In some African cultures, rituals surrounding the planting of crops or the first rains are performed to invoke blessings and ensure that the harvest will be abundant and resilient to environmental stresses (Gaillard & Mercer, 2013). These practices, while deeply rooted in spirituality, also function as a way of reinforcing community cohesion and preparedness, fostering collective action in the face of potential hazards.

Risk Perception and Adaptation Strategies: Local knowledge also includes a community's perception of risk and their strategies for adapting to environmental hazards. This form of knowledge reflects how people understand and interpret the threats posed by disasters, and how they adjust their behaviors and practices accordingly. In many cases, communities rely on past experiences to shape their current risk perceptions, which in turn informs their decision-making processes during times of crisis. For instance, the awareness of recurrent droughts might lead a community to adopt water-saving agricultural techniques or diversify their livelihoods to reduce vulnerability. In the case of cyclones in coastal regions, communities may recognize the patterns of storms and prepare by securing homes or relocating to higher ground (Campbell, 2009).

These forms of local knowledge—environmental awareness, historical practices, community rituals, and risk perception—form an interconnected system that enables communities to cope with, adapt to, and reduce the risks associated with natural hazards. The integration of these forms of knowledge into formal DRR frameworks can provide a more comprehensive, culturally relevant approach to disaster management, enhancing the resilience of vulnerable communities.

CASE STUDIES

The integration of traditional practices into disaster risk reduction (DRR) initiatives has been demonstrated through various successful case studies from different regions worldwide. These case studies showcase how indigenous knowledge and practices can complement formal disaster management strategies, providing culturally appropriate and locally effective solutions. The following examples illustrate how traditional practices have been used to enhance resilience to disasters, with particular focus on land-use practices, early warning systems, and cultural rituals.

Traditional Land-Use Practices and Soil Conservation

In many parts of the world, traditional land-use practices have been instrumental in reducing environmental hazards such as landslides and soil erosion, which are exacerbated by improper land management and deforestation. One notable example comes from the Andean communities of South America, where traditional agricultural practices such as *terracing* and *agroforestry* have been utilized for centuries. These methods help conserve soil and maintain water retention, reducing the likelihood of landslides and soil degradation. Research by Milla et al. (2013) highlights how the Quechua people of Peru use traditional terrace systems combined with indigenous knowledge of soil fertility to minimize erosion and enhance agricultural productivity. Similarly, in the mountainous regions of Nepal, local farming communities use crop rotation and the strategic planting of certain plant species to prevent soil erosion and landslides, a practice rooted in centuries of observation of local environmental dynamics (Berkes, 2018).

In Zimbabwe, traditional land-use practices such as *kontoma* (shallow basin farming) and the use of indigenous trees for windbreaks have been effective in mitigating soil erosion and enhancing crop yields, particularly in drought-prone areas (Mudenge, 2016). These practices have not only helped preserve the land but also contribute to local food security and sustainable agricultural systems.

Community-Led Early Warning Systems

Indigenous meteorological knowledge, which includes the observation of natural signs and patterns, plays a crucial role in the development of early warning systems in many communities. These systems often rely on indigenous knowledge of weather patterns, animal behavior, and seasonal changes to predict hazards such as storms, floods, or droughts. In the Pacific Islands, communities have used indigenous knowledge to develop community-led early warning systems for cyclones and tsunamis. For example, the people of Vanuatu and the Solomon Islands rely on the knowledge of elders who observe cloud formations, wind shifts, and the behavior of animals, such as the migration patterns of birds, to forecast impending storms (Campbell, 2009). These early warning systems are often complemented by modern scientific data, creating a hybrid system that is more effective and culturally appropriate.

In Zimbabwe, rural communities have traditionally relied on the observations of the skies and local wildlife to anticipate seasonal changes, such as the onset of rains or the arrival of droughts. Elders, who are skilled in interpreting these natural cues, often play a key role in disseminating early warnings to the community. These systems are particularly effective in areas where modern communication infrastructure is limited (Nyong et al., 2007). By integrating these traditional methods with formal meteorological data, Zimbabwean communities have improved their preparedness for extreme weather events, such as the floods caused by heavy rains or droughts.

Cultural Rituals Promoting Community Solidarity and Preparedness

Cultural rituals and community practices are integral to building social cohesion and ensuring collective action during disasters. Many communities engage in rituals that invoke spiritual protection or seek to ensure favorable environmental conditions. For example, in parts of sub-Saharan Africa, community-wide rituals are performed to ensure an abundant harvest or to appease spirits believed to control rainfall and crop fertility (Gaillard & Mercer, 2013). These rituals often involve prayers, dances, and the offering of food or other gifts to ancestral spirits. While these practices may be seen primarily as spiritual, they also serve a social function, bringing the community together and fostering a shared sense of purpose and preparedness.

In the Philippines, the Tagbanua indigenous people have rituals that promote disaster preparedness by strengthening community ties. One ritual includes the collective planting of mangrove trees in coastal areas, which not only serves to protect the environment but also reinforces community solidarity. These community-driven environmental practices are embedded in the cultural fabric of the people, ensuring that disaster preparedness becomes a shared responsibility (Campbell, 2009).

The emphasis on community participation through ritual activities enhances the collective resilience of the group, providing both emotional and practical support during times of crisis.

In Zimbabwe, the practice of communal labor and ceremonies during agricultural seasons fosters cooperation and preparedness for adverse weather events. Rituals associated with planting and harvesting often include community feasts and collective work, which help build social networks that can be mobilized in times of crisis (Mudenge, 2016). These practices, rooted in the belief that collective action and mutual support are essential to overcoming disasters, help strengthen the community's ability to cope with environmental shocks.

COLLABORATION AND INCLUSIVITY

Effective disaster risk reduction (DRR) requires collaboration across multiple levels of governance and across various sectors, including local communities, governmental organizations, and non-governmental organizations (NGOs). The success of DRR initiatives often hinges on how well these different stakeholders work together, sharing resources, expertise, and knowledge. This section explores the critical role of collaboration and inclusivity in enhancing disaster preparedness and response, highlighting the importance of mutual respect and understanding in fostering effective partnerships.

The Role of Local Communities in DRR Collaboration

Local communities are often the first responders to disasters, and their involvement in DRR efforts is crucial. Indigenous knowledge systems, which are deeply embedded in the cultural fabric of these communities, can offer valuable insights into local hazards, vulnerabilities, and coping mechanisms. However, the success of incorporating this knowledge into formal disaster management frameworks depends on the willingness of local communities to collaborate with external actors. This collaboration ensures that local knowledge is valued and integrated into official plans, while also ensuring that local populations receive the support and resources they need in times of crisis. The ability to share and validate indigenous knowledge with external stakeholders, such as NGOs and government agencies, enhances the relevance and effectiveness of disaster preparedness strategies (Mercer et al., 2010).

In Zimbabwe, community-based approaches to disaster management have highlighted the importance of inclusivity in disaster preparedness. Local knowledge is often shared during village meetings or through community groups, where elders, who hold the traditional wisdom, impart crucial information on environmental changes and disaster risks (Mudenge, 2016). When governmental bodies and NGOs recognize the value of this knowledge and collaborate with local leaders, it strengthens the community's resilience and ensures that disaster response strategies are culturally appropriate and effective.

The Role of Governmental Organizations and NGOs

Governmental organizations and NGOs play an essential role in facilitating the integration of local knowledge into national and regional DRR frameworks. By engaging with local communities and respecting their knowledge systems, these external actors can help build capacity for disaster preparedness while enhancing the overall resilience of communities. This collaborative approach is necessary for achieving a more comprehensive understanding of risks and ensuring the effectiveness of interventions.

The role of NGOs, in particular, has been crucial in many parts of the world where government capacity may be limited. NGOs often act as bridges between local communities and governmental agencies, helping to facilitate communication, advocate for the inclusion of local knowledge, and mobilize resources. For example, in the Philippines, NGOs have worked alongside indigenous communities to integrate traditional knowledge into the formal disaster response system. These organizations often organize workshops, training, and capacity-building programs that promote mutual understanding and respect between local communities and external actors (Gaillard & Mercer, 2013). By fostering partnerships based on trust and respect, NGOs contribute to the sustainability and effectiveness of disaster preparedness efforts.

In Zimbabwe, NGOs such as the Zimbabwe Red Cross have been active in building disaster resilience at the community level, often incorporating indigenous knowledge into their programming. Their collaboration with local communities helps to improve early warning systems, enhance local capacity for disaster management, and ensure that disaster preparedness activities are inclusive of all community members, including marginalized groups (Nyong et al., 2007).

The Importance of Mutual Respect and Understanding

Mutual respect and understanding between local communities and external stakeholders are essential for fostering effective collaboration. Local communities must feel that their knowledge and cultural practices are respected, not dismissed or undermined by formal disaster management structures. Similarly, governmental agencies and NGOs must approach community-based DRR efforts with humility, recognizing that local knowledge can provide insights that may not be available through formal scientific or technical data alone.

One important way to build mutual respect is through participatory decision-making, where local communities are involved in the planning and implementation of DRR initiatives. Participatory approaches, such as Participatory Rural Appraisal (PRA) and Community-Based Participatory Research (CBPR), emphasize the importance of including local voices in the development of disaster preparedness strategies (Chambers, 1994). These approaches help ensure that DRR programs are tailored to the specific needs and realities of the communities they aim to serve, leading to more effective and sustainable outcomes.

Additionally, the inclusion of local communities in DRR efforts promotes social capital and strengthens the social fabric, which is essential for community resilience. As communities work together to assess risks, plan responses, and implement mitigation strategies, they develop a collective sense of ownership and responsibility. This sense of ownership increases the likelihood of long-term success, as communities are more likely to adhere to and maintain initiatives that they have helped design (Gaillard & Mercer, 2013).

CHALLENGES AND OPPORTUNITIES

The integration of local knowledge into formal disaster risk reduction (DRR) frameworks presents both significant challenges and considerable opportunities. While there is increasing recognition of the value of indigenous knowledge in enhancing community resilience, obstacles remain in translating this knowledge into mainstream disaster management practices. These challenges often stem from power dynamics, insufficient recognition of local knowledge, and the need for capacity building. However, there are strategies that can overcome these barriers, leading to more inclusive and effective DRR approaches.

Challenges in Integrating Local Knowledge into Formal DRR Frameworks

One of the primary challenges in integrating local knowledge into formal DRR frameworks is the power dynamics between local communities and external actors, such as governmental agencies and international organizations. Often, formal knowledge systems, based on scientific data and quantitative models, dominate the decision-making processes, sidelining indigenous knowledge as anecdotal or irrelevant (Blaikie, 2006). This marginalization of local knowledge stems from an epistemological divide, where indigenous wisdom is often considered subjective and not scientifically verifiable (Mercer et al., 2010). In the context of Zimbabwe, where a history of colonialism has deeply influenced both the educational system and governance structures, these power imbalances persist, further complicating efforts to integrate local knowledge into national disaster management policies (Mudenge, 2016).

A related challenge is the lack of formal recognition and documentation of local knowledge. While many communities possess rich repositories of knowledge about risk management and adaptation, this knowledge is often passed down orally or through traditional practices, making it difficult to formalize and incorporate into official DRR plans. Without adequate recognition and institutional support, these knowledge systems remain underutilized, limiting their contribution to disaster preparedness and response (Manase, 2020). Furthermore, this lack of recognition may also result in the erosion of traditional practices, as younger generations may not fully appreciate the value of indigenous knowledge, particularly in the face of modernization and global influences (Mudenge, 2016).

In addition to these barriers, there is also a need for capacity building within both local communities and external organizations. For communities to effectively contribute to DRR, they must have the tools and resources to articulate their knowledge in ways that are understandable and applicable within formal frameworks. External stakeholders, such as NGOs and government bodies, must also be equipped with the skills to engage with and integrate local knowledge into their strategies. Without this capacity, efforts to incorporate indigenous knowledge are likely to be ineffective or superficial (Gumbo & Zirebwa, 2017).

Opportunities for Overcoming Challenges

Despite these challenges, there are numerous opportunities for integrating local knowledge into formal DRR frameworks. One of the most promising strategies is the promotion of participatory approaches that involve local communities in the planning, implementation, and evaluation of DRR initiatives. Participatory approaches, such as Participatory Rural Appraisal (PRA) and Community-Based Participatory Research (CBPR), emphasize the importance of including local voices and knowledge in the development of disaster risk reduction strategies (Chambers, 1994). These approaches ensure that local communities are not passive recipients of external aid but active participants in shaping their own disaster resilience.

In Zimbabwe, the adoption of participatory methodologies in disaster management has shown promising results. For example, community-based approaches to early warning systems and hazard mapping, informed by indigenous knowledge, have been successfully implemented in rural areas, where traditional practices such as the observation of environmental cues (e.g., changes in plant behavior or animal migration patterns) are integrated into scientific monitoring systems (Manase, 2020). These initiatives have not only improved the accuracy of early warning systems but also strengthened local ownership and engagement in DRR activities, fostering a sense of empowerment and responsibility among community members.

Another important opportunity lies in capacity building, both within local communities and among external stakeholders. Training programs that emphasize the value of indigenous knowledge and provide tools for its integration into formal frameworks are essential. This can involve educating government officials and NGOs about the importance of local knowledge in disaster risk reduction, as well as building the technical capacity of local communities to document and communicate their knowledge in ways that align with formal disaster management practices (Gumbo & Zirebwa, 2017). For example, using modern technologies, such as mobile apps or GIS mapping tools, can help communities document their knowledge in digital formats, making it easier to integrate into broader DRR frameworks.

Furthermore, fostering stronger partnerships between local communities, governmental organizations, and NGOs can help overcome power imbalances and create more inclusive disaster management processes. By engaging in mutual dialogue and respecting each other's knowledge systems, these stakeholders can co-create solutions that are more contextually appropriate and sustainable. This collaborative approach can lead to more effective and culturally sensitive DRR interventions that address both local vulnerabilities and broader societal challenges (Gaillard & Mercer, 2013).

CONCLUSION

The integration of local knowledge into formal Disaster Risk Reduction (DRR) frameworks represents a critical step toward enhancing community resilience in the face of increasing environmental hazards. Indigenous knowledge systems, rooted in centuries of local experience, offer valuable insights that are often overlooked in conventional disaster management strategies. However, this integration is not without its challenges. Power dynamics, the lack of recognition for local knowledge, and the need for capacity building among both local communities and external stakeholders are significant obstacles that hinder effective collaboration. Despite these challenges, the potential benefits of incorporating local knowledge into formal DRR frameworks far outweigh the barriers.

This approach requires a shift toward participatory models that recognize the expertise of local communities, empower them through capacity-building initiatives, and foster collaboration between indigenous knowledge holders and formal institutions. By addressing these challenges head-on, it is possible to overcome the barriers and create disaster management strategies that are not only more inclusive but also culturally appropriate and more effective in mitigating risks and reducing vulnerabilities. In doing so, disaster response and resilience-building efforts will become more holistic, grounded in both modern scientific knowledge and the wisdom of local traditions.

This paper explores the integration of local knowledge into formal DRR frameworks, highlighting its importance in fostering resilient communities, particularly in the context of Zimbabwe and other developing regions. Through this exploration, we aim to demonstrate how the synergy between traditional knowledge and formal disaster management strategies can create more sustainable and adaptive approaches to disaster risk reduction.

REFERENCES

Agrawal, A. (1995). Dismantling the divide between indigenous and scientific knowledge. *Development and Change*, *26*(3), 413–439. DOI: 10.1111/j.1467-7660.1995.tb00560.x

Berkes, F. (2018). *Sacred ecology* (4th ed.). Routledge.

Berkes, F., & Folke, C. (2002). *Back to the future: Ecosystem dynamics and local knowledge*. Cambridge University Press.

Blaikie, P. (2006). Risk and disaster reduction: A challenge for the social sciences. *International Journal of Risk Assessment and Management*, *6*(2-3), 137–145.

Campbell, J. R. (2009). Islandness: Vulnerability and resilience in Oceania. *Shima. The International Journal of Research into Island Cultures*, *3*(1), 85–97.

Chambers, R. (1994). The origins and practice of participatory rural appraisal. *World Development*, *22*(7), 953–969. DOI: 10.1016/0305-750X(94)90141-4

Chanza, N., & de Wit, A. (2016). Enhancing climate governance through indigenous knowledge: Case of the Shona in Zimbabwe. *Climate Policy*, *16*(4), 543–562. DOI: 10.1080/14693062.2015.1033674

Chirisa, I. (2021). Education 5.0 in Zimbabwe: Mapping the history and trajectory of University Education. *Zambia Journal of Science and Technology*, *16*(1), 22–35.

Folke, C., Carpenter, S. R., Walker, B., Scheffer, M., Chapin, T., & Rockström, J. (2010). Resilience thinking: Integrating resilience, adaptability and transformability. *Ecology and Society*, *15*(4), 20. DOI: 10.5751/ES-03610-150420

Gaillard, J. C., & Mercer, J. (2013). From knowledge to action: Bridging gaps in disaster risk reduction. *Progress in Human Geography*, *37*(1), 93–114. DOI: 10.1177/0309132512446717

Gaillard, J. C., & Mercer, J. (2022). Understanding and addressing root causes of vulnerability in disaster risk reduction. *Progress in Disaster Science*, *10*, 100203. DOI: 10.1016/j.pdisas.2022.100203

Gumbo, D., & Zirebwa, K. (2017). *Indigenous knowledge systems and climate change adaptation in Zimbabwe: Challenges and opportunities*. IGI Global.

Manase, S. (2020). *Community-based disaster risk management in Zimbabwe: Integrating local knowledge into national strategies*. IGI Global.

Mavhura, E. (2021). Indigenous knowledge and disaster risk reduction: The role of cultural beliefs in building resilience. *Sustainability, 13*(5), 2558. DOI: 10.3390/su13052558

Mavhura, E. (2021). Indigenous knowledge and disaster risk reduction: The role of cultural beliefs in building resilience. *Sustainability, 13*(5), 2558. DOI: 10.3390/su13052558

Mavhura, E., Manyena, B., & Collins, A. (2020). Drought vulnerability and risk perception: A case study from Zimbabwe. *Natural Hazards, 101*(1), 219–241. DOI: 10.1007/s11069-020-03854-5

McAdoo, B. G., Moore, A., & Baumwoll, J. (2009). Indigenous knowledge and the near field population response during the 2007 Solomon Islands tsunami. *Natural Hazards, 48*(1), 73–82. DOI: 10.1007/s11069-008-9249-z

Mercer, J., Kelman, I., Taranis, L., & Suchet-Pearson, S. (2010). Framework for integrating indigenous and scientific knowledge for disaster risk reduction. *Disasters, 34*(1), 214–239. DOI: 10.1111/j.1467-7717.2009.01126.x PMID: 19793324

Milla, A., Rojas, A., & Salazar, G. (2013). Traditional ecological knowledge and agricultural land-use practices in the Andes: A case study of the Quechua people of Peru. *Journal of Sustainable Agriculture, 37*(2), 221–234.

Mudenge, M. (2016). Indigenous knowledge systems and practices in Zimbabwe: Implications for sustainable development. *Indigenous Knowledge and Development Journal, 15*(1), 75–89.

Nguyen, P. T., Anderson, C., & Läderach, P. (2023). Leveraging indigenous knowledge to improve climate resilience: Insights from the Maasai pastoralists in Kenya. *Journal of Environmental Management, 324*, 116435. DOI: 10.1016/j.jenvman.2022.116435

Nyong, A., Adesina, F., & Elasha, B. O. (2007). The value of indigenous knowledge in climate change mitigation and adaptation strategies in the African Sahel. *Mitigation and Adaptation Strategies for Global Change, 12*(5), 787–797. DOI: 10.1007/s11027-007-9099-0

Nyong, A., Adesina, F., & Elasha, B. O. (2007). The value of indigenous knowledge in climate change mitigation and adaptation strategies in the African Sahel. *Mitigation and Adaptation Strategies for Global Change, 12*(5), 787–797. DOI: 10.1007/s11027-007-9099-0

Nyong, A., Naswa, P., & Tabopda, C. (2021). The importance of integrating indigenous knowledge in climate change adaptation and mitigation strategies in Africa. *Climate Change Review, 45*(3), 197–210.

Sudmeier-Rieux, K., Ash, N., Murti, R., & Dunning, E. (Eds.). (2019). *Ecosystems, livelihoods, and disasters: An integrated approach to disaster risk management*. Springer.

United Nations Office for Disaster Risk Reduction (UNDRR). (2023). *Global assessment report on disaster risk reduction 2023*. UNDRR., DOI: 10.18356/9789214030488

United Nations (UN). (2022). *The Sustainable Development Goals Report 2022*. United Nations., DOI: 10.18356/9789210017988

Chapter 7
Building Climate Resilience in the Health Sector Through Community-Based Adaptation Strategies

Naboshree Bhattacharya
https://orcid.org/0000-0001-5572-4098
Amity University, Ranchi, India

ABSTRACT

The global health landscape is increasingly shaped by the complex impacts of climate change, posing risks to human health and health systems. Building resilience in the health sector is a critical priority. This study explores the transformative potential of community-based adaptation (CBA) strategies in strengthening health systems' capacity to withstand climate-related shocks. CBA empowers communities to identify their vulnerabilities and develop context-specific resilience strategies. The study presents a comprehensive "Triple-A" framework for operationalizing CBA. The study examines the institutional and policy landscape shaping CBA initiatives. It identifies priorities to accelerate CBA uptake and strengthened institutions for mainstreaming CBA into health planning. This study makes a compelling case for CBA's transformative potential in building climate-resilient health systems. It offers a hopeful vision for a more resilient, equitable, and sustainable future for global health.

DOI: 10.4018/979-8-3693-9745-9.ch007

1. INTRODUCTION

Climate change poses significant challenges to human health and the health systems that support it. The Intergovernmental Panel on Climate Change (IPCC) has highlighted the wide-ranging impacts of climate change on health, including increased heat-related morbidity and mortality, the spread of infectious diseases, and the exacerbation of respiratory and cardiovascular illnesses (IPCC, 2022). The World Health Organization (WHO) emphasizes that the frequency and intensity of extreme weather events are expected to rise, necessitating the adaptation of health systems to ensure their resilience and ability to provide essential services to communities (WHO, 2021). The impacts of climate change on health are not evenly distributed, with vulnerable populations, such as the poor, elderly, and those with pre-existing health conditions, being disproportionately affected (Watts et al., 2019).

Building climate resilience in the health sector is crucial for minimizing the adverse effects of climate change on human health and ensuring the continuity of healthcare services during and after climate-related disasters (Ebi et al., 2021). Climate resilience refers to the capacity of health systems to anticipate, respond to, cope with, recover from, and adapt to climate-related stresses and shocks (WHO, 2015). Enhancing the resilience of health systems requires a multi-faceted approach that addresses the complex interactions between climate change, health, and socio-economic factors (Haines & Ebi, 2019).

Community-based adaptation (CBA) strategies have emerged as a promising approach to enhancing the resilience of health systems by engaging local communities in the planning, implementation, and monitoring of adaptation measures (McNamara & Buggy, 2017). CBA strategies recognize the importance of local knowledge, resources, and capacities in addressing the context-specific challenges posed by climate change (Reid, 2016). By involving communities in the adaptation process, CBA strategies can help to identify and prioritize local health needs, mobilize resources, and build capacity for effective and sustainable adaptation (Ayers & Forsyth, 2009).

The objectives of this chapter are to:

1. Provide an overview of the impacts of climate change on health systems and the need for building climate resilience.
2. Discuss the role of community-based adaptation strategies in enhancing the resilience of health systems.
3. Present case studies and best practices of successful community-based adaptation initiatives in the health sector.
4. Identify key challenges and opportunities for scaling up community-based adaptation in the health sector.

The scope of this chapter covers both the theoretical foundations and practical applications of community-based adaptation strategies in building climate resilience in the health sector. It draws upon recent research and case studies from various geographical contexts to provide a comprehensive understanding of the topic. The chapter aims to contribute to the growing body of knowledge on climate change adaptation in the health sector and to inform policy and practice in this field.

2. CLIMATE CHANGE AND HEALTH SYSTEM VULNERABILITIES

Climate change poses significant and far-reaching risks to human health and the health systems that support it. The direct and indirect health impacts of climate change are complex, multifaceted, and can exacerbate existing health inequities, particularly among marginalized and underserved communities (WHO, 2021). As the global climate crisis continues to unfold, it is becoming increasingly clear that the threats to human health and well-being are both imminent and severe, requiring urgent action to build resilience and adapt to the changing environmental conditions.

The direct health impacts of climate change are perhaps the most immediately visible and alarming. These impacts include increased morbidity and mortality from extreme weather events, such as heatwaves, floods, and storms (IPCC, 2022). As global temperatures continue to rise, the frequency and intensity of these events are expected to increase, putting more people at risk of injury, illness, and death. For example, extreme heat waves can cause heat stress, heat stroke, and exacerbate pre-existing health conditions, such as cardiovascular and respiratory diseases. Similarly, floods and storms can lead to drowning, physical trauma, and the spread of water-borne diseases, particularly in areas with inadequate infrastructure and sanitation

However, the indirect health impacts of climate change are equally concerning and may be even more pervasive in the long run. These impacts arise from changes in the environmental and social determinants of health, such as air pollution, water scarcity, food insecurity, and the spread of infectious diseases (Romanello et al., 2023). For example, rising temperatures and changing precipitation patterns can lead to more frequent and severe droughts, which can compromise food production and increase the risk of malnutrition, particularly in developing countries. At the same time, changes in temperature and humidity can alter the distribution and abundance of disease vectors, such as mosquitoes and ticks, leading to the spread of vector-borne diseases like malaria, dengue fever, and Lyme disease.

The health impacts of climate change are not evenly distributed, with marginalized and underserved communities being disproportionately affected (Berry et al., 2022). These communities often have limited access to healthcare, live in areas with poor infrastructure, and have pre-existing health conditions that make them more susceptible to climate-related risks (Smith et al., 2023). For example, low-income communities and communities of color are more likely to live in urban heat islands, where the lack of green space and the prevalence of heat-absorbing surfaces can amplify the effects of extreme heat. Similarly, indigenous communities and rural populations may be more dependent on natural resources for their livelihoods and may be more vulnerable to the impacts of climate change on food security and water availability

Climate change can also exacerbate existing health inequities and create new vulnerabilities, particularly among low-income, indigenous, and minority populations (Patz & Thomson, 2022). For example, climate-related disasters can disrupt access to healthcare services, particularly for those who are already underserved or have limited mobility at the same time, the economic impacts of climate change, such as job losses and reduced income, can further compromise the ability of vulnerable populations to access healthcare and maintain healthy living conditions

Given the vulnerabilities of health systems to climate change, there is a pressing need for adaptation and resilience-building strategies to protect population health (WHO, 2023). These strategies should aim to strengthen the capacity of health systems to anticipate, prepare for, and respond to climate-related risks (Bowen et al., 2022). This can involve a range of measures, such as improving disease surveillance and early warning systems, developing climate-resilient health infrastructure, and building the capacity of health workers to respond to climate-related health emergencies (Ebi & Hess, 2023).

One key strategy for building health system resilience is to invest in community-based adaptation (CBA) approaches that engage local communities in identifying and addressing their own climate-related health risks. CBA approaches recognize that communities have valuable knowledge and expertise about their local environments and are best placed to identify and prioritize their own adaptation needs. By empowering communities to take action on climate change, CBA approaches can help to build local capacity, foster social cohesion, and promote more equitable and sustainable outcomes.

However, building health system resilience to climate change is not a simple or straightforward task. It requires a concerted and coordinated effort across multiple sectors and levels of government, as well as sustained investment in research, innovation, and capacity building. It also requires a fundamental shift in how we think about and value health, recognizing that the health of people and the planet are inextricably linked and that protecting one requires protecting the other.

Ultimately, the challenges posed by climate change to human health and health systems are daunting, but they are not insurmountable. By taking urgent and ambitious action to reduce greenhouse gas emissions, build resilience, and adapt to the changing climate, we can help to protect the health and well-being of current and future generations. This will require leadership, collaboration, and a commitment to social justice and equity, but the stakes could not be higher. As the COVID-19 pandemic has shown, the health of our communities is a shared responsibility and a global imperative. In the face of the climate crisis, we must rise to this challenge and work together to build a healthier, more resilient, and more sustainable world for all.

3. COMMUNITY-BASED ADAPTATION (CBA) STRATEGIES

Community-Based Adaptation (CBA) represents a collaborative approach to climate resilience that emphasizes local participation and knowledge integration in developing adaptation strategies. According to Reid et al. (2019), CBA is defined as an equitable process that involves community members, researchers, and stakeholders working together to address climate-related challenges while recognizing the unique strengths each participant brings to the process.

The advantages of CBA in the health sector are particularly significant. McNamara and Buggy (2017) emphasize that this approach enables communities to identify their specific health vulnerabilities to climate change and develop locally appropriate solutions. CBA facilitates the development of climate-smart interventions that are both culturally appropriate and sustainable, as they are rooted in local understanding and ownership (Ensor et al., 2021). The approach has gained significant traction in developing countries, where it has proven effective in addressing climate change impacts in vulnerable communities (WHO, 2022).

A key strength of CBA lies in its integration of traditional knowledge with modern healthcare practices. For example, Kirkland et al. (2021) document how successful CBA initiatives have merged indigenous healing practices with contemporary medical approaches for adaptive management. This integration helps ensure that adaptation strategies are both scientifically sound and culturally appropriate, increasing their likelihood of adoption and long-term sustainability (Dasgupta et al., 2020).

Participatory approaches and community engagement are fundamental principles of CBA. Dodman and Mitlin (2023) emphasize that the methodology prioritizes equitable involvement of community members in all phases of the adaptation process, from initial planning to implementation and monitoring. This participatory nature empowers stakeholders to contribute their expertise and participate in decision-making processes, leading to more effective and sustainable outcomes (Ford et al., 2020). The approach has proven particularly valuable when working with vulnerable populations, as it ensures their voices and needs are central to the adaptation strategy (Sherman et al., 2022).

CBA's effectiveness in the health sector is enhanced through its emphasis on joining with communities as full and equal partners throughout the research and implementation process. This partnership approach has been particularly successful in addressing health challenges in vulnerable populations, where traditional top-down approaches have often fallen short (Williams et al., 2021). Through continuous community engagement and participatory monitoring and evaluation, CBA strategies can be adjusted and improved over time, ensuring their continued relevance and effectiveness in addressing climate-related health challenges (Ebi & Otmani del Barrio, 2023).

4. THE "TRIPLE-A" FRAMEWORK FOR CBA IMPLEMENTATION

The Triple-A framework represents a comprehensive approach to implementing community-based adaptation strategies for addressing climate-related health challenges. This framework, consisting of Awareness, Action, and Adaptation phases, provides a structured methodology for building climate resilience while ensuring strong community engagement and sustainable outcomes (WHO, 2021).

4.1 Phase 1: Awareness

The Awareness phase serves as the foundation of the Triple-A framework, emphasizing comprehensive understanding of climate-related health vulnerabilities. Climate and health vulnerability assessments (CHVAs) provide systematic methods for identifying population-specific risks and adaptive capacities (WHO, 2021). These assessments integrate both quantitative health data and qualitative community perspectives to create robust vulnerability profiles (Ebi et al., 2023). Community risk mapping exercises enable local stakeholders to identify and prioritize climate-related health risks based on their severity, likelihood, and community impact (Preston et al., 2022). This participatory approach ensures that subsequent adaptation strategies address the most pressing concerns while incorporating valuable local knowledge and perspectives (McNamara & Buggy, 2021).

The integration of scientific data with local knowledge during the Awareness phase ensures a comprehensive understanding of historical climate patterns, their health impacts, current vulnerabilities in health systems, and available resources and capacities. This phase also emphasizes documenting traditional knowledge about environmental changes and creating visual representations of vulnerable areas and resources (Ebi et al., 2023). The process helps build community ownership of the adaptation process from the outset while ensuring that interventions are grounded in both scientific evidence and local reality.

4.2 Phase 2: Action

The Action phase translates awareness into concrete interventions through carefully designed and implemented strategies. This phase focuses on developing practical solutions that address identified vulnerabilities while building on existing community strengths (Thomson et al., 2023). A key component is the development of early warning systems that integrate meteorological and health surveillance data, creating community-based monitoring networks and establishing clear communication protocols. Infrastructure strengthening includes climate-proofing health facilities, improving emergency response capabilities, and developing sustainable supply chains (WHO, 2023).

Capacity building programs form another crucial element of the Action phase, focusing on training health workers in climate-health relationships, developing community health volunteer networks, and strengthening local leadership capabilities (Ford et al., 2021). These programs ensure that communities have the knowledge and skills necessary to implement and maintain adaptation strategies effectively. The emphasis on local capacity building helps ensure the sustainability of interventions and reduces dependence on external support.

4.3 Phase 3: Adaptation

The Adaptation phase ensures the long-term effectiveness and sustainability of implemented strategies through systematic monitoring and continuous improvement processes (Sherman et al., 2023). This phase includes developing and implementing comprehensive monitoring and evaluation systems with clear key performance indicators, regular data collection protocols, and assessment of community engagement levels. Adaptive management practices enable regular review of strategy effectiveness and adjustment of approaches based on monitoring results (Dodman & Mitlin, 2022). Continuous improvement processes incorporate regular stakeholder feedback sessions, updates to strategies based on new climate projections, and integration of emerging best practices. This iterative approach ensures that adaptation strategies remain relevant and effective over time while strengthening community ownership of the process (Brooks et al., 2023). The success of these processes depends on maintaining strong stakeholder engagement and effective coordination among various participants.

The Triple-A framework has demonstrated particular effectiveness in developing countries where community-based approaches are essential for sustainable adaptation to climate change impacts on health systems (Ensor et al., 2023). Its success relies heavily on sustained community engagement throughout all phases, effective coordination among stakeholders, integration of local knowledge with scientific expertise, and long-term commitment to the process (Reid et al., 2022). This comprehensive approach ensures that climate adaptation strategies are not only technically sound but also socially acceptable and sustainable within the local context.

5. CASE STUDIES: SUCCESSFUL CBA INITIATIVES IN THE HEALTH SECTOR

Community-based adaptation (CBA) has emerged as a promising approach for building climate resilience within health systems, particularly in vulnerable communities. CBA initiatives prioritize local participation, knowledge integration, and capacity building to develop context-specific strategies for addressing climate-related health risks (Ensor et al., 2023). The following case studies provide an in-depth analysis of successful CBA projects in developing and developed countries, highlighting their unique strategies, outcomes, and lessons learned.

5.1 Examples from Developing Countries

A. Community-based surveillance and response systems in Ethiopia:

In the Somali Region of Ethiopia, where pastoralist communities face increasing health risks due to climate change, a community-based surveillance and response system was established to monitor and respond to climate-sensitive diseases (Taye et al., 2021). The initiative leveraged the strengths of traditional social networks and local knowledge to develop a culturally appropriate and sustainable system. Key strategies included:

i. Training local health workers and community volunteers in disease surveillance and reporting using mobile technologies
ii. Integrating traditional early warning indicators with scientific data to improve the accuracy and timeliness of disease detection
iii. Establishing community-based rapid response teams to investigate and contain outbreaks

The participatory approach fostered strong community ownership and trust, leading to high levels of engagement and data reporting. As a result, the system significantly improved disease detection rates, reduced response times, and minimized outbreak severity. The success of this initiative demonstrates the potential for CBA to strengthen health system resilience in resource-limited settings by empowering communities to take an active role in disease surveillance and response.

B. Climate-resilient health facilities and infrastructure in Bangladesh:

Bangladesh's coastal regions are highly vulnerable to climate change impacts, including sea-level rise, floods, and cyclones, which pose significant risks to health infrastructure and service delivery (Rahman et al., 2022). To address these challenges, a CBA project was implemented to build climate-resilient health facilities and infrastructure. The initiative employed a multi-stakeholder approach, involving collaboration between local communities, health authorities, and development partners. Key strategies included:

i. Conducting participatory vulnerability assessments to identify critical health infrastructure at risk
ii. Upgrading health facilities with climate-resilient features, such as elevated foundations, reinforced walls, and flood-resistant electrical systems
iii. Installing renewable energy systems, such as solar panels and biogas plants, to ensure uninterrupted power supply during extreme weather events
iv. Establishing community-based emergency response teams and protocols to ensure continuity of health services during disasters

The project's success was evident in the improved functionality and resilience of health facilities during subsequent flood and cyclone events. Community members reported increased trust and utilization of health services, as well as enhanced capacity to cope with climate-related health risks. This case study highlights the importance of investing in climate-resilient health infrastructure and engaging communities in the design and management of these assets.

5.2 Examples from Developed Countries

A. Integration of CBA into local health planning and policy in Canada:

In the province of Quebec, Canada, where extreme heat events are becoming more frequent and severe, a CBA approach was integrated into local health planning and policy to address the associated health risks (Gosselin et al., 2022). The initiative demonstrated the effectiveness of mainstreaming CBA into existing governance structures to ensure the sustainability and scalability of adaptation efforts. Key strategies included:

i. Conducting participatory risk assessments and vulnerability mapping to identify heat-sensitive populations and neighbourhoods
ii. Developing local heat action plans through collaboration between public health authorities, municipalities, and community organizations
iii. Implementing a range of heat adaptation measures, such as establishing cooling centers, enhancing urban green spaces, and improving access to air conditioning for vulnerable populations
iv. Integrating heat warning systems and emergency response protocols into local public health practice

The success of this initiative was evident in the increased awareness and adoption of heat adaptation measures at the community level, as well as the strengthened coordination between public health and municipal authorities. The integration of CBA into local health planning and policy frameworks ensured the long-term sustainability and institutionalization of adaptation efforts.

B. Public-private partnerships for health adaptation financing in the United States:

Financing climate adaptation in the health sector remains a significant challenge, particularly in resource-constrained settings. In California, a public-private partnership model was piloted to leverage diverse funding sources and expertise for health adaptation initiatives (Moser et al., 2021). The partnership involved collaboration between state agencies, health care providers, and private foundations. Key strategies included:

i. Establishing a dedicated health adaptation fund to pool resources from multiple sources
ii. Developing a participatory grant-making process to prioritize and fund community-driven adaptation projects
iii. Supporting a range of initiatives, such as the development of climate-smart health facilities, the implementation of early warning systems for climate-related health risks, and the creation of community resilience hubs
iv. Fostering knowledge exchange and capacity building among grantees to promote innovation and replication of successful adaptation strategies

The partnership model demonstrated the potential for mobilizing significant resources for health adaptation and fostering collaboration between diverse stakeholders. The funded projects contributed to building climate resilience in vulnerable communities and strengthening the adaptive capacity of local health systems. This case study highlights the importance of innovative financing mechanisms and multi-sectoral partnerships in scaling up CBA efforts in the health sector.

These case studies provide valuable insights into the diverse applications and benefits of CBA in the health sector. They demonstrate the effectiveness of CBA in building climate resilience in both developing and developed countries, and underscore the importance of community engagement, multi-stakeholder collaboration, and integration of local knowledge and scientific expertise (Ensor et al., 2023). As the health impacts of climate change continue to intensify, the lessons learned from these successful CBA initiatives can inform the design and implementation of future adaptation strategies, ensuring that health systems are better prepared to protect and promote population health in a changing climate.

6. ENABLING FACTORS AND CHALLENGES FOR CBA IMPLEMENTATION

The successful implementation of community-based adaptation (CBA) in the health sector relies on a complex interplay of enabling factors and challenges. These factors span across social, economic, institutional, and cultural dimensions, and their careful consideration is crucial for designing effective and sustainable CBA initiatives. This section provides a detailed analysis of the key enablers and barriers to CBA implementation, drawing on insights from both academic literature and practical experiences.

6.1 Multi-stakeholder Partnerships and Collaboration

Multi-stakeholder partnerships and collaboration are fundamental to the success of CBA initiatives. These partnerships involve the active engagement of diverse actors, including communities, local governments, health authorities, non-governmental organizations (NGOs), and academic institutions (Ensor et al., 2023). Effective partnerships are built on principles of trust, transparency, and shared decision-making, enabling the pooling of resources, expertise, and local knowledge for the development of context-specific and socially acceptable adaptation strategies.

However, establishing and maintaining effective partnerships can be challenging, particularly in settings with limited trust, power imbalances, or competing interests among stakeholders (Reid et al., 2022). For example, in some contexts, communities may have a history of mistrust towards government institutions or external actors, which can hinder their willingness to engage in collaborative adaptation efforts. Similarly, power imbalances between different stakeholder groups, such as between health authorities and local communities, can lead to the marginalization of community voices and priorities in decision-making processes.

To overcome these challenges, CBA initiatives must invest in building strong and equitable partnerships from the outset. This requires dedicating time and resources to stakeholder mapping, trust-building activities, and the development of clear partnership agreements that outline roles, responsibilities, and decision-making processes (Dodman & Mitlin, 2022). Partnerships should also prioritize the meaningful participation of marginalized groups, such as women, youth, and indigenous communities, to ensure that their needs and perspectives are adequately represented in adaptation planning and implementation.

6.2 Sustainable Financing Mechanisms

Sustainable financing mechanisms are essential for the long-term viability and scalability of CBA initiatives. CBA projects often require significant upfront investments in capacity building, infrastructure development, and community mobilization (Moser et al., 2021). Securing adequate and predictable funding sources can be a major challenge, particularly in resource-constrained settings where health budgets are already stretched thin.

Innovative financing models have emerged as potential solutions to this challenge. For example, public-private partnerships (PPPs) have been used to leverage private sector resources and expertise for CBA initiatives (Sherman et al., 2023). In these models, governments partner with private companies, foundations, or impact investors to co-finance and implement adaptation projects. PPPs can help to diversify funding sources, share risks, and promote innovation in adaptation approaches.

Community-based savings schemes and microfinance initiatives have also shown promise in mobilizing local resources for CBA (Ensor et al., 2023). These approaches enable communities to pool their savings and access small loans for adaptation investments, such as climate-resilient agriculture, water harvesting, or health infrastructure improvements. By building on existing community networks and financial practices, these initiatives can foster local ownership and sustainability of adaptation efforts.

However, ensuring the equitable distribution of adaptation funds and benefits among vulnerable communities remains a critical concern. CBA financing mechanisms must be designed with careful attention to issues of access, transparency, and accountability to prevent the capture of resources by elite groups or the exacerbation of existing inequalities (Reid et al., 2022).

6.3 Institutional and Policy Support

Institutional and policy support are key enabling factors for the integration of CBA into health systems. CBA approaches are most effective when they are mainstreamed into existing health planning, budgeting, and service delivery processes (WHO, 2021). This requires the development of supportive policies, guidelines, and institutional arrangements that recognize the value of community-driven adaptation and provide a conducive environment for its implementation.

At the national level, the integration of CBA into health policies and plans can help to ensure the allocation of resources and political support for adaptation initiatives. This may involve the development of specific CBA guidelines or the inclusion of CBA targets and indicators in national health strategies and monitoring frameworks (Ebi et al., 2023). The establishment of dedicated climate change and health units within health ministries can also help to coordinate and support CBA efforts across different levels of the health system.

At the local level, the decentralization of decision-making and resource allocation to sub-national health authorities and community-based organizations is critical for the effective implementation of CBA (Dodman & Mitlin, 2022). This requires the strengthening of local institutional capacities, including the training of health workers and community leaders in CBA approaches, and the establishment of mechanisms for community participation in health planning and budgeting processes.

However, many health systems face challenges in integrating CBA due to rigid bureaucratic structures, limited decentralization of decision-making, and competing priorities (Ensor et al., 2023). Health authorities may lack the flexibility and adaptability needed to respond to the dynamic and context-specific nature of CBA initiatives. Moreover, the short-term priorities of curative care and disease control often take precedence over the longer-term investments required for building adaptive capacity and resilience.

To overcome these challenges, CBA proponents must engage in sustained advocacy and capacity building efforts to build institutional buy-in and support for adaptation initiatives. This may involve the development of evidence-based arguments for the health benefits and cost-effectiveness of CBA, as well as the cultivation of champions and allies within health institutions and policy circles (WHO, 2021).

6.4 Social and Cultural Barriers to Adaptation

Addressing social and cultural barriers to adaptation is critical for the success of CBA initiatives. Climate change impacts and adaptation strategies are often mediated by social and cultural factors, such as gender norms, power dynamics, and traditional beliefs (McNamara & Buggy, 2021). These factors can shape individuals' and communities' perceptions of risk, their willingness to engage in adaptation actions, and their access to resources and decision-making processes.

For example, in many contexts, women and girls face disproportionate vulnerabilities to climate change impacts due to their roles in water collection, food production, and caregiving (Ensor et al., 2023). However, they may also face barriers to participating in adaptation decision-making and accessing resources due to gender-based discrimination and exclusion. Similarly, indigenous communities may possess valuable traditional knowledge and practices for coping with climate variability, but these may be overlooked or marginalized in formal adaptation planning processes.

CBA approaches must be sensitive to these social and cultural factors and actively engage communities in identifying and addressing potential barriers to adaptation. This requires investing in community mobilization, awareness-raising, and capacity building efforts that empower marginalized groups and foster inclusive decision-making processes (Reid et al., 2022). Participatory methods, such as community vulnerability assessments, scenario planning, and social mapping, can help to surface local knowledge, priorities, and capacities for adaptation.

However, overcoming deeply entrenched social and cultural barriers can be a complex and time-consuming process that requires sustained engagement and trust-building (Ensor et al., 2023). CBA initiatives must navigate sensitive issues of power, identity, and cultural change, and may face resistance from entrenched interests or traditional authorities. Building the capacity of local change agents, such as community health workers, women's groups, and youth leaders, can help to facilitate these processes and promote the adoption of new norms and practices.

In conclusion, the successful implementation of CBA in the health sector depends on a range of enabling factors and the ability to overcome various challenges. Effective multi-stakeholder partnerships, sustainable financing mechanisms, institutional and policy support, and the capacity to address social and cultural barriers are all critical elements for the success of CBA initiatives. While CBA holds significant potential for building climate resilience in vulnerable communities, it also requires careful planning, sustained engagement, and context-specific strategies to navigate the complex social, economic, and political landscapes in which it operates. By learning from the experiences of successful CBA initiatives and investing in the necessary enabling conditions, health systems can harness the power of community-driven adaptation to protect and promote population health in a changing climate.

7. MEASURING THE IMPACT OF CBA ON HEALTH SYSTEM RESILIENCE

Evaluating the impact of community-based adaptation (CBA) initiatives on health system resilience is a complex and multi-faceted endeavor that requires a comprehensive approach. It involves assessing various dimensions of CBA interventions, from their immediate effects on health outcomes and service delivery to their long-term contributions to community adaptive capacity and resilience. This section delves deeper into the key considerations and methods for measuring the impact of CBA on health system resilience, highlighting the importance of using both quantitative and qualitative indicators, employing participatory approaches, and sharing lessons learned to inform future adaptation efforts.

7.1 Quantitative and Qualitative Indicators of Success

To comprehensively assess the impact of CBA initiatives, it is crucial to employ a combination of quantitative and qualitative indicators. Quantitative indicators provide objective and measurable data on the outputs and outcomes of CBA interventions. These may include metrics such as the number of health facilities upgraded to be climate-resilient, the percentage of health workers trained in climate change and health, and the incidence rates of climate-sensitive diseases. Quantitative indicators allow for the tracking of progress over time, the comparison of results across different contexts, and the identification of areas for improvement.

However, relying solely on quantitative indicators may not capture the full complexity and nuances of CBA impacts. Qualitative indicators, on the other hand, provide valuable insights into the subjective experiences, perceptions, and social dynamics that shape the implementation and outcomes of CBA initiatives. These may include community perceptions of the relevance and effectiveness of CBA interventions, changes in knowledge and attitudes related to climate change and health, and the inclusion and empowerment of marginalized groups in adaptation decision-making. Qualitative indicators allow for a deeper understanding of the context-specific factors that influence the success of CBA initiatives and the ways in which they contribute to health system resilience.

To ensure that the selected indicators are relevant, meaningful, and owned by local stakeholders, it is important to employ participatory approaches in their development and measurement. Community-based monitoring and evaluation (M&E) methods, such as participatory rural appraisal (PRA) and most significant change (MSC) techniques, can help to engage communities in the process of defining, collecting, and analyzing data on CBA impacts. These approaches not only enhance the validity and reliability of the indicators but also foster local ownership and capacity building for ongoing adaptation efforts.

7.2 Evaluating Improvements in Health Outcomes and Service Delivery

A primary goal of CBA initiatives in the health sector is to improve health outcomes and service delivery in the face of climate change impacts. To evaluate these improvements, a multi-disciplinary approach that combines epidemiological, health systems, and social science research methods is necessary.

Epidemiological studies can help to assess changes in the burden of climate-sensitive health outcomes, such as vector-borne diseases, malnutrition, and heat stress. These studies may involve the analysis of routine health surveillance data, the conduct of cross-sectional surveys, or the use of quasi-experimental designs to compare health outcomes between communities with and without CBA interventions. For example, a study in Ethiopia found that the implementation of a community-based early warning system for malaria, coupled with the distribution of insecticide-treated bed nets and the training of community health workers, led to a significant reduction in malaria incidence and mortality rates in intervention villages compared to control villages.

Health systems research can provide valuable insights into the effectiveness, efficiency, and equity of CBA-related health services and interventions. This may involve the assessment of health facility readiness, the quality of care provided, the utilization of services by different population groups, and the cost-effectiveness of CBA investments. Participatory action research methods, such as community scorecards and social audits, can help to capture community perspectives on the accessibility, acceptability, and responsiveness of health services. For instance, a study in Cambodia used community scorecards to assess the quality of health services in villages affected by climate change, revealing gaps in the availability of essential medicines, the skills of health workers, and the cleanliness of health facilities. The findings were used to develop action plans for improving health service delivery and strengthening community-health system linkages.

Social science research methods, such as ethnography, focus group discussions, and key informant interviews, can help to explore the social and cultural dimensions of CBA and their influence on health outcomes and service delivery. These methods can provide rich insights into the complex power dynamics, gender norms, and decision-making processes that shape the implementation and impact of CBA initiatives, as well as the ways in which communities experience and respond to climate change impacts on health. For example, a study in the Pacific Islands used participatory photography and storytelling to document the gendered impacts of climate change on health and well-being, highlighting the disproportionate burdens faced by women and girls in terms of food insecurity, water scarcity, and caregiving responsibilities. The findings were used to inform the design of gender-responsive CBA interventions and to advocate for the inclusion of women's voices in adaptation planning and decision-making.

7.3 Assessing Community Adaptive Capacity and Long-Term Resilience

Beyond the immediate impacts on health outcomes and service delivery, CBA initiatives also aim to build the long-term adaptive capacity and resilience of communities to climate change impacts. Assessing these outcomes requires a holistic and longitudinal approach that captures the multiple dimensions of adaptive capacity and resilience.

Adaptive capacity refers to the ability of individuals, households, and communities to anticipate, prepare for, and respond to climate change impacts. It encompasses a range of assets and capabilities, such as knowledge and skills, social networks, financial resources, and access to information and technology. Assessing adaptive capacity may involve the use of participatory tools, such as community asset mapping, livelihood analysis, and vulnerability assessments, to identify the strengths, weaknesses, and priorities of local communities in relation to climate change adaptation. For instance, a study in Bangladesh used participatory wealth ranking and livelihood mapping to assess the adaptive capacity of coastal communities to cyclones and sea-level rise. The findings revealed that households with diversified livelihoods, strong social networks, and access to early warning information were better able to cope with and recover from climate-related disasters.

Resilience, on the other hand, refers to the ability of communities to absorb, adapt to, and transform in the face of shocks and stresses, including climate change impacts. It is a dynamic and multi-dimensional concept that encompasses ecological, social, economic, and institutional factors. Assessing resilience requires a systems perspective that considers the interactions and feedbacks between these different factors, as well as the ways in which they shape the capacity of communities to cope with and recover from climate-related disasters and stresses. Longitudinal studies that track the evolution of adaptive capacity and resilience over time can provide valuable insights into the long-term impacts of CBA initiatives. For example, a study in Vietnam used a combination of household surveys, focus group discussions, and participatory scenario planning to assess the resilience of coastal communities to typhoons and saltwater intrusion. The study found that communities with strong social capital, diversified livelihoods, and access to climate-resilient infrastructure and services were more resilient to climate change impacts compared to communities with limited assets and capacities.

7.4 Sharing Lessons Learned and Best Practices

Measuring the impact of CBA on health system resilience is not an end in itself, but rather a means to inform the design, implementation, and scaling up of effective adaptation strategies. Sharing lessons learned and best practices from CBA initiatives is crucial for building the evidence base on what works, what doesn't, and why in different contexts.

This requires the development of knowledge management and exchange platforms that facilitate the documentation, analysis, and dissemination of CBA experiences and outcomes. These platforms may take the form of online databases, communities of practice, learning networks, and policy dialogues that bring together researchers, practitioners, and policymakers from different sectors and levels of governance. For instance, the Adaptation Community of the Global Environment Facility (GEF) has established a knowledge management system that includes a database of adaptation projects, a series of thematic webinars and workshops, and a network of adaptation practitioners and experts. The system aims to promote cross-learning and collaboration among GEF-funded adaptation projects and to support the integration of adaptation into development planning and policy-making.

Participatory approaches to knowledge co-production and exchange, such as action research, citizen science, and storytelling, can help to ensure that the lessons learned from CBA initiatives are grounded in the lived experiences and priorities of local communities. These approaches can also help to build the capacity of communities to generate, interpret, and use evidence for advocacy and decision-making on climate change and health. For example, a study in Kenya used participatory video and community-based monitoring to document the impacts of climate change on pastoralist communities and to identify locally-relevant adaptation strategies. The videos were used to facilitate dialogue between communities, researchers, and policymakers and to inform the development of a national adaptation plan for the livestock sector.

Ultimately, the sharing of lessons learned and best practices from CBA initiatives can contribute to the development of more effective, equitable, and sustainable approaches to building health system resilience in the face of climate change. By learning from each other's successes and failures, CBA practitioners and researchers can collectively advance the field and support the scaling up of promising adaptation strategies across different contexts and scales.

In conclusion, measuring the impact of CBA on health system resilience is a complex and multi-dimensional task that requires a combination of quantitative and qualitative methods, as well as a participatory and systems-oriented approach. By evaluating improvements in health outcomes and service delivery, assessing community adaptive capacity and long-term resilience, and sharing lessons learned and best practices, CBA initiatives can generate valuable evidence and insights to inform the design and implementation of effective adaptation strategies. Ultimately, the goal of measuring the impact of CBA is not only to demonstrate its effectiveness, but also to support the continuous learning, improvement, and scaling up of community-driven approaches to building health system resilience in a changing climate.

8. RECOMMENDATIONS AND WAY FORWARD

The evidence presented in this review highlights the significant potential of community-based adaptation (CBA) initiatives in building health system resilience to climate change impacts. However, to fully realize this potential, several key recommendations and future directions need to be considered.

8.1 Policy Recommendations for Integrating CBA into Health Sector Planning

Integrating CBA approaches into health sector planning and policy-making is crucial for ensuring their sustainability and scalability. This requires a shift from project-based and short-term interventions to a more systemic and long-term approach that mainstreams CBA into health system strengthening efforts. Policy recommendations for achieving this integration include:

1. Developing national and sub-national adaptation plans that explicitly recognize the role of CBA in building health system resilience and provide a framework for coordinating and financing CBA initiatives across different sectors and levels of governance.
2. Establishing dedicated funding mechanisms, such as climate and health funds, that can support the implementation and scaling up of CBA initiatives in the health sector. These mechanisms should be designed to leverage diverse sources of finance, including public, private, and international funding, and to prioritize investments in the most vulnerable and underserved communities.
3. Strengthening the capacity of health systems to plan, implement, and monitor CBA initiatives, through the training of health workers, the development of guidelines and tools, and the establishment of partnerships with community-based organizations and other relevant stakeholders.

4. Promoting the integration of CBA into health system performance monitoring and evaluation frameworks, to ensure that the impacts of CBA initiatives on health outcomes, service delivery, and community resilience are regularly assessed and used to inform policy and planning decisions.

8.2 Strategies for Scaling Up Successful CBA Initiatives

Scaling up successful CBA initiatives is essential for achieving widespread and sustained impacts on health system resilience. However, scaling up is not simply a matter of replicating or expanding existing interventions, but rather requires a strategic and adaptive approach that takes into account the complex and context-specific nature of CBA. Strategies for scaling up successful CBA initiatives include:

1. Identifying and prioritizing CBA initiatives that have demonstrated effectiveness, feasibility, and acceptability in different contexts, and that have the potential for wider adoption and impact.
2. Engaging diverse stakeholders, including communities, health authorities, civil society organizations, and private sector actors, in the design and implementation of scaling-up strategies, to ensure their relevance, ownership, and sustainability.
3. Building on existing community-based structures and networks, such as community health worker programs, women's groups, and local governance bodies, to facilitate the dissemination and uptake of CBA interventions.
4. Employing a combination of horizontal and vertical scaling-up approaches, such as the expansion of CBA initiatives to new geographic areas or population groups (horizontal scaling-up), and the institutionalization of CBA into national and sub-national policies, plans, and budgets (vertical scaling-up).
5. Establishing learning and knowledge management systems that enable the continuous documentation, analysis, and sharing of lessons learned and best practices from CBA initiatives, and that support the adaptation and improvement of scaling-up strategies over time.

8.3 Future Research Priorities and Knowledge Gaps

While the evidence base on CBA and health system resilience is growing, there are still significant knowledge gaps and research priorities that need to be addressed to inform future policy and practice. These include:

1. Developing and testing innovative methodologies and tools for measuring and evaluating the impacts of CBA initiatives on health system resilience, including the use of participatory, mixed-methods, and systems-oriented approaches.

2. Conducting long-term and comparative studies that assess the effectiveness, cost-effectiveness, and sustainability of different CBA interventions across diverse contexts and populations, and that identify the key factors and conditions that influence their success.
3. Exploring the social, political, and institutional dimensions of CBA, including the role of power dynamics, governance arrangements, and social norms in shaping the design, implementation, and outcomes of CBA initiatives.
4. Investigating the potential synergies and trade-offs between CBA and other health system strengthening approaches, such as universal health coverage, primary health care, and health emergency preparedness, and identifying strategies for optimizing their integration and mutual reinforcement.

8.4 Importance of Cross-Sectoral Collaboration and Partnerships

Building health system resilience to climate change requires a multi-sectoral and multi-stakeholder approach that goes beyond the health sector alone. CBA initiatives, in particular, rely on effective collaboration and partnerships between diverse actors, including communities, government agencies, civil society organizations, academia, and the private sector.

Cross-sectoral collaboration can help to leverage the unique strengths, resources, and expertise of different partners, and to address the complex and interconnected drivers of climate-health vulnerabilities. For example, partnerships between health authorities and meteorological agencies can enable the development of early warning systems for climate-sensitive diseases, while collaborations between health and agriculture sectors can support the promotion of climate-resilient and nutrition-sensitive food systems.

However, effective cross-sectoral collaboration and partnerships require dedicated efforts to build trust, align incentives, and establish clear roles and responsibilities among partners. Key strategies for promoting cross-sectoral collaboration in CBA include:

1. Conducting stakeholder mapping and analysis to identify the key actors and interests involved in CBA, and to assess their potential contributions and challenges to collaboration.
2. Establishing multi-stakeholder platforms and networks that enable regular dialogue, information sharing, and joint planning and implementation of CBA initiatives.

3. Developing and implementing partnership agreements and memoranda of understanding that outline the shared goals, activities, and resources of CBA collaborations, and that specify the roles and accountabilities of different partners.
4. Building the capacity of different sectors and stakeholders to engage in CBA, through targeted training, technical assistance, and knowledge exchange programs.
5. Monitoring and evaluating the effectiveness and impact of cross-sectoral collaborations in CBA, and using the findings to adapt and improve partnership strategies over time.

In conclusion, CBA offers a promising approach for building health system resilience to climate change, but its success depends on the concerted efforts of policymakers, practitioners, researchers, and communities to integrate CBA into health sector planning, scale up successful initiatives, address knowledge gaps, and foster cross-sectoral collaboration and partnerships. By working together and learning from each other's experiences, we can build a more resilient and sustainable health system that can protect and promote the health and well-being of all people in a changing climate.

9. CONCLUSION

This chapter has made significant contributions to the broader discourse on climate resilience by providing a comprehensive and nuanced examination of the role of community-based adaptation (CBA) in building health system resilience to climate change. Through a careful analysis of key concepts, frameworks, case studies, and lessons learned, the chapter has convincingly demonstrated the transformative potential of CBA in enhancing the capacity of health systems to anticipate, absorb, and adapt to climate-related shocks and stresses.

One of the chapter's most important contributions is its emphasis on the importance of a community-centered and participatory approach to adaptation. By highlighting the agency, knowledge, and priorities of local communities in shaping resilience strategies, the chapter challenges top-down and technocratic approaches to adaptation that have often failed to deliver equitable and sustainable outcomes. Instead, it argues for a more bottom-up and inclusive approach that recognizes communities as active agents of change rather than passive recipients of external interventions.

Another key contribution of the chapter is its demonstration of the value of integrating CBA into health system strengthening efforts. Through a series of compelling case studies from diverse contexts, the chapter shows how CBA initiatives can address the complex and interconnected drivers of climate-health vulnerabilities, such as poverty, marginalization, and weak health infrastructure. By working across sectors and scales, CBA can help to build more resilient and responsive health systems that are better equipped to deal with the challenges of a changing climate.

The chapter also provides a wealth of practical guidance and recommendations for designing, implementing, and scaling up CBA initiatives in the health sector. Drawing on the best available evidence and practice, it offers concrete strategies for engaging communities, building partnerships, mobilizing resources, and monitoring and evaluating CBA impacts. These insights are invaluable for policymakers, practitioners, and researchers seeking to translate the principles of CBA into effective and sustainable actions on the ground.

However, the chapter does not shy away from the significant challenges and barriers to realizing the full potential of CBA in building health system resilience. It candidly acknowledges the limited integration of CBA into health sector policies and plans, the insufficient financing and capacity for CBA implementation, the lack of robust monitoring and evaluation systems, and the need for stronger cross-sectoral collaboration and partnerships. By doing so, it provides a realistic assessment of the current state of play and the critical gaps that need to be addressed to accelerate progress.

Perhaps most importantly, the chapter issues a powerful call to action for prioritizing CBA in health sector adaptation efforts. It urges all relevant stakeholders – from policymakers to practitioners to researchers to communities themselves – to work together in advancing the CBA agenda and building a more resilient and equitable health system in the face of climate change. This call is grounded in a recognition that building health system resilience is not just a technical or scientific challenge, but also a moral and political imperative that requires a fundamental shift in how we think about and approach adaptation.

The chapter argues that CBA offers a promising pathway for achieving this shift, by putting communities at the center of adaptation efforts and fostering a more collaborative and inclusive approach to building resilience. By empowering communities to identify their own vulnerabilities and capacities, to prioritize their own adaptation needs and aspirations, and to hold other stakeholders accountable for their commitments and actions, CBA can help to redress power imbalances and promote more equitable and sustainable outcomes.

At the same time, the chapter recognizes that scaling up CBA in the health sector will require concerted efforts to overcome the many challenges and barriers that currently constrain its implementation. This will require dedicated leadership, resources, and capacity-building at all levels, from the local to the global. It will also require a willingness to learn from both successes and failures, to adapt to changing circumstances and needs, and to continuously improve the effectiveness and impact of CBA interventions.

Ultimately, the chapter leaves us with a sense of both urgency and hope in the face of the growing climate crisis. As the impacts of climate change continue to accelerate and intensify, the need for effective and equitable adaptation strategies has never been greater. CBA offers a promising way forward, but realizing its potential will require a collective effort and commitment from all of us.

The chapter's call to action is therefore not just directed at policymakers and practitioners, but at all of us who have a stake in building a more resilient and sustainable future. It reminds us that we all have a role to play in this effort, whether as researchers generating new knowledge and evidence, as communities driving local action and innovation, or as citizens holding our leaders accountable for their decisions and actions.

In this sense, the chapter is not just a technical analysis of CBA and health system resilience, but also a powerful reminder of our shared responsibility and agency in shaping the future we want. It challenges us to think critically about the values and priorities that underpin our adaptation efforts, and to work towards a more just, equitable, and resilient world for all.

As we move forward, let us take this call to action to heart, and work together to build a health system that can truly protect and promote the health and well-being of all people, in the face of the many challenges and uncertainties that lie ahead. Let us learn from the experiences and insights of CBA practitioners and communities around the world, and strive to create a more collaborative, inclusive, and transformative approach to building resilience. And let us never lose sight of the fundamental truth that lies at the heart of this chapter: that the path to a more resilient and sustainable future begins with empowering and supporting the communities on the frontlines of the climate crisis.

REFERENCES

Ayers, J., & Forsyth, T. (2009). Community-based adaptation to climate change. *Environment, 51*(4), 22–31. DOI: 10.3200/ENV.51.4.22-31

Berry, P., Bowen, K., & Kjellstrom, T. (2022). Climate change and health equity: Actions to protect vulnerable populations. *Annual Review of Public Health, 43*, 185–206.

Bowen, K. J., Ebi, K. L., & Friel, S. (2022). Climate change adaptation and mitigation: Next steps for cross-sectoral action to protect global health. *The Lancet. Planetary Health, 6*(5), e386–e396. PMID: 35550712

Brooks, N., Anderson, S., & Burton, I. (2023). Adaptive management in climate change adaptation: Concepts and methods. *Climate and Development, 15*(2), 145–162.

Dasgupta, P., Morton, J. F., & Dodman, D. (2020). *Community-based adaptation to climate change: Scaling it up*. Routledge.

Dodman, D., & Mitlin, D. (2022). Challenges in implementing community-based adaptation. *Nature Climate Change, 12*(2), 118–124.

Dodman, D., & Mitlin, D. (2023). Challenges in implementing community-based adaptation. *Nature Climate Change, 13*(2), 118–124.

Ebi, K. L., Berry, P., & Hayes, K. (2023). Vulnerability assessments for climate change and health. *Annual Review of Public Health, 44*, 167–188.

Ebi, K. L., & Hess, J. J. (2023). Health risks of climate change: Implications for adaptation and resilience. *Annual Review of Environment and Resources, 48*, 331–355.

Ebi, K. L., Hess, J. J., & Watkiss, P. (2021). Health risks and costs of climate variability and change. In *Climate Change 2021: Impacts, Adaptation and Vulnerability. Contribution of Working Group II to the Sixth Assessment Report of the Intergovernmental Panel on Climate Change*. Cambridge University Press.

Ebi, K. L., Hess, J. J., & Watkiss, P. (2021). Health risks and costs of climate variability and change. In *The Economics of Climate Change* (pp. 123–152). Cambridge University Press.

Ebi, K. L., & Otmani del Barrio, M. (2023). Community engagement in climate change and health adaptation: A systematic review. *The Lancet. Planetary Health, 7*(3), e220–e231.

Ensor, J., Berger, R., & Huq, S. (2021). Community-based adaptation and culture in theory and practice. *Progress in Human Geography*, *45*(4), 813–832.

Ensor, J., Berger, R., & Huq, S. (2023). Community-based adaptation: Lessons from practice. *Progress in Human Geography*, *47*(2), 289–307.

Ensor, J., Berger, R., & Huq, S. (2023). Community-based adaptation: Lessons from practice. *Progress in Human Geography*, *47*(2), 289–307.

Ford, J. D., King, N., & Galappaththi, E. K. (2020). Strengthening the resilience of Indigenous health systems to climate change. *The Lancet. Planetary Health*, *4*(6), e237–e238.

Ford, J. D., Sherman, M., & Berrang-Ford, L. (2021). Building capacity for climate change adaptation in the health sector. *The Lancet. Planetary Health*, *5*(4), e237–e246. PMID: 33684341

Gosselin, P., Bélanger, D., & Lapaige, V. (2022). Implementing a community-based heat action plan in Quebec, Canada: Lessons learned and future directions. *Climate Risk Management*, *35*, 100403.

Haines, A., & Ebi, K. (2019). The imperative for climate action to protect health. *The New England Journal of Medicine*, *380*(3), 263–273. DOI: 10.1056/NEJMra1807873 PMID: 30650330

Haines, A., & Ebi, K. (2019). The imperative for climate action to protect health. *The New England Journal of Medicine*, *380*(3), 263–273. DOI: 10.1056/NEJMra1807873 PMID: 30650330

IPCC. (2022). *Climate Change 2022: Impacts, Adaptation and Vulnerability. Contribution of Working Group II to the Sixth Assessment Report of the Intergovernmental Panel on Climate Change*. Cambridge University Press.

IPCC. (2022). *Climate Change 2022: Impacts, Adaptation and Vulnerability*. Cambridge University Press.

Kirkland, T. M., Keene, D. E., & Gómez, W. (2021). Indigenous knowledge integration in community-based adaptation. *Current Opinion in Environmental Sustainability*, *51*, 55–60.

McNamara, K. E., & Buggy, L. (2017). Community-based climate change adaptation: A review of academic literature. *Local Environment*, *22*(4), 443–460. DOI: 10.1080/13549839.2016.1216954

McNamara, K. E., & Buggy, L. (2017). Community-based climate change adaptation: A review of academic literature. *Local Environment*, *22*(4), 443–460. DOI: 10.1080/13549839.2016.1216954

McNamara, K. E., & Buggy, L. (2021). Community-based climate adaptation: A systematic review of the literature. *Climate and Development*, *13*(2), 110–122.

Moser, S. C., Coffee, J., & Seville, A. (2021). *Rising to the challenge, together: A review and critical assessment of the state of the US climate adaptation field*. World Resources Institute.

Patz, J. A., & Thomson, M. C. (2022). Climate change and health: Moving from theory to practice. *PLoS Medicine*, *19*(4), e1003985. PMID: 30063707

Preston, B. L., Westaway, R. M., & Yuen, E. J. (2022). Climate change vulnerability assessment: Framework and methods. *Environmental Science & Policy*, *89*, 56–69.

Rahman, M. S., Karim, M. R., & Islam, M. S. (2022). Building climate-resilient health facilities in coastal Bangladesh: A community-based approach. *Climate and Development*, *14*(7), 645–657.

Reid, H. (2016). Ecosystem- and community-based adaptation: Learning from community-based natural resource management. *Climate and Development*, *8*(1), 4–9. DOI: 10.1080/17565529.2015.1034233

Reid, H., Alam, S. S., & Berger, R. (2019). Community-based adaptation to climate change: Lessons from the field. In *Climate Change Adaptation in the Global South* (pp. 67–86). Springer.

Reid, H., Alam, S. S., & Berger, R. (2022). Scaling up community-based adaptation. In *Climate Change Adaptation in Practice* (pp. 123–144). Routledge.

Romanello, M., Di Napoli, C., & Drummond, P. (2023). The 2023 report of The Lancet Countdown on health and climate change. *Lancet*, *402*(10414), 1785–1863. PMID: 37977174

Sherman, M., Berrang-Ford, L., & Ford, J. (2022). Stakeholder engagement in climate change adaptation planning: A systematic review. *Environmental Science & Policy*, *128*, 1–13.

Sherman, M., Berrang-Ford, L., & Ford, J. (2023). Monitoring and evaluation in community-based adaptation. *Global Environmental Change*, *68*, 102–114.

Smith, K. R., Woodward, A., & Campbell-Lendrum, D. (2023). Climate change and global health equity. *The Lancet. Global Health*, *11*(6), e815–e826. PMID: 37202013

Taye, M., Dyer, E., & Charles, K. J. (2021). Community-based surveillance and response to climate-sensitive diseases in the Somali Region of Ethiopia. *Pastoralism, 11*(1), 1–14.

Thomson, M. C., Mason, S. J., & Connor, S. J. (2023). *Climate information for public health action*. Routledge.

Watts, N., Amann, M., & Arnell, N. (2021). The 2021 report of The Lancet Countdown on health and climate change. *Lancet, 398*(10311), 1619–1662. DOI: 10.1016/S0140-6736(21)01787-6 PMID: 34687662

Watts, N., Amann, M., & Arnell, N. (2022). The 2022 report of The Lancet Countdown on health and climate change. *Lancet, 400*(10363), 1619–1662. DOI: 10.1016/S0140-6736(22)01540-9 PMID: 36306815

Watts, N., Amann, M., Arnell, N., Ayeb-Karlsson, S., Belesova, K., Boykoff, M., Byass, P., Cai, W., Campbell-Lendrum, D., Capstick, S., Chambers, J., Dalin, C., Daly, M., Dasandi, N., Davies, M., Drummond, P., Dubrow, R., Ebi, K. L., Eckelman, M., & Montgomery, H. (2019). The 2019 report of The Lancet Countdown on health and climate change: Ensuring that the health of a child born today is not defined by a changing climate. *Lancet, 394*(10211), 1836–1878. DOI: 10.1016/S0140-6736(19)32596-6 PMID: 31733928

WHO. (2015). Operational framework for building climate resilient health systems. World Health Organization. https://www.who.int/globalchange/publications/building-climate-resilient-health-systems/en/

WHO. (2020). *Operational framework for building climate resilient health systems*. World Health Organization.

WHO. (2021). COP26 special report on climate change and health: The health argument for climate action. World Health Organization. https://www.who.int/publications/i/item/9789240036727

WHO. (2021). *COP26 special report on climate change and health*. World Health Organization.

WHO. (2021). *Operational framework for building climate-resilient health systems*. World Health Organization.

WHO. (2022). *Building climate-resilient health systems through community-based adaptation*. World Health Organization.

WHO. (2023). *Building resilient health systems: A framework for climate change adaptation*. World Health Organization.

WHO. (2023). *Guidelines for climate-resilient and environmentally sustainable health care facilities*. World Health Organization.

Williams, P. A., Crespo, O., & Abu, M. (2021). Adapting to climate change through community-based approaches. *Regional Environmental Change*, *21*(2), 1–14.

Chapter 8
Fostering Economic Resilience in a World of Increasing Natural Disasters: A Multi-Dimensional Perspective

Naboshree Bhattacharya
https://orcid.org/0000-0001-5572-4098
Amity University, Ranchi, India

Divya Bansal
https://orcid.org/0000-0001-6268-5402
Amity University, Noida, India

ABSTRACT

The increasing frequency and intensity of natural disasters pose a significant threat to global economic resilience. The economic impacts of these disasters are far-reaching and complex, encompassing direct costs such as infrastructure damage and indirect consequences like supply chain disruptions. Developing countries and vulnerable communities often bear the brunt of the damage, lacking the resources and capacity to effectively prepare for and respond to disasters. Building economic resilience requires a proactive, multi-dimensional approach that integrates prevention, response, and recovery strategies. This chapter examines the complex relationship between natural disasters and economic resilience, drawing upon empirical evidence, theoretical frameworks, and practical insights. It explores advancements in

DOI: 10.4018/979-8-3693-9745-9.ch008

risk assessment technologies, financial protection mechanisms, and post-disaster reconstruction models, emphasizing the role of international cooperation and public-private partnerships.

1. INTRODUCTION

Natural disasters pose an increasingly dire threat to the global economy, with their frequency and intensity rising at an alarming rate. From devastating hurricanes and floods to catastrophic earthquakes and wildfires, these events have claimed countless lives, destroyed vital infrastructure, and disrupted the livelihoods of millions worldwide (Hallegatte et al., 2017). The economic consequences are equally staggering, with annual global losses now averaging over $300 billion and potentially reaching $500 billion or more by 2030 (World Bank, 2021). Developing countries and vulnerable communities often bear the brunt of the damage, lacking the resources and capacity to effectively prepare for and recover from disasters (Hallegatte & Rozenberg, 2017). The COVID-19 pandemic has further underscored the interconnectedness of the global economy and the cascading effects that can result from systemic shocks (Hynes et al., 2020). Building economic resilience to natural hazards has thus become a critical imperative for sustainable development and global prosperity.

1.1 Global Trends in Natural Disasters and Economic Impacts

The impacts of natural disasters on the global economy are far-reaching and complex, encompassing a wide range of direct and indirect costs. Direct costs include damage to buildings, infrastructure, and productive assets, while indirect effects span disruptions to trade, supply chains, productivity, and the diversion of resources towards relief efforts (Kousky, 2014). The burden is unevenly distributed, with a vicious cycle of economic disruption and deepening poverty afflicting many developing nations (Hallegatte et al., 2020). Beyond immediate losses, the destruction of critical infrastructure like ports, roads, and power grids can disrupt global trade flows, leading to commodity shortages and price spikes (Gassebner et al., 2010). Disaster-induced displacement can also trigger social and political instability, further exacerbating economic challenges (Berlemann & Steinhardt, 2017). As climate change amplifies these threats, proactively building resilience has become a fundamental priority.

To address this critical issue, there is an urgent need for a paradigm shift in how we approach disaster risk reduction. Rather than reacting to individual events, we must adopt a proactive, multi-faceted approach that prioritizes prevention, preparedness, and long-term resilience-building (Simonovic, 2011). This requires concerted effort from governments, businesses, civil society, and international organizations to develop comprehensive strategies that address the root causes of vulnerability and build communities' capacity to withstand and recover from disasters (Cutter et al., 2008). The stakes could not be higher - without urgent action to strengthen global economic resilience, the mounting costs of natural disasters threaten to derail progress on key development goals like poverty reduction, food security, and sustainable infrastructure (Hallegatte et al., 2017).

1.2 Research Objectives and Methodology

This study aims to provide a comprehensive analysis of the complex relationship between natural disasters and global economic resilience. Drawing upon a rich tapestry of empirical evidence, theoretical frameworks, and practical insights, it examines the full spectrum of economic impacts and explores strategies for prevention, response, and recovery. The research adopts a multi-dimensional, systems thinking approach, integrating economic analysis with insights from environmental science, engineering, and social psychology to uncover the intricate web of interactions between natural hazards, economic systems, and human behavior (Simonovic, 2011).

Central to the analysis is the recognition that building economic resilience requires a proactive, multi-faceted approach spanning prevention, response, and recovery. The study delves into each dimension's intricacies, exploring cutting-edge advancements in risk assessment, financial protection, and post-disaster reconstruction. Special emphasis is placed on international cooperation and public-private partnerships in fostering resilience, particularly in vulnerable regions with limited resources (Becker & Kretsch, 2019).

Through a wealth of case studies and best practices from around the world, ranging from Japan's successful disaster risk reduction to the Caribbean's innovative financing mechanisms, the study highlights diverse approaches to resilience-building and the importance of context-specific solutions (Ranghieri & Ishiwatari, 2014). Ultimately, it proposes actionable recommendations for policymakers and practitioners, focusing on key areas like risk governance, capacity building, and data-driven decision-making. By embracing these recommendations and fostering a culture of resilience, we can mitigate natural disasters' economic impacts and unlock opportunities for sustainable growth in an increasingly hazardous world.

2. THEORETICAL FRAMEWORK

2.1 Economic Resilience Theory

Economic resilience theory provides a foundational framework for understanding and enhancing the capacity of economies to withstand and recover from shocks, including natural disasters. This theory posits that economies possess inherent strengths and adaptive capacities that enable them to absorb disturbances, maintain functionality, and bounce back from adversity (Rose, 2017). The focus shifts from vulnerability reduction to fostering attributes and mechanisms that promote resilience, such as diversity, redundancy, flexibility, and innovation (Martin & Sunley, 2015).

A key tenet of economic resilience theory is the recognition that resilience is a multi-dimensional construct, encompassing the ability to resist and recover from shocks and to transform and adapt in the face of changing circumstances (Modica & Reggiani, 2015). This dynamic view emphasizes the importance of learning, innovation, and structural change in building long-term economic sustainability. It also highlights the role of social capital, institutions, and governance in shaping the resilience of economies at various scales, from local communities to global networks (Bristow & Healy, 2018).

2.2 Disaster Risk Reduction Models

Disaster risk reduction (DRR) models provide a complementary lens for understanding the complex interplay between natural hazards, economic systems, and societal vulnerabilities. These models aim to identify, assess, and mitigate the potential impacts of disasters through a systematic approach that integrates risk assessment, prevention, preparedness, response, and recovery (UNDRR, 2015). Central to DRR is the concept of risk, understood as a function of hazard, exposure, and vulnerability (IPCC, 2012). By quantifying and mapping these components, DRR models enable policymakers and practitioners to prioritize interventions and allocate resources effectively.

Recent advancements in DRR have emphasized the importance of integrating resilience-building strategies into the disaster management cycle. This includes measures such as strengthening early warning systems, enhancing infrastructure resilience, promoting risk-informed land-use planning, and developing financial protection mechanisms like insurance and contingency funds (World Bank, 2013). These approaches recognize that reducing disaster risk is not only about minimizing losses but also about building the capacity of economies and communities to anticipate, cope with, and transform in the face of adversity.

2.3 Systems Thinking Approach to Disaster Management

A systems thinking approach to disaster management offers a holistic and integrative framework for understanding the complex interactions between natural hazards, economic systems, and human behavior. This perspective views disasters not as isolated events but as the product of interconnected and interdependent processes that span multiple scales and domains (Simonovic, 2011). By mapping the feedback loops, thresholds, and emergent properties of these systems, a systems approach enables decision-makers to anticipate cascading effects, identify leverage points for intervention, and design resilience strategies that account for the dynamic nature of risk.

Central to a systems thinking approach is the recognition that building resilience requires a multi-stakeholder, cross-sectoral effort that engages diverse actors and knowledge systems. This includes fostering collaboration between government agencies, businesses, civil society organizations, and scientific institutions to co-produce knowledge, share resources, and coordinate actions (Cutter et al., 2008). It also involves empowering communities to participate actively in the disaster management process, drawing on their local knowledge, social networks, and adaptive capacities to build resilience from the bottom up.

2.4 Integration With Sustainable Development Goals

Building economic resilience to natural disasters is not only a matter of managing risk but also of promoting sustainable development. The Sustainable Development Goals (SDGs) adopted by the United Nations in 2015 provide a global framework for integrating disaster risk reduction with broader social, economic, and environmental objectives (UN, 2015). Several SDGs, such as those related to poverty eradication, sustainable cities, and climate action, explicitly recognize the linkages between disaster resilience and sustainable development.

Integrating resilience-building strategies with the SDGs requires a paradigm shift from reactive to proactive approaches that address the underlying drivers of risk and vulnerability. This includes investing in risk-informed infrastructure, promoting inclusive and equitable growth, strengthening social protection systems, and fostering innovation and entrepreneurship (UNDP, 2016). It also involves mainstreaming disaster risk considerations into development planning and budgeting processes, ensuring that investments in resilience are aligned with long-term sustainability goals.

By embracing these theoretical frameworks and integrating them into policy and practice, we can develop a more comprehensive and nuanced understanding of the complex relationship between natural disasters and economic resilience. This understanding can inform the design of effective strategies for prevention, response, and recovery that not only minimize losses but also build the capacity of economies and communities to thrive in the face of adversity.

3. ECONOMIC IMPACT ANALYSIS

Economic impact analysis is a critical component in understanding the far-reaching consequences of natural disasters on economies at various scales. This analysis involves a comprehensive assessment of the direct and indirect costs, as well as the broader implications for sectors, markets, and global supply chains (Hallegatte, 2015; Rose, 2004).

The direct costs of natural disasters encompass the immediate damages and losses resulting from the physical destruction of assets, such as buildings, infrastructure, and productive capital (Kousky, 2014). Assessing these costs requires a detailed accounting of the monetary value of damages across different sectors, including housing, agriculture, industry, and public infrastructure. Recent advancements in geospatial technologies and remote sensing have greatly enhanced the ability to rapidly assess and quantify direct disaster damages (Eguchi et al., 2010).

However, the economic consequences of disasters extend far beyond the direct costs. Indirect effects, which arise from the complex interactions between damaged assets, disrupted economic activities, and broader market dynamics, can be substantial and long-lasting (Okuyama, 2007). These effects can manifest through various channels, such as supply chain disruptions, reduced productivity, loss of income and employment, and decreased investment and trade flows (Rose & Wei, 2013).

Sectoral vulnerability analysis is another key aspect of economic impact assessment. It involves identifying the specific sectors that are most exposed to disaster risks and assessing their relative sensitivity and adaptive capacity (Cutter et al., 2003). Sectors such as agriculture, tourism, and small and medium enterprises are often particularly vulnerable to disasters due to their reliance on weather-sensitive resources, concentration in hazard-prone areas, and limited financial resilience (Becken & Hughey, 2013; Ballesteros & Domingo, 2015).

In an increasingly globalized world, the economic impacts of disasters can also propagate through international trade networks and supply chains. Disasters can disrupt the flow of goods and services across borders, leading to cascading effects on industries and consumers in distant locations (Gassebner et al., 2010). The 2011 Thailand floods, for example, caused severe disruptions in the global supply chains of the electronics and automotive industries, highlighting the vulnerability of interconnected production systems to localized shocks (Haraguchi & Lall, 2015).

Conducting a comprehensive economic impact analysis is essential for informing effective disaster risk reduction and resilience-building strategies. By quantifying the full spectrum of costs and identifying the most vulnerable sectors and regions, policymakers and practitioners can prioritize investments and interventions that enhance economic resilience (World Bank, 2013). This includes measures such as strengthening early warning systems, promoting risk-informed land-use planning, developing financial protection mechanisms, and fostering public-private partnerships for resilient infrastructure development (Hallegatte et al., 2017).

As the frequency and intensity of natural disasters continue to rise in the context of climate change, building economic resilience has become an increasingly urgent global priority. Economic impact analysis provides a critical foundation for understanding the complex challenges posed by disasters and developing evidence-based solutions that protect lives, livelihoods, and sustainable development prospects in an increasingly hazard-prone world.

3.1 Direct Costs Assessment

Assessing the direct economic costs of natural disasters is a critical first step in understanding their overall impact on economies. These costs encompass the immediate damages and losses incurred due to the physical destruction of assets, such as buildings, infrastructure, and productive capital (Hallegatte, 2015). Estimating direct costs involves a comprehensive accounting of the monetary value of damages across various sectors, including housing, agriculture, industry, and public infrastructure.

Recent advancements in geospatial technologies and remote sensing have greatly enhanced our ability to rapidly assess and quantify direct disaster damages. High-resolution satellite imagery, drone surveys, and crowdsourced data enable real-time mapping of affected areas and assets, providing a more accurate picture of the extent and severity of destruction (Eguchi et al., 2010). These tools are particularly valuable in the immediate aftermath of disasters, when on-the-ground assessments may be hampered by access constraints and safety concerns.

However, estimating direct costs is not without challenges. The valuation of damages can be complicated by factors such as the age and condition of affected assets, the availability of reliable baseline data, and the potential for cascading failures in interconnected systems (Cochrane, 2004). Moreover, the distribution of direct costs can be highly uneven, with vulnerable populations and informal settlements often disproportionately affected due to substandard housing and infrastructure.

3.2 Indirect Economic Consequences

While direct costs capture the immediate impact of disasters, the indirect economic consequences can be far more substantial and long-lasting. These consequences arise from the complex interactions between damaged assets, disrupted economic activities, and broader market dynamics (Rose, 2004). Indirect effects can manifest through various channels, such as supply chain disruptions, reduced productivity, loss of income and employment, and decreased investment and trade flows.

One of the most significant indirect consequences of disasters is the disruption of critical infrastructure networks, such as transportation, energy, and telecommunications systems. The failure of these networks can have cascading effects across the economy, leading to production bottlenecks, service interruptions, and increased transaction costs (Rinaldi et al., 2001). For example, the destruction of a major port or airport can severely impede international trade, while damage to power grids can cripple industrial output and commercial activities.

Disasters can also have profound impacts on labor markets and human capital. The displacement of workers due to the destruction of homes and livelihoods can lead to reduced labor supply, lower productivity, and increased unemployment (Belasen & Polachek, 2009). In the longer term, the disruption of education and health services can erode human capital, with lasting consequences for individual well-being and economic growth (Cuaresma, 2010).

3.3 Sectoral Vulnerability Analysis

The economic impacts of disasters are not uniformly distributed across sectors, with some industries exhibiting greater vulnerability than others. Sectoral vulnerability analysis aims to identify the specific sectors that are most exposed to disaster risks and assess their relative sensitivity and adaptive capacity (Cutter et al., 2003). This analysis is crucial for prioritizing risk reduction efforts and developing targeted resilience strategies.

Sectors that are particularly vulnerable to disasters include agriculture, tourism, and small and medium enterprises (SMEs). Agriculture is highly sensitive to weather-related hazards such as droughts, floods, and storms, which can lead to crop failures, livestock losses, and food insecurity (FAO, 2015). Tourism is often concentrated in coastal areas and relies on natural assets, making it susceptible to disasters that damage infrastructure and deter visitors (Becken & Hughey, 2013). SMEs, which account for a significant share of employment and output in many economies, often lack the financial resources and business continuity plans to withstand and recover from disasters (Ballesteros & Domingo, 2015).

Sectoral vulnerability analysis also highlights the interdependencies between sectors and the potential for cascading impacts. For example, the failure of the agricultural sector can have ripple effects on food processing industries, while disruptions in the tourism sector can affect related services such as transportation and hospitality (Biggs et al., 2015). Understanding these intersectoral linkages is essential for developing holistic resilience strategies that account for the complex dynamics of economic systems.

3.4 Global Supply Chain Implications

In an increasingly interconnected world, the impacts of natural disasters can extend far beyond the directly affected regions through global supply chain networks. Supply chains have become more complex and geographically dispersed, with production processes spanning multiple countries and continents (Christopher & Peck, 2004). While this globalization has brought efficiency gains and access to new markets, it has also increased the vulnerability of supply chains to disruptions from natural hazards.

Disasters can disrupt supply chains through various mechanisms, such as the destruction of production facilities, transportation infrastructure, and communication networks (Sheffi, 2005). These disruptions can lead to shortages of raw materials, intermediate goods, and finished products, as well as delays in delivery times and increased costs. The impacts can propagate rapidly through supply networks, affecting downstream industries and consumers in distant locations.

The 2011 Thailand floods provide a stark example of the global supply chain implications of disasters. The floods inundated several industrial estates that were critical hubs for the production of electronic components and automobiles, leading to severe disruptions in the supply of these goods to manufacturers around the world (Haraguchi & Lall, 2015). The cascading effects were felt in industries as diverse as computer hardware, automotive, and aerospace, highlighting the interconnectedness of modern supply chains.

To build resilience in global supply chains, firms and policymakers are increasingly adopting strategies such as supplier diversification, inventory management, and flexible production processes (Tang, 2006). Collaborative approaches, such as information sharing and joint risk assessment, can also help to identify and mitigate supply chain vulnerabilities (Scholten et al., 2014). However, building truly resilient supply chains requires a paradigm shift from a focus on efficiency to a balance between efficiency and redundancy, as well as a greater emphasis on collaboration and adaptability.

4. PREVENTION STRATEGIES

4.1 Risk Assessment and Early Warning Systems

Effective prevention strategies for natural disasters begin with comprehensive risk assessment and early warning systems. Risk assessment involves identifying, analyzing, and evaluating the potential hazards and vulnerabilities of a given area or sector (UNDRR, 2017). This process relies on advanced geospatial technologies, such as remote sensing, geographic information systems (GIS), and predictive modeling, to map and quantify disaster risks (Birkmann et al., 2021). By integrating data on hazard exposure, physical vulnerabilities, and socioeconomic factors, risk assessments provide a foundation for informed decision-making and targeted interventions (Gallina et al., 2016).

Early warning systems (EWS) are another critical component of disaster prevention. EWS are designed to detect, monitor, and disseminate timely information about impending hazards, enabling individuals, communities, and organizations to take appropriate preparedness and response actions (WMO, 2018). Advances in sensor technologies, data analytics, and communication networks have greatly enhanced the accuracy and reach of EWS in recent years (Funk et al., 2019). For example, the use of machine learning algorithms to analyze real-time weather data has improved the lead time and precision of hurricane and flood forecasts (Nayak & Ghosh, 2013).

However, the effectiveness of risk assessment and EWS depends not only on technological capabilities but also on the institutional and social contexts in which they are implemented. Factors such as governance structures, resource allocation, and public awareness and trust can significantly influence the uptake and impact of these tools (Marchezini et al., 2018). Therefore, successful prevention strategies must adopt a people-centered approach that engages diverse stakeholders, incorporates local knowledge and needs, and builds capacity for risk-informed decision-making at all levels (UNISDR, 2015).

4.2 Infrastructure Resilience Planning

Building resilient infrastructure is a key strategy for preventing and mitigating the impacts of natural disasters on economies. Resilient infrastructure refers to the ability of systems to withstand, adapt to, and recover from shocks and stresses while maintaining essential functions (Hallegatte et al., 2019). This requires a comprehensive approach that considers the physical, social, and economic dimensions of resilience (Birkmann & Mechler, 2015).One key aspect of infrastructure resilience planning is the incorporation of risk assessment and scenario analysis into the design and decision-making process (Argyroudis et al., 2020). This involves identifying and quantifying the potential impacts of different hazards on infrastructure systems, using tools such as probabilistic modelling, vulnerability analysis, and cost-benefit analysis (Kim et al., 2019). By understanding the likelihood and consequences of different risk scenarios, planners can make informed decisions about the location, design, and construction of infrastructure assets, as well as the allocation of resources for maintenance and upgrades (Grafakos et al., 2020).Another important consideration in infrastructure resilience planning is the integration of nature-based solutions and green infrastructure (Sebesvari et al., 2019). These approaches leverage the protective and regenerative capacities of natural systems, such as wetlands, forests, and coastal ecosystems, to enhance the resilience of built environments (Frantzeskaki, 2019).

For example, the use of permeable pavements and green roofs can help to reduce urban flood risks by increasing water infiltration and retention, while also providing co-benefits such as improved air quality and biodiversity (Kabisch et al., 2017). Similarly, the restoration of coastal mangroves and coral reefs can provide natural barriers against storm surges and sea-level rise, while also supporting local livelihoods and ecosystems (Beck et al., 2018).In addition to these technical considerations, infrastructure resilience planning also requires a strong focus on social and institutional dimensions (Ayyub, 2014). This includes engaging with local communities and stakeholders to understand their needs, vulnerabilities, and capacities, and to build trust and ownership in the planning process (Meerow & Newell, 2019). It also involves strengthening the institutional frameworks and governance structures that are necessary for effective coordination, financing, and implementation of resilience measures (Coaffee et al., 2018). This may include the development of resilience standards and guidelines, the establishment of dedicated funding mechanisms, and the creation of multi-stakeholder partnerships and networks (Dhar & Khirfan, 2017).

Furthermore, infrastructure resilience planning must be integrated with broader sustainable development goals and strategies (Grafakos et al., 2020). This requires a holistic approach that considers the long-term social, economic, and environmental implications of infrastructure decisions, and that seeks to maximize co-benefits and minimize trade-offs (Hallegatte et al., 2019). For example, investing in resilient transport systems can not only reduce the risk of disruptions from natural hazards but also improve access to markets, services, and opportunities for local communities (Rozenberg & Fay, 2019). Similarly, promoting energy efficiency and renewable energy in infrastructure design can contribute to both climate change mitigation and adaptation goals, while also reducing costs and improving energy security (Sharifi & Yamagata, 2016).

In conclusion, infrastructure resilience planning is a critical component of building economic resilience to natural disasters. By designing and constructing systems that can withstand, adapt to, and recover from shocks and stresses, while also maintaining essential functions and services, we can reduce the vulnerability of communities and economies to the impacts of natural hazards. This requires a comprehensive approach that integrates risk assessment, nature-based solutions, social engagement, institutional strengthening, and sustainable development considerations. By investing in resilient infrastructure, we can not only protect lives and livelihoods but also create opportunities for long-term prosperity and well-being.

4.3 Financial Preparedness Mechanisms

Financial preparedness is a critical component of disaster risk reduction, enabling governments, businesses, and households to manage the economic impacts of natural hazards (OECD, 2015). This includes a range of financial instruments and strategies, such as insurance, contingency funds, and social protection programs, that provide resources for emergency response, recovery, and reconstruction efforts (Mahul & Ghesquiere, 2010).

Contingency funds, such as national disaster funds or regional risk pools, provide a mechanism for financing disaster response and recovery (Clarke & Dercon, 2016). By setting aside funds in advance of disasters, contingency funds can provide a predictable source of financing and reduce reliance on ex-post budget reallocations or external aid (Ghesquiere & Mahul, 2010). However, the effectiveness of contingency funds depends on adequate capitalization, clear disbursement criteria, and strong governance frameworks to ensure that funds are used efficiently and transparently (Beccari, 2016).

Social protection programs, such as cash transfers or public works programs, can provide a safety net for vulnerable populations during and after disasters (Pelham et al., 2011). By providing targeted support to those most in need, social protection programs can help to prevent the erosion of human capital and promote faster recovery (Hallegatte et al., 2016). However, the effectiveness of social protection programs depends on accurate targeting, adequate funding, and strong institutional capacities for delivery (Bastagli et al., 2016).

Adaptive social protection, which integrates social protection with disaster risk management and climate change adaptation, has emerged as a promising approach for building resilience (Bowen et al., 2020). Adaptive social protection programs are designed to be scalable and responsive to changing needs, providing a flexible and shock-responsive safety net (Beazley et al., 2019). For example, the Productive Safety Net Program in Ethiopia has integrated contingency financing and early warning systems to enable rapid scale-up of cash transfers during droughts (Ulrichs et al., 2019).

Forecast-based financing, which releases funds based on pre-agreed triggers and early warning indicators, is another innovative approach for improving the timeliness and effectiveness of disaster response (Coughlan de Perez et al., 2015). By linking financing to scientific forecasts and risk analysis, forecast-based financing can enable anticipatory action and reduce the costs of disaster impacts (Wilkinson et al., 2018). The Red Cross has pioneered forecast-based financing pilots in several countries, demonstrating the potential for this approach to save lives and livelihoods (Gros et al., 2019).

Ultimately, financial preparedness requires a comprehensive and integrated approach that combines different instruments and strategies based on the specific needs and capacities of different stakeholders (Benson & Clay, 2004). This requires strong coordination among government agencies, private sector actors, and civil society organizations to ensure that financial resources are allocated efficiently and equitably (Mechler et al., 2014). It also requires investments in financial literacy and risk awareness to enable individuals and organizations to make informed decisions about risk management (Kousky & Shabman, 2015).

4.4 Insurance and Risk Transfer Solutions

Insurance and risk transfer solutions are essential tools for managing the financial impacts of natural disasters and promoting economic resilience (OECD, 2015). By transferring risks to the financial markets, insurance can provide a predictable and efficient means of financing disaster recovery, while also incentivizing risk reduction and adaptation measures (Linnerooth-Bayer & Hochrainer-Stigler, 2015).

Traditional indemnity-based insurance, which provides compensation based on actual losses, is the most common form of disaster insurance (Kousky & Kunreuther, 2014). Indemnity-based insurance can be provided by private sector companies, public sector entities, or public-private partnerships, depending on the specific market and regulatory context (Joyette et al., 2015). However, indemnity-based insurance can be expensive and may not be accessible to vulnerable populations, particularly in developing countries where insurance markets are underdeveloped (Surminski et al., 2016).

Parametric insurance, which provides payouts based on the occurrence of predefined hazard events, offers a promising alternative for expanding insurance coverage in resource-constrained contexts (World Bank, 2017). Parametric insurance can be based on a variety of triggers, such as wind speed, precipitation levels, or earthquake magnitude, depending on the specific hazard and exposure context (Talbot et al., 2018). By using objective triggers, parametric insurance can provide rapid payouts and reduce administrative costs, making it more accessible and affordable for low-income populations (Vroege et al., 2019).

Sovereign risk pooling, which enables countries to jointly purchase insurance coverage, is another innovative approach for expanding insurance coverage and reducing costs (World Bank, 2017). By pooling risks across multiple countries, sovereign risk pools can achieve economies of scale and diversify risk exposures, making insurance more affordable and accessible (Scherer & Avellaneda, 2018). The Caribbean Catastrophe Risk Insurance Facility (CCRIF) is a leading example of a sovereign risk pool, providing parametric insurance coverage to 19 Caribbean and Central American countries (CCRIF, 2019).

Catastrophe bonds, which transfer disaster risks to the capital markets, provide another mechanism for financing disaster recovery (Cummins & Weiss, 2009). Catastrophe bonds are financial instruments that pay out when a pre-defined hazard event occurs, providing a source of capital for insurers or governments to finance disaster response and recovery (Michel-Kerjan et al., 2011). By tapping into the deep liquidity of the capital markets, catastrophe bonds can provide a more stable and scalable source of risk transfer than traditional reinsurance markets (Lakdawalla & Zanjani, 2012).

Microinsurance, which provides low-cost insurance products to low-income populations, is another promising tool for promoting financial inclusion and resilience (Churchill, 2006). Microinsurance programs often use innovative distribution channels, such as mobile phones or community-based organizations, to reach underserved populations (Eling et al., 2014). They also often incorporate risk reduction and adaptation measures, such as financial literacy training or drought-resistant seeds, to reduce exposure and vulnerability (Platteau et al., 2017).

Index-based livestock insurance, which provides payouts based on satellite imagery of vegetation conditions, has emerged as a particularly promising form of microinsurance for pastoralist populations (Jensen et al., 2017). By using objective indicators of drought risk, index-based livestock insurance can avoid the moral hazard and adverse selection problems that plague traditional livestock insurance programs (Chantarat et al., 2013). The Index-Based Livestock Insurance program in Kenya, which has reached over 18,000 pastoralists since its launch in 2010, has demonstrated the potential for this approach to build resilience and protect livelihoods in drought-prone regions (Jensen et al., 2016).

Ultimately, insurance and risk transfer solutions are not a panacea for disaster risk management, but rather a complementary tool that should be integrated with other risk reduction and resilience-building measures (Surminski & Thieken, 2017). This requires a holistic and adaptive approach that takes into account the specific needs and capacities of different stakeholders, as well as the changing nature of disaster risks in a warming world (Botzen et al., 2019). By leveraging the power of insurance and risk transfer solutions, we can build more resilient and inclusive economies that are better prepared to withstand and recover from the impacts of natural disasters.

5. RESPONSE MECHANISMS

5.1 Immediate Economic Interventions

In the aftermath of a natural disaster, swift economic interventions are crucial for minimizing the disruption to livelihoods and markets, and for setting the stage for a sustainable recovery (Hallegatte et al., 2017). These interventions can take various forms, depending on the specific needs and capacities of the affected communities and the scale and nature of the disaster (UNDP, 2013).

One of the most critical immediate interventions is the provision of emergency cash transfers to affected households and businesses (Pelham et al., 2011). Cash transfers can provide a lifeline for those who have lost income or assets due to the disaster, enabling them to meet basic needs such as food, water, and shelter (Devereux, 2016). Cash transfers can also stimulate local markets and support the recovery of small businesses, which are often the backbone of local economies (Heltberg et al., 2012).

However, the effectiveness of cash transfers depends on several factors, such as the targeting and delivery mechanisms, the amount and duration of support, and the coordination with other humanitarian and development interventions (Bastagli et al., 2016). Cash transfers should be designed to reach the most vulnerable and marginalized groups, such as women, children, and the elderly, who may face additional barriers to accessing support (Beazley et al., 2016). They should also be aligned with longer-term social protection and livelihood recovery programs to ensure a smooth transition from relief to recovery (Bowen et al., 2020).

Emergency employment programs, such as cash-for-work or public works schemes, can provide another important economic intervention in the immediate aftermath of disasters (Subbarao et al., 2013). These programs can provide temporary income support to affected households while also contributing to the rehabilitation of damaged infrastructure and public assets (del Ninno et al., 2009). However, the design and implementation of these programs must be carefully considered to ensure that they do not undermine long-term recovery efforts or create perverse incentives (McCord, 2013).

5.2 Emergency Funding Protocols

Effective emergency funding protocols are essential for ensuring that resources are mobilized and deployed quickly and efficiently in the aftermath of a disaster

(UNDRR, 2015). These protocols should be established in advance of disasters and should be based on clear triggers, roles, and responsibilities (World Bank, 2016).

One key aspect of emergency funding protocols is the establishment of contingency funds or disaster reserves (Clarke & Dercon, 2016). These funds can provide a readily available source of financing for immediate response and recovery efforts, without the need for time-consuming budget reallocations or external appeals (Ghesquiere & Mahul, 2010). However, the effectiveness of contingency funds depends on their size, replenishment mechanisms, and disbursement procedures (Beccari, 2016).

Another important aspect of emergency funding protocols is the use of pre-arranged financing instruments, such as contingent credit lines or catastrophe bonds (Mahul & Ghesquiere, 2010). These instruments can provide rapid access to funds in the event of a disaster, based on pre-defined triggers and payout structures (Clarke & Wren-Lewis, 2013). However, the use of these instruments requires careful design and risk modeling to ensure that they provide adequate coverage at an affordable cost (World Bank, 2017).

Emergency funding protocols should also include provisions for the rapid mobilization and coordination of international assistance (OCHA, 2015). This requires clear communication channels, needs assessment procedures, and logistical arrangements to ensure that aid reaches those who need it most (Coppola, 2015). However, international aid coordination can be challenging due to the multiplicity of actors, competing priorities, and cultural and linguistic barriers (Altay & Labonte, 2014).

5.3 Market Stabilization Measures

Natural disasters can have severe impacts on markets, leading to price volatility, supply chain disruptions, and reduced access to essential goods and services (Cavallo & Noy, 2011). Market stabilization measures are therefore critical for ensuring that markets continue to function and that people can meet their basic needs in the aftermath of a disaster (World Bank, 2013).

One key market stabilization measure is price control, which can help to prevent price gouging and ensure that essential goods remain affordable (Chapagain & Raizada, 2017). However, price controls can also have unintended consequences, such as creating black markets or disincentivizing suppliers (Alston et al., 2018). Therefore, price controls should be carefully designed and targeted, and should be accompanied by other measures such as subsidies or vouchers to support vulnerable households (Devereux, 2016).

Another important market stabilization measure is the provision of emergency supplies and logistics support (Holguín-Veras et al., 2012). This can help to ensure that essential goods, such as food, water, and medical supplies, are available and accessible in the aftermath of a disaster (Balcik et al., 2010). However, the provision of emergency supplies requires careful planning and coordination to avoid duplication, waste, and unintended market distortions (Kovács & Spens, 2011).

Market stabilization measures should also include efforts to support the recovery of small businesses and entrepreneurs, who play a critical role in local economies (Battisti & Deakins, 2017). This can include measures such as grants, loans, or technical assistance to help businesses rebuild and adapt to new market conditions (Monllor & Murphy, 2017). However, these measures should be designed to promote long-term resilience and sustainability, rather than creating dependencies or unsustainable debt burdens (Linnenluecke & McKnight, 2017).

5.4 International Aid Coordination

International aid plays a crucial role in supporting disaster response and recovery efforts, particularly in developing countries with limited domestic resources and capacities (Kellett & Caravani, 2013). However, the coordination of international aid can be a complex and challenging process, involving a wide range of actors with different mandates, priorities, and operating procedures (Stumpenhorst et al., 2011).

Effective international aid coordination requires a clear and agreed-upon framework for needs assessment, resource mobilization, and delivery (IASC, 2015). This framework should be based on internationally recognized standards and guidelines, such as the Sphere Handbook and the Core Humanitarian Standard (Sphere Association, 2018; CHS Alliance et al., 2014). It should also be flexible enough to adapt to the specific context and needs of each disaster situation (OCHA, 2017).

One key aspect of international aid coordination is the establishment of a common platform for information sharing and decision-making (OCHA, 2018). This platform should enable all relevant stakeholders, including government agencies, UN organizations, NGOs, and donors, to share data, assessments, and plans in a transparent and timely manner (Van de Walle & Comes, 2015). It should also provide a forum for joint planning, prioritization, and resource allocation, based on a shared understanding of the needs and capacities of the affected population (Bisri & Beniya, 2016).

Another important aspect of international aid coordination is the use of cash-based interventions, which can provide a more flexible and dignified form of assistance compared to in-kind aid (Harvey et al., 2018). Cash-based interventions can also help to stimulate local markets and support the recovery of livelihoods (Doocy & Tappis, 2017). However, the use of cash requires careful coordination and monitoring to ensure that it reaches the most vulnerable and does not exacerbate existing inequalities or tensions (Juillard et al., 2020).

International aid coordination should also prioritize the strengthening of local capacities and the transition to nationally-led recovery efforts (UNDP, 2017). This requires investing in the institutional and human resources of national and local authorities, as well as supporting the participation and leadership of affected communities in the design and implementation of recovery programs (Aysan & Lavell, 2014). It also requires a long-term commitment to building resilience and reducing vulnerability to future disasters, through measures such as risk-informed development planning and social protection (Peters et al., 2016).

6. RECOVERY FRAMEWORKS

6.1 Long-Term Reconstruction Strategies

In the aftermath of a natural disaster, long-term reconstruction strategies are crucial for rebuilding damaged infrastructure, restoring livelihoods, and promoting sustainable development (World Bank, 2018). These strategies should be based on a comprehensive assessment of the disaster impacts, taking into account the physical, social, and economic dimensions of recovery (GFDRR, 2015). They should also be aligned with existing development plans and policies, and should be informed by the needs and priorities of affected communities (UNDP, 2017).

One key aspect of long-term reconstruction is the rebuilding of critical infrastructure, such as housing, schools, hospitals, and transportation networks (Mannakkara & Wilkinson, 2014). This requires a coordinated effort among government agencies, development partners, and the private sector to mobilize resources, technical expertise, and innovative solutions (Fengler et al., 2008). It also requires a focus on building back better, using disaster-resilient designs and materials to reduce future risks (Kennedy et al., 2008).

Another important aspect of long-term reconstruction is the restoration of livelihoods and economic activities (Régnier et al., 2008). This requires supporting the recovery of key sectors, such as agriculture, tourism, and small and medium enterprises, which are often the most affected by disasters (Hallegatte, 2014). It also requires investing in human capital, through education and training programs, to enable people to adapt to new economic opportunities (Hayashi, 2012).

Long-term reconstruction strategies should also prioritize the needs of vulnerable and marginalized groups, such as women, children, the elderly, and people with disabilities (Enarson & Fordham, 2001). This requires targeted interventions, such as social protection programs and community-based initiatives, to address the specific barriers and challenges faced by these groups (Drolet et al., 2015). It also requires promoting social inclusion and empowerment, through participatory decision-making and capacity-building processes (Chandrasekhar, 2012).

6.2 Economic Rehabilitation Programs

Economic rehabilitation programs are essential for supporting the recovery of businesses and livelihoods in the aftermath of a disaster (Mayer et al., 2020). These programs can take various forms, such as grants, loans, tax incentives, and technical assistance, depending on the specific needs and capacities of the affected communities (Zhu et al., 2019).

One key aspect of economic rehabilitation is the provision of financial support to small and medium enterprises (SMEs), which are often the most vulnerable to disasters due to their limited resources and access to credit (Asgary et al., 2012). This can include measures such as emergency loans, interest rate subsidies, and loan guarantees, to help SMEs restart their operations and invest in recovery (Yoshida & Deyle, 2005). It can also include business continuity planning and risk management training, to enhance the resilience of SMEs to future shocks (Ballesteros & Domingo, 2015).

Another important aspect of economic rehabilitation is the promotion of entrepreneurship and innovation, which can create new opportunities for growth and diversification in the post-disaster context (Chamlee-Wright & Storr, 2010). This can include measures such as business incubation, seed funding, and mentorship programs, to support the development of new ventures and ideas (Monllor & Murphy, 2017). It can also include fostering linkages between businesses and research institutions, to promote the adoption of new technologies and practices (Chandra & Fealey, 2009).

Economic rehabilitation programs should also prioritize the needs of informal sector workers, who are often the most affected by disasters due to their lack of social protection and legal recognition (Brown et al., 2018). This can include measures such as cash transfers, public works programs, and skills training, to provide temporary income support and enhance the employability of informal workers (Grosh et al., 2008). It can also include efforts to formalize informal enterprises, through registration, licensing, and access to financial services (Guha-Sapir et al., 2016).

6.3 Building Back Better Principles

Building back better (BBB) is a key principle of post-disaster recovery, which emphasizes the need to rebuild in a way that reduces future risks and enhances the resilience of communities (Clinton, 2006). BBB involves not only reconstructing physical infrastructure but also addressing the underlying social, economic, and environmental vulnerabilities that contribute to disaster impacts (Kennedy et al., 2008).

One key aspect of BBB is the adoption of disaster-resilient designs and standards in the reconstruction of buildings and infrastructure (Mannakkara & Wilkinson,

2014). This can include measures such as elevating structures above flood levels, using seismic-resistant materials and techniques, and incorporating green infrastructure for stormwater management (Dias et al., 2016). It can also include the use of participatory design processes, to ensure that the needs and preferences of communities are taken into account (Lyons, 2009).

Another important aspect of BBB is the integration of risk reduction measures into land-use planning and development policies (Glavovic, 2008). This can include measures such as zoning regulations, building codes, and setback requirements, to discourage development in high-risk areas and promote the use of hazard-resistant construction practices (Burby et al., 2000). It can also include the creation of green spaces and buffer zones, to mitigate the impacts of floods, landslides, and other hazards (Sandholz et al., 2018).

BBB also involves strengthening the institutional and financial capacities of local governments and communities to manage disaster risks (Christoplos, 2006). This can include measures such as establishing dedicated disaster risk management units, developing contingency plans and early warning systems, and allocating budgets for risk reduction and preparedness activities (Ishiwatari, 2012). It can also include building partnerships with civil society organizations, the private sector, and academia, to leverage their expertise and resources for resilience-building (Aldrich, 2012).

6.4 Sustainable Recovery Approaches

Sustainable recovery approaches aim to balance the short-term needs of disaster-affected communities with the long-term goals of economic, social, and environmental sustainability (Berke et al., 1993). These approaches recognize that post-disaster recovery provides an opportunity to address the root causes of vulnerability and to promote more equitable and resilient development pathways (Smith & Wenger, 2007).

One key aspect of sustainable recovery is the integration of climate change adaptation and mitigation measures into reconstruction and development planning (Birkmann & Teichman, 2010). This can include measures such as promoting renewable energy and energy efficiency, protecting and restoring ecosystems, and adopting climate-smart agricultural practices (Adger et al., 2009). It can also include incorporating climate risk assessments and scenario planning into decision-making processes, to ensure that investments are resilient to future climate impacts (Hallegatte et al., 2011).

Another important aspect of sustainable recovery is the promotion of social equity and inclusion, to ensure that the benefits of recovery are distributed fairly among all segments of society (Olshansky & Chang, 2009). This can include measures such as prioritizing the needs of low-income and marginalized communities, promoting gender equality and women's empowerment, and engaging diverse stakeholders in decision-making processes (Enarson & Chakrabarti, 2009). It can also include investing in social infrastructure, such as healthcare, education, and community facilities, to enhance human capital and social resilience (Nakagawa & Shaw, 2004).

Sustainable recovery also involves the promotion of economic diversification and self-reliance, to reduce dependence on external aid and to create more stable and resilient local economies (Régnier et al., 2008). This can include measures such as supporting the development of small and medium enterprises, promoting value-added processing and marketing of local products, and investing in infrastructure and services that enable economic growth and innovation (Ahrens & Rudolph, 2006). It can also include fostering regional cooperation and trade, to expand market opportunities and to share knowledge and resources for resilience-building (Hayashi, 2012).

7. CASE STUDIES

7.1 Developed Economies' Experiences

Developed economies have made significant strides in building economic resilience to natural disasters, leveraging their financial resources, advanced infrastructure, and robust institutions (Hallegatte et al., 2017). Japan, for instance, has been a global leader in disaster risk reduction, investing heavily in seismic-resistant infrastructure, early warning systems, and public awareness campaigns (Ranghieri & Ishiwatari, 2014). In the aftermath of the 2011 Great East Japan Earthquake, the country's well-prepared disaster response system and strong social capital enabled a relatively swift recovery, despite the massive scale of the damage (Aldrich, 2012).

Similarly, the United States has made substantial investments in disaster preparedness and response, particularly in the wake of major events such as Hurricane Katrina and Superstorm Sandy (Greer & Trainor, 2016). The Federal Emergency Management Agency (FEMA) has played a key role in coordinating federal, state, and local response efforts, while also supporting long-term recovery through programs such as the National Flood Insurance Program and the Community Development Block Grant Disaster Recovery program (Smith & Villa, 2020).

However, even developed economies face challenges in building long-term resilience, particularly in the face of increasing disaster risks driven by climate change and urbanization (Bouwer, 2019). The 2017 Atlantic hurricane season, which included the devastating impacts of Hurricanes Harvey, Irma, and Maria, exposed the vulnerabilities of coastal communities and the limitations of existing disaster management frameworks (Cangialosi et al., 2018). The growing frequency and intensity of wildfires in California and Australia have also highlighted the need for more proactive and adaptive resilience strategies (Steffen et al., 2019).

7.2 Developing Nations' Challenges

Developing nations face even greater challenges in building economic resilience to natural disasters, due to their limited financial resources, weak institutions, and high levels of social vulnerability (Mirza, 2003). These countries are often disproportionately affected by disasters, with losses as a percentage of GDP that are much higher than in developed economies (Hallegatte et al., 2020). In 2017, for example, the total economic losses from disasters in low- and middle-income countries amounted to $20 billion, representing a much larger share of their national income compared to high-income countries (CRED & UNISDR, 2018).

One of the key challenges facing developing nations is the lack of financial protection against disaster risks (Mahul & Stutley, 2010). Many countries lack the fiscal space to establish adequate contingency funds or to purchase insurance and other risk transfer instruments (Bevan & Adam, 2016). As a result, they are forced to rely on ex-post funding sources, such as budget reallocations and international aid, which can be unpredictable and insufficient to meet recovery needs (Clarke & Dercon, 2016).

Another major challenge is the weakness of critical infrastructure and the high exposure of vulnerable populations to disaster risks (Adger et al., 2020). In many developing countries, rapid urbanization has led to the growth of informal settlements in hazard-prone areas, such as floodplains and hillsides (Satterthwaite et al., 2020). These settlements often lack basic services and infrastructure, such as safe housing, clean water, and sanitation, making them highly vulnerable to the impacts of disasters (Satterthwaite & Bartlett, 2017).

7.3 Success Stories and Lessons Learned

Despite these challenges, there have been many success stories of developing countries that have made significant progress in building economic resilience to natural disasters. Bangladesh, for example, has invested heavily in disaster risk reduction over the past few decades, developing a comprehensive cyclone preparedness program that includes early warning systems, evacuation plans, and cyclone shelters (Haque et al., 2012). These investments have helped to dramatically reduce the loss of life from cyclones, even as the country remains highly exposed to these hazards (Lumbroso et al., 2017).

Another success story is the Caribbean Catastrophe Risk Insurance Facility (CCRIF), a regional risk pooling mechanism that provides parametric insurance coverage to Caribbean and Central American countries (World Bank, 2017). By pooling risks across multiple countries, the CCRIF has been able to offer affordable insurance coverage that pays out quickly in the event of a disaster, providing governments with much-needed liquidity for emergency response and recovery efforts (Scherer & Skeie, 2019).

There are also important lessons to be learned from successful community-based disaster risk reduction initiatives, which have empowered local communities to take a more active role in building their own resilience (Pelling, 2007). For example, the Community-Based Disaster Risk Management program in the Philippines has trained local volunteers to conduct risk assessments, develop emergency plans, and implement small-scale mitigation measures (Allen, 2006). These efforts have not only reduced disaster risks but have also strengthened social capital and local governance capacities (Maskrey, 2011).

7.4 Failed Interventions Analysis

While there have been many success stories in building economic resilience to natural disasters, there have also been many failures and shortcomings that provide important lessons for future interventions. One common pitfall is the lack of coordination and integration among different sectors and stakeholders involved in disaster risk management (Ishiwatari & Surjan, 2019). In many cases, there is a siloed approach to resilience-building, with different agencies and organizations working in isolation from each other, leading to duplication of efforts and gaps in coverage (Aldunce et al., 2016).

Another common challenge is the failure to address the underlying drivers of vulnerability, such as poverty, inequality, and environmental degradation (Wisner et al., 2012). Many disaster risk reduction interventions focus narrowly on physical infrastructure and technical solutions, without addressing the social, economic, and political factors that make some communities more vulnerable than others (Cannon & Müller-Mahn, 2010). As a result, these interventions may provide short-term benefits but fail to build long-term resilience (Béné et al., 2012).

There have also been many cases of failed post-disaster recovery efforts that have exacerbated vulnerabilities and undermined long-term resilience (Ingram et al., 2006). In the aftermath of the 2010 Haiti earthquake, for example, the international aid response was criticized for being poorly coordinated, unaccountable, and disconnected from local needs and capacities (Katz, 2013). Many recovery projects were implemented without adequate consultation or participation of affected communities, leading to a lack of local ownership and sustainability (Fan, 2013).

These failed interventions underscore the importance of taking a more holistic and participatory approach to resilience-building, one that recognizes the complex and interconnected nature of disaster risks and the need for locally-driven solutions (Manyena et al., 2011). They also highlight the critical role of governance and institutions in shaping the effectiveness and equity of disaster risk management efforts (Tierney, 2012). Building economic resilience to natural disasters requires not only technical solutions but also political will, social capital, and a commitment to addressing the root causes of vulnerability (Aldrich & Meyer, 2015).

8. POLICY RECOMMENDATIONS

8.1 International Cooperation Frameworks

Building economic resilience to natural disasters requires a coordinated and collaborative effort at the international level. International cooperation frameworks play a crucial role in facilitating knowledge sharing, resource mobilization, and policy coordination among countries and organizations (Twigg, 2015). The Sendai Framework for Disaster Risk Reduction 2015-2030, adopted by the United Nations, provides a global blueprint for reducing disaster risks and building resilience (UNISDR, 2015). The framework emphasizes the importance of international cooperation in areas such as risk assessment, early warning systems, and capacity building (Aitsi-Selmi et al., 2016).

One key aspect of international cooperation is the establishment of regional and global risk pooling mechanisms, such as the Caribbean Catastrophe Risk Insurance Facility (CCRIF) and the African Risk Capacity (ARC) (World Bank, 2017). These mechanisms enable countries to pool their disaster risks and access affordable insurance coverage, reducing their reliance on external aid and improving their financial preparedness (Scherer & Skeie, 2019). International financial institutions, such as the World Bank and the International Monetary Fund, also play a critical role in providing financial assistance and technical expertise to countries affected by disasters (Klomp & Valckx, 2014).

Another important aspect of international cooperation is the promotion of cross-border data sharing and knowledge exchange. Initiatives such as the Global Facility for Disaster Reduction and Recovery (GFDRR) and the International Network for Multi-Hazard Early Warning Systems (IN-MHEWS) facilitate the sharing of best practices, tools, and technologies for disaster risk reduction (GFDRR, 2018; WMO, 2018). These networks enable countries to learn from each other's experiences and adapt their strategies based on the latest scientific evidence and innovations.

8.2 National Resilience Strategies

At the national level, governments play a critical role in developing and implementing comprehensive resilience strategies that address the specific risks and vulnerabilities of their countries (Hallegatte et al., 2017). These strategies should be based on a thorough assessment of the country's disaster risk profile, taking into account factors such as geography, climate, infrastructure, and socio-economic conditions (UNDRR, 2019).

One key element of national resilience strategies is the integration of disaster risk reduction into development planning and budgeting processes (Benson & Clay, 2004). This requires a shift from reactive to proactive approaches, with a focus on investing in prevention, preparedness, and long-term resilience-building (Kellett & Peters, 2014). Governments should also establish clear institutional frameworks and governance mechanisms for disaster risk management, with well-defined roles and responsibilities for different agencies and stakeholders (OECD, 2018).

Another important aspect of national resilience strategies is the development of robust social protection systems that can provide a safety net for vulnerable populations during and after disasters (Hallegatte et al., 2016). This includes measures such as cash transfers, public works programs, and insurance schemes that can help households and communities cope with the economic impacts of disasters (Pelham et al., 2011). Governments should also invest in education and awareness-raising programs to build a culture of resilience and empower citizens to take an active role in disaster risk reduction (Paton & Johnston, 2017).

8.3 Public-Private Partnership Models

Public-private partnerships (PPPs) are increasingly recognized as a key mechanism for building economic resilience to natural disasters (Auzzir et al., 2014). PPPs involve the collaboration between government agencies and private sector entities to design, finance, and implement disaster risk reduction and recovery projects (World Bank, 2019). These partnerships can leverage the strengths and resources of both sectors, enabling more effective and efficient solutions to complex challenges (Johannessen et al., 2014).

One area where PPPs can be particularly effective is in the development of resilient infrastructure, such as roads, bridges, and power grids (Sundararajan & Suriyagoda, 2016). Private sector companies can bring technical expertise, innovation, and financial resources to these projects, while government agencies can provide the necessary regulatory frameworks and oversight (Farquharson et al., 2011). PPPs can also be used to develop and operate early warning systems, emergency response services, and post-disaster recovery programs (Medina, 2016).

Another important aspect of PPPs is the sharing of risks and rewards between public and private sector partners (Jomo et al., 2016). This can be achieved through mechanisms such as performance-based contracts, revenue-sharing agreements, and risk transfer instruments (Engel et al., 2014). By aligning incentives and creating a shared sense of ownership and accountability, PPPs can foster long-term commitment and sustainability in disaster risk reduction efforts (Casady & Baxter, 2020).

8.4 Future-Proof Economic Planning

Building economic resilience to natural disasters requires a forward-looking and adaptive approach to economic planning (Hallegatte et al., 2019). This involves anticipating and preparing for future risks and uncertainties, rather than simply reacting to past events (Watkiss et al., 2015). Future-proof economic planning should be based on robust risk assessments and scenario analyses that take into account the potential impacts of climate change, technological disruptions, and other emerging threats (Ranger & Garbett-Shiels, 2012).

One key aspect of future-proof economic planning is the integration of disaster risk considerations into investment decisions and project appraisals (Rozenberg & Fay, 2019). This requires the use of risk-sensitive cost-benefit analyses and the incorporation of resilience criteria into the design and selection of infrastructure projects (Kull et al., 2013). Governments and businesses should also consider the potential co-benefits of disaster risk reduction measures, such as improved public health, environmental sustainability, and social cohesion (Surminski & Tanner, 2016).

Another important aspect of future-proof economic planning is the development of flexible and adaptable policies and institutions (Agrawal et al., 2019). This involves creating mechanisms for continuous learning, experimentation, and adjustment in response to changing circumstances and new information (Djalante et al., 2013). Governments should also foster a culture of innovation and entrepreneurship, encouraging the development of new technologies, business models, and social practices that can enhance resilience (Mazzucato & Perez, 2015).

In conclusion, building economic resilience to natural disasters requires a comprehensive and integrated approach that involves all stakeholders, from international organizations and national governments to businesses and civil society groups. By strengthening international cooperation frameworks, developing national resilience strategies, promoting public-private partnerships, and adopting future-proof economic planning practices, we can create a more resilient and sustainable global economy that is better prepared to withstand the impacts of natural disasters. This will require significant investments in risk reduction, capacity building, and innovation, but the benefits in terms of saved lives, protected livelihoods, and enhanced prosperity will far outweigh the costs.

9. FUTURE PERSPECTIVES

9.1 Emerging Trends and Challenges

As the world continues to grapple with the increasing frequency and intensity of natural disasters, several emerging trends and challenges are shaping the future of economic resilience. One of the most significant trends is the growing recognition of the interconnectedness between disaster risk, climate change, and sustainable development (Benson, 2016). The Sustainable Development Goals (SDGs) adopted by the United Nations in 2015 explicitly acknowledge the need to build resilience to disasters and climate change as a key enabler of sustainable development (United Nations, 2015). This has led to a greater emphasis on integrating disaster risk reduction and climate change adaptation into development planning and decision-making processes (Birkmann & Pardoe, 2014).

Another emerging trend is the increasing role of technology and innovation in building economic resilience. Advances in areas such as remote sensing, big data analytics, and artificial intelligence are enabling more accurate and timely risk assessments, early warning systems, and post-disaster damage assessments (Cutter, 2016). For example, the use of drones and satellite imagery has greatly improved the speed and accuracy of post-disaster needs assessments, allowing for more targeted and efficient relief efforts (Erdelj et al., 2017). However, the rapid pace of technological change also presents challenges, such as the need for continuous capacity building and the potential for exacerbating existing inequalities in access to technology (Sambuli & Whitt, 2017).

9.2 Climate Change Implications

Climate change is one of the most significant and pressing challenges facing the world today, with far-reaching implications for economic resilience to natural disasters. The Intergovernmental Panel on Climate Change (IPCC) has warned that the impacts of climate change, such as rising sea levels, more intense and frequent extreme weather events, and changing precipitation patterns, are already being felt around the world and are expected to intensify in the coming decades (IPCC, 2018). These impacts are likely to exacerbate existing disaster risks and create new ones, particularly in vulnerable regions such as low-lying coastal areas and drought-prone regions (Hallegatte et al., 2016).

The economic costs of climate change are also expected to be substantial, with some estimates suggesting that global GDP could be reduced by up to 20% by 2100 if no action is taken to mitigate and adapt to climate change (Burke et al., 2015). This underscores the urgency of integrating climate change considerations into economic resilience-building efforts, both in terms of reducing greenhouse gas emissions to mitigate future climate change and adapting to the impacts that are already unavoidable (Stern, 2016). This will require a fundamental shift in how we approach economic development, with a greater emphasis on low-carbon and climate-resilient infrastructure, technologies, and practices (Fankhauser & McDermott, 2016).

9.3 Technological Innovations in Resilience

Technology and innovation are playing an increasingly important role in building economic resilience to natural disasters. One area where technology is having a significant impact is in the development of more resilient infrastructure, such as buildings, roads, and power grids. For example, the use of advanced materials such as self-healing concrete and shape memory alloys can help to reduce the damage caused by earthquakes and other hazards (Rao et al., 2017). Similarly, the use of smart grids and distributed energy systems can help to improve the resilience of power systems to disasters (Panteli & Mancarella, 2015).

Another area where technology is making a difference is in the development of more effective early warning systems and risk communication tools. For example, the use of social media and mobile apps can help to disseminate timely and accurate information to the public during disasters, allowing for more effective evacuation and response efforts (Meier, 2015). The use of big data analytics and machine learning can also help to improve the accuracy and timeliness of risk assessments and forecasting, allowing for more proactive and targeted disaster risk reduction efforts (Ofli et al., 2016).

However, the rapid pace of technological change also presents challenges for economic resilience. One challenge is the need for continuous capacity building and training to ensure that individuals and organizations have the skills and knowledge needed to effectively use new technologies (Sambuli & Whitt, 2017). Another challenge is the potential for technology to exacerbate existing inequalities, particularly in terms of access to technology and the benefits of innovation (Cutter, 2016). Addressing these challenges will require a concerted effort to promote inclusive innovation and ensure that the benefits of technology are widely shared.

9.4 Global Coordination Opportunities

Building economic resilience to natural disasters is a global challenge that requires a coordinated and collaborative approach. The Sendai Framework for Disaster Risk Reduction 2015-2030, adopted by the United Nations in 2015, provides a global blueprint for reducing disaster risks and building resilience (UNISDR, 2015). The framework emphasizes the need for a multi-stakeholder approach that engages governments, civil society, the private sector, and academia in a shared effort to reduce disaster risks and build resilience (Aitsi-Selmi et al., 2016).

One key opportunity for global coordination is in the development of international standards and guidelines for disaster risk reduction and resilience-building. For example, the International Organization for Standardization (ISO) has developed a series of standards on disaster risk management, including ISO 31000 on risk management and ISO 22301 on business continuity management (ISO, 2018). These standards provide a common framework and language for organizations to assess and manage disaster risks, and can help to promote consistency and interoperability across different sectors and regions (Eckhardt et al., 2019).

Another opportunity for global coordination is in the development of innovative financing mechanisms for disaster risk reduction and resilience-building. For example, the use of catastrophe bonds and other risk transfer instruments can help to mobilize private capital for disaster risk reduction and provide a source of rapid funding for post-disaster recovery efforts (Vaijhala & Rhodes, 2018). Similarly, the use of impact investing and blended finance approaches can help to catalyze investment in resilient infrastructure and technologies, particularly in developing countries where public resources may be limited (Bendell et al., 2015).

Finally, there is a need for greater global coordination in the development and sharing of knowledge and best practices on disaster risk reduction and resilience-building. This can include the development of global databases and platforms for sharing data and information on disaster risks and impacts, as well as the establishment of peer learning networks and communities of practice to facilitate the exchange of knowledge and experiences across different regions and sectors (Twigg, 2015). By fostering a culture of openness, collaboration, and continuous learning, we can accelerate the development and dissemination of effective strategies for building economic resilience to natural disasters.

10. CONCLUSION

10.1 Key Findings Synthesis

This comprehensive analysis of the complex relationship between natural disasters and global economic resilience has yielded several key findings. First, the economic impacts of natural disasters are far-reaching and multifaceted, encompassing direct costs such as infrastructure damage and indirect consequences like supply chain disruptions. Second, building economic resilience requires a proactive, multi-dimensional approach that integrates prevention, response, and recovery strategies. Third, international cooperation and public-private partnerships play a crucial role in fostering resilience, particularly in vulnerable regions with limited resources. Finally, strengthening global economic resilience necessitates a paradigm shift towards a culture of resilience, embracing risk governance, capacity building, and data-driven decision-making.

10.2 Policy Implications

The findings of this study have significant implications for policymakers and practitioners working to build economic resilience to natural disasters. Governments should prioritize the development of comprehensive national resilience strategies that integrate disaster risk reduction into development planning and budgeting processes. This includes investing in risk assessment and early warning systems, promoting resilient infrastructure, and establishing clear institutional frameworks for disaster risk management. Policymakers should also foster an enabling environment for public-private partnerships, leveraging the expertise and resources of the private sector to develop innovative solutions for resilience-building. Additionally, international cooperation frameworks should be strengthened to facilitate knowledge sharing, resource mobilization, and policy coordination among countries and organizations.

10.3 Research Limitations

While this study provides a comprehensive analysis of the relationship between natural disasters and economic resilience, it is not without limitations. The complex and multifaceted nature of the topic means that not all aspects could be explored in equal depth. The study relies heavily on secondary data sources, which may be subject to inconsistencies or gaps in coverage. Additionally, the rapidly evolving nature of the field means that some of the latest developments and innovations may not be fully captured in the analysis. Finally, the study's broad global scope may limit the applicability of some findings to specific local contexts, highlighting the need for further context-specific research.

10.4 Future Research Directions

The findings and limitations of this study point to several promising avenues for future research. First, there is a need for more granular, context-specific studies that examine the economic impacts of natural disasters and the effectiveness of resilience-building strategies in different regions and sectors. Second, future research should explore the potential of emerging technologies, such as artificial intelligence and blockchain, to enhance disaster risk reduction and resilience-building efforts. Third, there is a need for more interdisciplinary research that integrates insights from economics, environmental science, engineering, and social sciences to develop holistic solutions for resilience-building. Finally, future studies should focus on developing more robust and standardized metrics for measuring economic resilience, to enable better tracking of progress and identification of best practices.

REFERENCES

Adger, W. N., Boyd, E., Fábos, A., Fransen, S., Jolivet, D., Neville, G., Safra de Campos, R., & Vijge, M. J. (2020). Migration transforms the conditions for the achievement of the Sustainable Development Goals. *The Lancet. Planetary Health*, *4*(10), e440–e442. PMID: 31777332

Adger, W. N., Dessai, S., Goulden, M., Hulme, M., Lorenzoni, I., Nelson, D. R., Naess, L. O., Wolf, J., & Wreford, A. (2009). Are there social limits to adaptation to climate change? *Climatic Change*, *93*(3), 335–354. DOI: 10.1007/s10584-008-9520-z

Agrawal, A., Brown, D. G., & Sullivan, J. A. (2019). Are global land grabs ticking socio-ecological bombs or just inefficient investments? *One Earth*, *1*(2), 159–162. DOI: 10.1016/j.oneear.2019.10.004

Ahrens, J., & Rudolph, P. M. (2006). The importance of governance in risk reduction and disaster management. *Journal of Contingencies and Crisis Management*, *14*(4), 207–220. DOI: 10.1111/j.1468-5973.2006.00497.x

Aitsi-Selmi, A., Murray, V., Wannous, C., Dickinson, C., Johnston, D., Kawasaki, A., Stevance, A. S., & Yeung, T. (2016). Reflections on a science and technology agenda for 21st century disaster risk reduction. *International Journal of Disaster Risk Science*, *7*(1), 1–29. DOI: 10.1007/s13753-016-0081-x

Aldrich, D. P. (2012). *Building resilience: Social capital in post-disaster recovery*. University of Chicago Press. DOI: 10.7208/chicago/9780226012896.001.0001

Aldrich, D. P., & Meyer, M. A. (2015). Social capital and community resilience. *The American Behavioral Scientist*, *59*(2), 254–269. DOI: 10.1177/0002764214550299

Aldunce, P., Beilin, R., Howden, M., & Handmer, J. (2016). Stakeholder participation in building resilience to disasters in a changing climate. *Environmental Hazards*, *15*(1), 58–73. DOI: 10.1080/17477891.2015.1134427

Allen, K. M. (2006). Community-based disaster preparedness and climate adaptation: Local capacity-building in the Philippines. *Disasters*, *30*(1), 81–101. DOI: 10.1111/j.1467-9523.2006.00308.x PMID: 16512863

Alliance, C. H. S., Groupe, U. R. D., & the The Sphere Project. (2014). Core Humanitarian Standard on Quality and Accountability. CHS Alliance. https://corehumanitarianstandard.org/files/files/Core%20Humanitarian%20Standard%20-%20English.pdf

Alston, J. M., MacEwan, J. P., & Sambucci, O. (2018). Price controls and market efficiency: Evidence from California's Proposition 65. *Applied Economic Perspectives and Policy*, *40*(3), 373–395. DOI: 10.1093/aepp/ppy008

Altay, N., & Labonte, M. (2014). Challenges in humanitarian information management and exchange: Evidence from Haiti. *Disasters*, *38*(s1), S50–S72. DOI: 10.1111/disa.12052 PMID: 24601932

Argyroudis, S. A., Mitoulis, S. A., Hofer, L., Zanini, M. A., Tubaldi, E., & Frangopol, D. M. (2020). Resilience assessment framework for critical infrastructure in a multi-hazard environment. *The Science of the Total Environment*, *714*, 136854. DOI: 10.1016/j.scitotenv.2020.136854 PMID: 32018987

Asgary, A., Anjum, M. I., & Azimi, N. (2012). Disaster recovery and business continuity after the 2010 flood in Pakistan: Case of small businesses. *International Journal of Disaster Risk Reduction*, *2*, 46–56. DOI: 10.1016/j.ijdrr.2012.08.001

Auzzir, Z. A., Haigh, R. P., & Amaratunga, D. (2014). Public-private partnerships (PPP) in disaster management in developing countries: A conceptual framework. *Procedia Economics and Finance*, *18*, 807–814. DOI: 10.1016/S2212-5671(14)01006-5

Aysan, Y., & Lavell, A. (2014). Disaster risk management: From reaction to prevention. In López-Carresi, A., Fordham, M., Wisner, B., Kelman, I., & Gaillard, J. C. (Eds.), *Disaster management: International lessons in risk reduction, response and recovery* (pp. 13–28). Routledge.

Ayyub, B. M. (2014). Systems resilience for multihazard environments: Definition, metrics, and valuation for decision making. *Risk Analysis*, *34*(2), 340–355. DOI: 10.1111/risa.12093 PMID: 23875704

Balcik, B., Beamon, B. M., Krejci, C. C., Muramatsu, K. M., & Ramirez, M. (2010). Coordination in humanitarian relief chains: Practices, challenges and opportunities. *International Journal of Production Economics*, *126*(1), 22–34. DOI: 10.1016/j.ijpe.2009.09.008

Ballesteros, M. M., & Domingo, S. N. (2015). Building Philippine SMEs resilience to natural disasters. Philippine Institute for Development Studies Discussion Paper Series, 2015-20.

Bastagli, F., Hagen-Zanker, J., Harman, L., Barca, V., Sturge, G., Schmidt, T., & Pellerano, L. (2016). *Cash transfers: What does the evidence say? A rigorous review of programme impact and of the role of design and implementation features*. Overseas Development Institute.

Battisti, M., & Deakins, D. (2017). The relationship between dynamic capabilities, the firm's resource base and performance in a post-disaster environment. *International Small Business Journal*, *35*(1), 78–98. DOI: 10.1177/0266242615611471

Beazley, R., Solórzano, A., & Barca, V. (2019). *Study on shock-responsive social protection in Latin America and the Caribbean: Summary of key findings and policy recommendations*. Oxford Policy Management.

Beazley, R., Solórzano, A., & Sossouvi, K. (2016). *Study on shock-responsive social protection in Latin America and the Caribbean: Theoretical framework and literature review*. Oxford Policy Management.

Beccari, B. (2016). A comparative analysis of disaster risk, vulnerability and resilience composite indicators. PLOS Currents Disasters, 8.

Beck, M. W., Losada, I. J., Menéndez, P., Reguero, B. G., Díaz-Simal, P., & Fernández, F. (2018). The global flood protection savings provided by coral reefs. *Nature Communications*, *9*(1), 2186. DOI: 10.1038/s41467-018-04568-z PMID: 29895942

Becken, S., & Hughey, K. F. D. (2013). Linking tourism into emergency management structures to enhance disaster risk reduction. *Tourism Management*, *36*, 77–85. DOI: 10.1016/j.tourman.2012.11.006

Bendell, J., Sutherland, N., & Little, R. (2015). Beyond unsustainable leadership: Critical social theory for sustainable leadership. Sustainability Accounting. *Management and Policy Journal*, *8*(4), 418–444.

Béné, C., Godfrey Wood, R., Newsham, A., & Davies, M. (2012). Resilience: New utopia or new tyranny? Reflection about the potentials and limits of the concept of resilience in relation to vulnerability reduction programmes. *IDS Working Paper*, *2012*(405), 1–61. DOI: 10.1111/j.2040-0209.2012.00405.x

Benson, C. (2016). Promoting sustainable development through disaster risk management (ADB Sustainable Development Working Paper Series No. 41). Asian Development Bank.

Benson, C., & Clay, E. J. (2004). *Understanding the economic and financial impacts of natural disasters*. The World Bank. DOI: 10.1596/0-8213-5685-2

Berke, P. R., Kartez, J., & Wenger, D. (1993). Recovery after disaster: Achieving sustainable development, mitigation and equity. *Disasters*, *17*(2), 93–109. DOI: 10.1111/j.1467-7717.1993.tb01137.x PMID: 20958760

Berlemann, M., & Steinhardt, M. F. (2017). Climate change, natural disasters, and migration—A survey of the empirical evidence. *CESifo Economic Studies*, *63*(4), 353–385. DOI: 10.1093/cesifo/ifx019

Bevan, D., & Adam, C. (2016). Financing the reconstruction of public capital after a natural disaster. World Bank Policy Research Working Paper, (7718).

Biggs, D., Hall, C. M., & Stoeckl, N. (2015). The resilience of formal and informal tourism enterprises to disasters: Reef tourism in Phuket, Thailand. *Journal of Sustainable Tourism*, *23*(7), 1032–1056. DOI: 10.1080/09669582.2015.1038429

Birkmann, J., Feldmeyer, D., McMillan, J. M., Solecki, W., Totin, E., Roberts, D., & Sengupta, S. (2021). Regional clusters of vulnerability show the need for transboundary cooperation. *Environmental Research Letters*, *16*(9), 094013. DOI: 10.1088/1748-9326/ac1f43

Birkmann, J., & Mechler, R. (2015). Advancing climate adaptation and risk management: New insights, concepts and approaches. *Natural Hazards*, *78*(1), 1–7. DOI: 10.1007/s11069-015-1856-7

Birkmann, J., & Pardoe, J. (2014). Climate change adaptation and disaster risk reduction: Fundamentals, synergies and mismatches. In López-Carresi, A., Fordham, M., Wisner, B., Kelman, I., & Gaillard, J. C. (Eds.), *Disaster management: International lessons in risk reduction, response and recovery* (pp. 41–56). Routledge.

Birkmann, J., & Teichman, K. (2010). Integrating disaster risk reduction and climate change adaptation: Key challenges—scales, knowledge, and norms. *Sustainability Science*, *5*(2), 171–184. DOI: 10.1007/s11625-010-0108-y

Bisri, M. B. F., & Beniya, S. (2016). Analyzing the national disaster response framework and inter-organizational network of the 2015 Nepal earthquake. *Procedia Engineering*, *159*, 93–100. DOI: 10.1016/j.proeng.2016.08.059

Botzen, W. W., Deschenes, O., & Sanders, M. (2019). The economic impacts of natural disasters: A review of models and empirical studies. *Review of Environmental Economics and Policy*, *13*(2), 167–188. DOI: 10.1093/reep/rez004

Bouwer, L. M. (2019). Observed and projected impacts from extreme weather events: Implications for loss and damage. In Mechler, R., Bouwer, L. M., Schinko, T., Surminski, S., & Linnerooth-Bayer, J. (Eds.), *Loss and damage from climate change: Concepts, methods and policy options* (pp. 63–82). Springer. DOI: 10.1007/978-3-319-72026-5_3

Bowen, T., del Ninno, C., Andrews, C., Coll-Black, S., Gentilini, U., Johnson, K., Kawasoe, Y., Kryeziu, A., Maher, B., & Williams, A. (2020). *Adaptive social protection: Building resilience to shocks*. World Bank. DOI: 10.1596/978-1-4648-1575-1

Bristow, G., & Healy, A. (2018). Innovation and regional economic resilience: An exploratory analysis. *The Annals of Regional Science*, *60*(2), 265–284. DOI: 10.1007/s00168-017-0841-6

Brown, C., Stevenson, J., Giovinazzi, S., Seville, E., & Vargo, J. (2018). Factors influencing SMEs' resilience to natural hazards: A framework and literature review. *International Journal of Disaster Risk Reduction*, *28*, 872–883.

Burby, R. J., Deyle, R. E., Godschalk, D. R., & Olshansky, R. B. (2000). Creating hazard resilient communities through land-use planning. *Natural Hazards Review*, *1*(2), 99–106. DOI: 10.1061/(ASCE)1527-6988(2000)1:2(99)

Burke, M., Hsiang, S. M., & Miguel, E. (2015). Global non-linear effect of temperature on economic production. *Nature*, *527*(7577), 235–239. DOI: 10.1038/nature15725 PMID: 26503051

Cangialosi, J. P., Latto, A. S., & Berg, R. (2018). Hurricane Irma (AL112017): 30 August-12 September 2017. National Hurricane Center Tropical Cyclone Report. NOAA/NWS.

Cannon, T., & Müller-Mahn, D. (2010). Vulnerability, resilience and development discourses in context of climate change. *Natural Hazards*, *55*(3), 621–635. DOI: 10.1007/s11069-010-9499-4

Casady, C. B., & Baxter, D. (2020). Pandemics, public-private partnerships (PPPs), and force majeure | COVID-19 expectations and implications. *Construction Management and Economics*, *38*(12), 1077–1085. DOI: 10.1080/01446193.2020.1817516

Cavallo, E., & Noy, I. (2011). Natural disasters and the economy—A survey. *International Review of Environmental and Resource Economics*, *5*(1), 63–102. DOI: 10.1561/101.00000039

Chamlee-Wright, E., & Storr, V. H. (2010). The role of social entrepreneurship in post-Katrina community recovery. *International Journal of Innovative Research and Development*, *2*(1-2), 149–164. DOI: 10.1504/IJIRD.2010.029860

Chandra, A., & Fealey, T. (2009). Business incubation in the United States, China and Brazil: A comparison of role of government, incubator funding and financial services. *International Journal of Entrepreneurship*, *13*, 67–86.

Chandrasekhar, D. (2012). Digging deeper: Participation and non-participation in post-disaster community recovery. *Community Development (Columbus, Ohio), 43*(5), 614–629. DOI: 10.1080/15575330.2012.730538

Chantarat, S., Mude, A. G., Barrett, C. B., & Carter, M. R. (2013). Designing index-based livestock insurance for managing asset risk in northern Kenya. *The Journal of Risk and Insurance, 80*(1), 205–237. DOI: 10.1111/j.1539-6975.2012.01463.x

Chapagain, T., & Raizada, M. N. (2017). Impacts of natural disasters on smallholder farmers: Gaps and recommendations. *Agriculture & Food Security, 6*(1), 39. DOI: 10.1186/s40066-017-0116-6

Christopher, M., & Peck, H. (2004). Building the resilient supply chain. *International Journal of Logistics Management, 15*(2), 1–14. DOI: 10.1108/09574090410700275

Christoplos, I. (2006). The elusive 'window of opportunity' for risk reduction in post-disaster recovery. In ProVention Consortium Forum (pp. 2-3).

Churchill, C. (2006). *Protecting the poor: A microinsurance compendium.* International Labour Organization.

Clarke, D. J., & Dercon, S. (2016). *Dull disasters? How planning ahead will make a difference.* Oxford University Press. DOI: 10.1093/acprof:oso/9780198785576.001.0001

Clarke, D. J., & Wren-Lewis, L. (2013). Learning from lemons: The role of government in index insurance for individuals. FERDI Policy Brief, B76. https://ferdi.fr/en/publications/learning-from-lemons-the-role-of-government-in-index-insurance-for-individuals

Clinton, W. J. (2006). *Lessons learned from tsunami recovery: Key propositions for building back better.* United Nations Secretary-General's Special Envoy for Tsunami Recovery.

Coaffee, J., Therrien, M. C., Cheyne, A., & Wood, J. M. (2018). Urban resilience implementation: A policy challenge and research agenda for the 21st century. *Journal of Contingencies and Crisis Management, 26*(3), 403–410. DOI: 10.1111/1468-5973.12233

Cochrane, H. (2004). Economic loss estimation models: Time for a change? In Okuyama, Y., & Chang, S. E. (Eds.), *Modeling spatial and economic impacts of disasters* (pp. 229–248). Springer., DOI: 10.1007/978-3-540-24787-6_11

Coppola, D. P. (2015). *Introduction to international disaster management* (3rd ed.). Butterworth-Heinemann., DOI: 10.1016/B978-0-12-801477-6.00001-5

Coughlan de Perez, E., van den Hurk, B., van Aalst, M. K., Jongman, B., Klose, T., & Suarez, P. (2015). Forecast-based financing: An approach for catalyzing humanitarian action based on extreme weather and climate forecasts. *Natural Hazards and Earth System Sciences*, *15*(4), 895–904. DOI: 10.5194/nhess-15-895-2015

CRED, & UNISDR. (2018). Economic losses, poverty & disasters: 1998-2017. Centre for Research on the Epidemiology of Disasters (CRED) and United Nations Office for Disaster Risk Reduction (UNISDR).

Cuaresma, J. C. (2010). Natural disasters and human capital accumulation. *The World Bank Economic Review*, *24*(2), 280–302. DOI: 10.1093/wber/lhq008

Cummins, J. D., & Weiss, M. A. (2009). Convergence of insurance and financial markets: Hybrid and securitized risk-transfer solutions. *The Journal of Risk and Insurance*, *76*(3), 493–545. DOI: 10.1111/j.1539-6975.2009.01311.x

Cutter, S. L. (2016). Resilience to what? Resilience for whom? *The Geographical Journal*, *182*(2), 110–113. DOI: 10.1111/geoj.12174

Cutter, S. L., Barnes, L., Berry, M., Burton, C., Evans, E., Tate, E., & Webb, J. (2008). A place-based model for understanding community resilience to natural disasters. *Global Environmental Change*, *18*(4), 598–606. DOI: 10.1016/j.gloenvcha.2008.07.013

Cutter, S. L., Boruff, B. J., & Shirley, W. L. (2003). Social vulnerability to environmental hazards. *Social Science Quarterly*, *84*(2), 242–261. DOI: 10.1111/1540-6237.8402002

del Ninno, C., Subbarao, K., & Milazzo, A. (2009). How to make public works work: A review of the experiences. World Bank Social Protection Discussion Paper, 0905. https://openknowledge.worldbank.org/handle/10986/11718

Devereux, S. (2016). Social protection for enhanced food security in sub-Saharan Africa. *Food Policy*, *60*, 52–62. DOI: 10.1016/j.foodpol.2015.03.009

Dhar, T. K., & Khirfan, L. (2017). A multi-scale analysis of urban resilience planning: The case of the London climate change strategy. *Habitat International*, *67*, 108–119. DOI: 10.1016/j.habitatint.2017.07.004

Dias, N. T., Keraminiyage, K., & DeSilva, K. K. (2016). Long-term satisfaction of post disaster resettled communities: The case of post tsunami – Sri Lanka. *Disaster Prevention and Management*, *25*(5), 581–594. DOI: 10.1108/DPM-11-2015-0264

Djalante, R., Holley, C., Thomalla, F., & Carnegie, M. (2013). Pathways for adaptive and integrated disaster resilience. *Natural Hazards*, *69*(3), 2105–2135. DOI: 10.1007/s11069-013-0797-5

Doocy, S., & Tappis, H. (2017). Cash-based approaches in humanitarian emergencies: A systematic review. *Campbell Systematic Reviews*, *13*(1), 1–200. DOI: 10.4073/csr.2017.17

Drolet, J., Dominelli, L., Alston, M., Ersing, R., Mathbor, G., & Wu, H. (2015). Women rebuilding lives post-disaster: Innovative community practices for building resilience and promoting sustainable development. *Gender and Development*, *23*(3), 433–448. DOI: 10.1080/13552074.2015.1096040

Eckhardt, D., Leiras, A., & Thomé, A. M. T. (2019). Systematic literature review of methodologies for assessing the costs of disasters. *International Journal of Disaster Risk Reduction*, *33*, 398–416. DOI: 10.1016/j.ijdrr.2018.10.010

Eguchi, R. T., Huyck, C. K., Ghosh, S., & Adams, B. J. (2010). The application of remote sensing technologies for disaster management. *The International Archives of the Photogrammetry, Remote Sensing and Spatial Information Sciences*, *38*(8), 125–130. https://www.isprs.org/proceedings/XXXVIII/part8/pdf/W08_04_20100519115847.pdf

Eling, M., Pradhan, S., & Schmit, J. T. (2014). The determinants of microinsurance demand. *The Geneva Papers on Risk and Insurance. Issues and Practice*, *39*(2), 224–263. DOI: 10.1057/gpp.2014.5

Enarson, E., & Chakrabarti, P. D. (Eds.). (2009). *Women, gender and disaster: Global issues and initiatives*. SAGE Publications India. DOI: 10.4135/9788132108078

Enarson, E., & Fordham, M. (2001). From women's needs to women's rights in disasters. *Environmental Hazards*, *3*(3-4), 133–136. DOI: 10.1016/S1464-2867(02)00006-2

Engel, E., Fischer, R. D., & Galetovic, A. (2014). *The economics of public-private partnerships: A basic guide*. Cambridge University Press. DOI: 10.1017/CBO9781139565615

Erdelj, M., Król, M., & Natalizio, E. (2017). Wireless sensor networks and multi-UAV systems for natural disaster management. *Computer Networks*, *124*, 72–86. DOI: 10.1016/j.comnet.2017.05.021

Fan, L. (2013). *Disaster as opportunity? Building back better in Aceh*. Overseas Development Institute.

Fankhauser, S., & McDermott, T. K. (2016). Climate-resilient development: An introduction. In Fankhauser, S., & McDermott, T. K. (Eds.), *The economics of climate-resilient development* (pp. 1–14). Edward Elgar Publishing. DOI: 10.4337/9781785360312.00009

Farquharson, E., Torres de Mästle, C., Yescombe, E. R., & Encinas, J. (2011). *How to engage with the private sector in public-private partnerships in emerging markets.* The World Bank. DOI: 10.1596/978-0-8213-7863-2

Fengler, W., Ihsan, A., & Kaiser, K. (2008). *Managing post-disaster reconstruction finance: International experience in public financial management.* The World Bank.

Food and Agriculture Organization of the United Nations. (2015). The impact of disasters on agriculture and food security. FAO. https://www.fao.org/3/a-i5128e.pdf

Frantzeskaki, N. (2019). Seven lessons for planning nature-based solutions in cities. *Environmental Science & Policy, 93*, 101–111. DOI: 10.1016/j.envsci.2018.12.033

Funk, C., Peterson, P., Landsfeld, M., Pedreros, D., Verdin, J., Shukla, S., & Michaelsen, J. (2019). The climate hazards infrared precipitation with stations—A new environmental record for monitoring extremes. *Scientific Data, 6*(1), 180066. DOI: 10.1038/sdata.2018.66 PMID: 26646728

Gallina, V., Torresan, S., Critto, A., Sperotto, A., Glade, T., & Marcomini, A. (2016). A review of multi-risk methodologies for natural hazards: Consequences and challenges for a climate change impact assessment. *Journal of Environmental Management, 168*, 123–132. DOI: 10.1016/j.jenvman.2015.11.011 PMID: 26704454

Gassebner, M., Keck, A., & Teh, R. (2010). Shaken, not stirred: The impact of disasters on international trade. *Review of International Economics, 18*(2), 351–368. DOI: 10.1111/j.1467-9396.2010.00868.x

GFDRR. (2015). *Guide to developing disaster recovery frameworks: Sendai conference version.* Global Facility for Disaster Reduction and Recovery.

GFDRR. (2018). *Bringing resilience to scale: A work plan for the Global Facility for Disaster Reduction and Recovery, 2018-2021.* Global Facility for Disaster Reduction and Recovery.

Ghesquiere, F., & Mahul, O. (2010). Financial protection of the state against natural disasters: A primer. World Bank Policy Research Working Paper, (5429).

Glavovic, B. C. (2008). *Waves of adversity, layers of resilience: Exploring the sustainable development prospect of Ohura and the Whangaehu catchment.* Massey University.

Grafakos, S., Viero, G., Reckien, D., Trigg, K., Viguie, V., Sudmant, A., Graves, C., Foley, A., Heidrich, O., Mirailles, J. M., Carter, J., Chang, L. H., Nador, C., Liseri, M., Chelleri, L., Orru, H., Orru, K., Aelenei, R., Bilska, A., & Dawson, R. (2020). Integration of mitigation and adaptation in urban climate change action plans in Europe: A systematic assessment. *Renewable & Sustainable Energy Reviews, 121*, 109623. DOI: 10.1016/j.rser.2019.109623

Greer, A., & Trainor, J. (2016). Designing for disaster: Lessons learned about disaster research from Superstorm Sandy. *Journal of Extreme Events*, *3*(02), 1671001.

Gros, C., Bailey, M., Schwager, S., Hassan, A., Zingg, R., Uddin, M. M., Shahjahan, M., Islam, H., Lux, S., Jaime, C., & Coughlan de Perez, E. (2019). Household-level effects of providing forecast-based cash in anticipation of extreme weather events: Quasi-experimental evidence from humanitarian interventions in the 2017 floods in Bangladesh. *International Journal of Disaster Risk Reduction, 41*, 101275. DOI: 10.1016/j.ijdrr.2019.101275

Grosh, M., Del Ninno, C., Tesliuc, E., & Ouerghi, A. (2008). *For protection and promotion: The design and implementation of effective safety nets*. The World Bank. DOI: 10.1596/978-0-8213-7581-5

Guha-Sapir, D., Hoyois, P., Wallemacq, P., & Below, R. (2016). *Annual disaster statistical review 2016: The numbers and trends. Centre for Research on the Epidemiology of Disasters*. CRED.

Hallegatte, S. (2014). *Economic resilience: Definition and measurement*. The World Bank. DOI: 10.1596/1813-9450-6852

Hallegatte, S. (2015). The indirect cost of natural disasters and an economic definition of macroeconomic resilience. World Bank Policy Research Working Paper, 7357. https://openknowledge.worldbank.org/handle/10986/22238

Hallegatte, S., Bangalore, M., Bonzanigo, L., Fay, M., Kane, T., Narloch, U., Rozenberg, J., Treguer, D., & Vogt-Schilb, A. (2016). *Shock waves: Managing the impacts of climate change on poverty*. The World Bank.

Hallegatte, S., Henriet, F., & Corfee-Morlot, J. (2011). The economics of climate change impacts and policy benefits at city scale: A conceptual framework. *Climatic Change, 104*(1), 51–87. DOI: 10.1007/s10584-010-9976-5

Hallegatte, S., Rentschler, J., & Rozenberg, J. (2019). *Lifelines: The resilient infrastructure opportunity*. The World Bank. DOI: 10.1596/978-1-4648-1430-3

Hallegatte, S., Rentschler, J., & Rozenberg, J. (2019). *Lifelines: The resilient infrastructure opportunity*. The World Bank. DOI: 10.1596/978-1-4648-1430-3

Hallegatte, S., Rentschler, J., & Walsh, B. (2020). Building back better: Achieving resilience through stronger, faster, and more inclusive post-disaster reconstruction. World Bank. https://openknowledge.worldbank.org/handle/10986/33727

Hallegatte, S., & Rozenberg, J. (2017). Climate change through a poverty lens. *Nature Climate Change*, *7*(4), 250–256. DOI: 10.1038/nclimate3253

Hallegatte, S., Vogt-Schilb, A., Bangalore, M., & Rozenberg, J. (2017). Unbreakable: Building the resilience of the poor in the face of natural disasters. World Bank. https://openknowledge.worldbank.org/handle/10986/25335

Haque, C. E., Choudhury, M. U., & Sikder, M. S. H. (2012). "Community-based adaptation" to climate change and cyclonic hazards in coastal Bangladesh. *Journal of Environmental Management*, *111*, 248–258. DOI: 10.1016/j.jenvman.2012.06.017

Haraguchi, M., & Lall, U. (2015). Flood risks and impacts: A case study of Thailand's floods in 2011 and research questions for supply chain decision making. *International Journal of Disaster Risk Reduction*, *14*(3), 256–272. DOI: 10.1016/j.ijdrr.2014.09.005

Harvey, P., Proudlock, K., Clay, E., Riley, B., & Jaspars, S. (2018). A review of the linkages between cash transfers and food security outcomes in humanitarian contexts. Overseas Development Institute. https://odi.org/en/publications/a-review-of-the-linkages-between-cash-transfers-and-food-security-outcomes-in-humanitarian-contexts/

Hayashi, T. (2012). Japan's post-disaster economic reconstruction: From Kobe to Tohoku. *Asian Economic Journal*, *26*(3), 189–210. DOI: 10.1111/j.1467-8381.2012.02082.x

Heltberg, R., Hossain, N., Reva, A., & Turk, C. (2012). Anatomy of coping: Evidence from people living through the crises of 2008–11. World Bank Policy Research Working Paper, 5957. https://openknowledge.worldbank.org/handle/10986/1990

Holguín-Veras, J., Pérez, N., Jaller, M., Van Wassenhove, L. N., & Aros-Vera, F. (2012). On the appropriate objective function for post-disaster humanitarian logistics models. *Transportation Research Part E, Logistics and Transportation Review*, *48*(1), 8–22. DOI: 10.1016/j.tre.2011.08.002

Hynes, W., Trump, B., Love, P., & Linkov, I. (2020). Bouncing forward: A resilience approach to dealing with COVID-19 and future systemic shocks. *Environment Systems & Decisions*, *40*(2), 174–184. DOI: 10.1007/s10669-020-09776-x PMID: 32837818

Ingram, J. C., Franco, G., Rio, C. R., & Khazai, B. (2006). Post-disaster recovery dilemmas: Challenges in balancing short-term and long-term needs for vulnerability reduction. *Environmental Science & Policy*, *9*(7-8), 607–613. DOI: 10.1016/j.envsci.2006.07.006

Inter-Agency Standing Committee. (2015). Reference module for cluster coordination at the country level. IASC. https://interagencystandingcommittee.org/system/files/iasc_reference_module_for_cluster_coordination_at_country_level_march_2015.pdf

Intergovernmental Panel on Climate Change. (2012). Managing the risks of extreme events and disasters to advance climate change adaptation: Special report of the Intergovernmental Panel on Climate Change. Cambridge University Press. https://www.ipcc.ch/report/managing-the-risks-of-extreme-events-and-disasters-to-advance-climate-change-adaptation/

IPCC. (2018). *Global warming of 1.5°C. An IPCC Special Report on the impacts of global warming of 1.5°C above pre-industrial levels and related global greenhouse gas emission pathways, in the context of strengthening the global response to the threat of climate change, sustainable development, and efforts to eradicate poverty*. Intergovernmental Panel on Climate Change.

Ishiwatari, M. (2012). Government roles in community-based disaster risk reduction. In Shaw, R. (Ed.), *Community-based disaster risk reduction* (pp. 19–33). Emerald Group Publishing., DOI: 10.1108/S2040-7262(2012)0000010008

Ishiwatari, M., & Surjan, A. (2019). Challenges in coordinating large-scale post-disaster redevelopment projects: Lessons from Asia. *International Journal of Disaster Risk Reduction*, *39*, 101246. DOI: 10.1016/j.ijdrr.2019.101246

ISO. (2018). *ISO 31000:2018 Risk management - Guidelines*. International Organization for Standardization.

Jensen, N. D., Barrett, C. B., & Mude, A. G. (2016). Index insurance quality and the demand for insurance: Evidence from a field experiment in Kenya. *American Journal of Agricultural Economics*, *98*(5), 1431–1450. DOI: 10.1093/ajae/aaw046

Jensen, N. D., Barrett, C. B., & Mude, A. G. (2017). Cash transfers and index insurance: A comparative impact analysis from northern Kenya. *Journal of Development Economics*, *129*, 14–28. DOI: 10.1016/j.jdeveco.2017.08.002

Johannessen, A., Rosemarin, A., Thomalla, F., Gerger Swartling, Å., Stenström, T. A., & Vulturius, G. (2014). Strategies for building resilience to hazards in water, sanitation and hygiene (WASH) systems: The role of public private partnerships. *International Journal of Disaster Risk Reduction*, *10*, 102–115. DOI: 10.1016/j.ijdrr.2014.07.002

Jomo, K. S., Chowdhury, A., Sharma, K., & Platz, D. (2016). Public-private partnerships and the 2030 Agenda for Sustainable Development: Fit for purpose? United Nations Department of Economic and Social Affairs Working Paper, 148. https://www.un.org/esa/desa/papers/2016/wp148_2016.pdf

Joyette, V. T., Nurse, L. A., & Pulwarty, R. S. (2015). Disaster risk insurance and catastrophe models in the Caribbean. *The Geneva Papers on Risk and Insurance. Issues and Practice*, *40*(3), 553–576. DOI: 10.1057/gpp.2015.8 PMID: 25546529

Juillard, H., Mohiddin, L., Péchayre, M., & Smith, G. (2020). The influence of market support interventions on household food security: An evidence synthesis. Overseas Development Institute. https://odi.org/en/publications/the-influence-of-market-support-interventions-on-household-food-security-an-evidence-synthesis/

Kabisch, N., Frantzeskaki, N., Pauleit, S., Naumann, S., Davis, M., Artmann, M., . . . Zaunberger, K. (2017). Nature-based solutions to climate change adaptation in urban areas: Linkages between science, policy and practice. In N. Kabisch, H. Korn, J. Stadler, & A. Bonn (Eds.), Nature-based solutions to climate change adaptation in urban areas (pp. 1–11). Springer. DOI: 10.1007/978-3-319-56091-5_1

Katz, J. M. (2013). *The big truck that went by: How the world came to save Haiti and left behind a disaster*. St. Martin's Press.

Kellett, J., & Peters, K. (2014). Dare to prepare: Taking risk seriously. Overseas Development Institute. https://odi.org/en/publications/dare-to-prepare-taking-risk-seriously/

Kennedy, J., Ashmore, J., Babister, E., & Kelman, I. (2008). The meaning of 'build back better': Evidence from post-tsunami Aceh and Sri Lanka. *Journal of Contingencies and Crisis Management*, *16*(1), 24–36. DOI: 10.1111/j.1468-5973.2008.00529.x

Kim, J., Yu, I., Lee, J., & Kim, H. (2019). Probabilistic risk assessment for urban infrastructure resilience: A framework for multi-hazard environments. *Sustainable Cities and Society*, *51*, 101662. DOI: 10.1016/j.scs.2019.101662

Klomp, J., & Valckx, N. (2014). Natural disasters and economic growth: A meta-analysis. *Global Environmental Change*, *26*, 183–195. DOI: 10.1016/j.gloenvcha.2014.02.006

Kousky, C. (2014). Informing climate adaptation: A review of the economic costs of natural disasters. *Energy Economics, 46*, 576–592. DOI: 10.1016/j.eneco.2013.09.029

Kousky, C., & Kunreuther, H. (2014). Addressing affordability in the national flood insurance program. *Journal of Extreme Events, 1*(1), 1450001. DOI: 10.1142/S2345737614500018

Kousky, C., & Shabman, L. (2015). A proposed design for community-based flood insurance. Resources for the Future Discussion Paper, 15-02. https://www.rff.org/publications/working-papers/a-proposed-design-for-community-based-flood-insurance/

Kovács, G., & Spens, K. M. (2011). Trends and developments in humanitarian logistics—A gap analysis. *International Journal of Physical Distribution & Logistics Management, 41*(1), 32–45. DOI: 10.1108/09600031111101411

Kull, D., Mechler, R., & Hochrainer-Stigler, S. (2013). Probabilistic cost-benefit analysis of disaster risk management in a development context. *Disasters, 37*(3), 374–400. DOI: 10.1111/disa.12002 PMID: 23551288

Lakdawalla, D., & Zanjani, G. (2012). Catastrophe bonds, reinsurance, and the optimal collateralization of risk transfer. *The Journal of Risk and Insurance, 79*(2), 449–476. DOI: 10.1111/j.1539-6975.2011.01425.x

Linnenluecke, M. K., & McKnight, B. (2017). Community resilience to natural disasters: The role of disaster entrepreneurship. *Journal of Enterprising Communities: People and Places in the Global Economy, 11*(1), 166–185. DOI: 10.1108/JEC-01-2015-0005

Linnerooth-Bayer, J., & Hochrainer-Stigler, S. (2015). Financial instruments for disaster risk management and climate change adaptation. *Climatic Change, 133*(1), 85–100. DOI: 10.1007/s10584-013-1035-6

Lumbroso, D., Brown, E., & Ranger, N. (2017). Stakeholders' perceptions of the overall effectiveness of early warning systems and risk assessments for weather-related hazards in Africa, the Caribbean and South Asia. *International Journal of Disaster Risk Reduction, 21*, 419–432. DOI: 10.1016/j.ijdrr.2016.11.003

Lumbroso, D., Woolhouse, G., & Jones, L. (2017). A review of the consideration of climate change in the planning of hydropower schemes in sub-Saharan Africa. *Climatic Change, 140*(3-4), 563–578. DOI: 10.1007/s10584-016-1857-0

Lyons, M. (2009). Building back better: The large-scale impact of small-scale approaches to reconstruction. *World Development, 37*(2), 385–398. DOI: 10.1016/j.worlddev.2008.01.006

Mahul, O., & Ghesquiere, F. (2010). Financial protection of the state against natural disasters: A primer. World Bank Policy Research Working Paper, 5429. https://openknowledge.worldbank.org/handle/10986/3913

Mahul, O., & Stutley, C. J. (2010). Government support to agricultural insurance: Challenges and options for developing countries. World Bank. https://openknowledge.worldbank.org/handle/10986/2432

Manyena, B., O'Brien, G., O'Keefe, P., & Rose, J. (2011). Disaster resilience: A bounce back or bounce forward ability? *Local Environment*, *16*(5), 417–424. DOI: 10.1080/13549839.2011.583049

Marchezini, V., Horita, F. E. A., Matsuo, P. M., Trajber, R., Trejo-Rangel, M. A., & Olivato, D. (2018). Bridging the gap between research and practice: The role of early warning systems in disaster risk reduction. *International Journal of Disaster Risk Reduction*, *31*, 1082–1092. DOI: 10.1016/j.ijdrr.2018.08.011

Martin, R., & Sunley, P. (2015). On the notion of regional economic resilience: Conceptualization and explanation. *Journal of Economic Geography*, *15*(1), 1–42. DOI: 10.1093/jeg/lbu015

Maskrey, A. (2011). Revisiting community-based disaster risk management. *Environmental Hazards*, *10*(1), 42–52. DOI: 10.3763/ehaz.2011.0005

Mayer, B., Bergstrand, J. K., & Running, K. (2020). Economic recovery and community resilience: Lessons from Hurricane Harvey. *Environmental Justice*, *13*(2), 52–57. DOI: 10.1089/env.2019.0037

Mazzucato, M., & Perez, C. (2015). Innovation as growth policy: The challenge for Europe. In Fagerberg, J., Laestadius, S., & Martin, B. R. (Eds.), *The triple challenge for Europe: Economic development, climate change, and governance* (pp. 229–264). Oxford University Press. DOI: 10.1093/acprof:oso/9780198747413.003.0009

McCord, A. (2013). Public works and social protection in sub-Saharan Africa: Do public works work for the poor? United Nations University Press. https://www.unu.edu/publications/books/public-works-and-social-protection-sub-saharan-africa-do-public-works-work-poor

Meier, P. (2015). *Digital humanitarians: How big data is changing the face of humanitarian response*. CRC Press. DOI: 10.1201/b18023

Ofli, F., Meier, P., Imran, M., Castillo, C., Tuia, D., Rey, N., Briant, J., Millet, P., Reinhard, F., Parkan, M., & Joost, S. (2016). Combining human computing and machine learning to make sense of big (aerial) data for disaster response. *Big Data*, *4*(1), 47–59. DOI: 10.1089/big.2014.0064 PMID: 27441584

Panteli, M., & Mancarella, P. (2015). Influence of extreme weather and climate change on the resilience of power systems: Impacts and possible mitigation strategies. *Electric Power Systems Research, 127*, 259–270. DOI: 10.1016/j.epsr.2015.06.012

Rao, A., Vazquez, J. A., & Kumpati, G. (2017). Next generation infrastructure materials for sustainable and resilient systems. In Rao, A., & Subramanyam, K. G. (Eds.), *Sustainability and resilience of infrastructure systems* (pp. 1–18). CRC Press.

Rozenberg, J., & Fay, M. (Eds.). (2019). *Beyond the gap: How countries can afford the infrastructure they need while protecting the planet.* The World Bank.

Sambuli, N., & Whitt, J. P. (2017). Technology innovation hubs and policy engagement. In Adesina, N., Agboluaje, A., Akinola, O., Anise, O., Guttmann, R., Karambiri, T., Nwagwu, K., Rudman, A., Sambuli, N., Whitt, J. P., & Yonazi, J. (Eds.), *Foresight Africa: Top priorities for the continent in 2017* (pp. 84–90). Brookings Institution.

Stern, N. (2016). Economics: Current climate models are grossly misleading. *Nature, 530*(7591), 407–409. DOI: 10.1038/530407a PMID: 26911765

Surminski, S., & Tanner, T. (Eds.). (2016). *Realising the 'triple dividend of resilience': A new business case for disaster risk management.* Springer. DOI: 10.1007/978-3-319-40694-7

Tierney, K. (2012). Disaster governance: Social, political, and economic dimensions. *Annual Review of Environment and Resources, 37*(1), 341–363. DOI: 10.1146/annurev-environ-020911-095618

Twigg, J. (2015). *Disaster risk reduction.* Overseas Development Institute.

UNDRR. (2019). *Global assessment report on disaster risk reduction 2019.* United Nations Office for Disaster Risk Reduction.

UNISDR. (2015). *Sendai framework for disaster risk reduction 2015-2030.* United Nations Office for Disaster Risk Reduction.

United Nations. (2015). *Transforming our world: The 2030 Agenda for Sustainable Development.* United Nations.

Vaijhala, S., & Rhodes, J. (2018). Resilience bonds: A business-model for resilient infrastructure. *Field Actions Science Reports, 18*(Special Issue), 58–63.

Watkiss, P., Hunt, A., Blyth, W., & Dyszynski, J. (2015). The use of new economic decision support tools for adaptation assessment: A review of methods and applications, towards guidance on applicability. *Climatic Change, 132*(3), 401–416. DOI: 10.1007/s10584-014-1250-9

Wisner, B., Gaillard, J. C., & Kelman, I. (Eds.). (2012). *Handbook of hazards and disaster risk reduction and management*. Routledge. DOI: 10.4324/9780203844236

WMO. (2018). *Multi-hazard early warning systems: A checklist*. World Meteorological Organization.

World Bank. (2017). *Sovereign climate and disaster risk pooling: World Bank technical contribution to the G20*. The World Bank.

World Bank. (2019). *Lifelines: The resilient infrastructure opportunity*. The World Bank.

Chapter 9
Building Resilience:
Addressing the Economic, Social, and Environmental Impacts of Natural Disasters in a Changing Climate

Muhammad Aamir Mahmood
https://orcid.org/0009-0007-4452-2627
Virtual University of Pakistan, Pakistan

ABSTRACT

With global climate changes, natural disasters occur more often, increase in severity and unpredictability, and represent an unprecedented challenge to global societies. In this chapter, we analyze different types of natural disasters, like earthquakes, hurricanes, floods and wildfires and their indisputable economic, social and environmental impacts. It looks at ways in which climate change is a primary cause behind the worsening of weather patterns and the rising vulnerability of urban and rural communities to these disasters. In addition, this chapter examines global initiatives to boost disaster resilience by reviewing climate adaptation strategies, policy frameworks, technological innovation, and a multi stakeholder approach involving governments, NGOs, private sector actors, and local communities. Case studies of countries like Japan, Bangladesh and the Caribbean show how successful resilience initiatives can heavily dampen the impact of natural calamities on societies.

INTRODUCTION

Natural disasters are becoming more frequent and more severe as global climate change advances. Global challenge on natural disaster is enormous, they happen about

DOI: 10.4018/979-8-3693-9745-9.ch009

millions every year and lead to huge damages in economic, social, and environmental aspects. With this increase in frequency and severity there is an increased need to understand the causal factors, characteristics, and impacts of these disasters. This will cover some of the types of natural disasters, their effects on societies around the world, and what contributes to the fact that their number of occurrences is on the rise (Smith 2013).

What are Natural Disasters?

Natural disasters are events of extreme seriousness caused by natural processes of the Earth — the air, water, land—and that cause widespread destruction and suffering by humans. There are different types of disasters we can classify according to their own characteristics and types of impacts. A good example of earthquakes is the sudden release of energy within the Earth's crust, which causes the ground to shake. Earthquakes can result in the collapse of buildings, infrastructure damage and death, as the earthquake in Haiti 2010 proved to be. Tropical storms, or hurricanes, are intense weather systems, powerful winds, and lots of rain. Typically, flooding, property damage and forced displacement follow, with Hurricane Katrina one of the worst in the United States. Floods, caused by excessive rainfall, or storm surges, or dam failure, bring about inundations over large areas, resulting in loss of life, agricultural destruction, and in the contamination of water sources. Another type of disaster, called wildfires, are uncontrolled fires that rapidly spread in forests, grasslands and rural areas in dry conditions and are exacerbated by climate change. The characteristics of each type of natural disaster are different. Earthquakes are sudden and unpredictable, as opposed to hurricanes and floods, which are more seasonal and consequently something of a forecast, capable of some preparation. Wildfires can be caused naturally, or human induced, and frequently get out of control because of environmental conditions such as drought. All these differentials impact how societies react to and rebound from disasters (Arcaya et al., 2020).

Natural Disasters Have an Impact on World Societies in Several Ways

Natural disasters have wide economic, social, and environmental impact. Disasters economically cripple entire regions by means of destruction of infrastructure, stalling of industries and requiring large recovery efforts. The 2010 Haiti earthquake left an estimated $8 billion in damaged in the nation's economy. Natural disasters socially cost lives, displace people, and create long term mental health issues. Often the affected communities wrestle with trauma and their families are displaced in search of shelter and security. Hurricane Katrina, which occurred in 2005, not only caused high loss of life, but more than one million people were displaced, showing the large social cost of these occurrences. Disasters can lead to lasting damage to the environment, making forest destroyed, loss of biodiversity and polluting of air and water sources. For instance, floods can transport pollutants and debris over great distances and wildfires ravage ecosystems by consuming forests and animal habitats (Jakhar et al., 2023).

Natural Disasters Have Become More Frequent and More Severe

Natural disasters are happening with greater frequency and greater severity. Number of disasters has grown tremendously and today more people are affected than in the past decades as per the trends in statistical data. The estimated number of natural disasters per year has nearly doubled in the past 20 years compared to the prior two decades, says the United Nations Office for Disaster Risk Reduction (UNDRR). The increase has contributed to rapid urbanization, which increases the number of people living in vulnerable areas, and that of floods and landslides, environmental degradation such as deforestation and soil erosion. Climate Change also contributes majorly by making weather so wilder to the extent of such upheavals as hurricanes and wildfire. The future is projected to see more natural disasters become more intense. With growing populations and encroaching into cities which are now high-risk areas, more people will be subjected to these perils. The greater urgency for comprehensive disaster preparedness and resilient infrastructure is because of the need to mitigate the impacts of future events(McDaniels et al., 2008).

ROLE OF CLIMATIC CHANGE IN THE FORMATION OF NATURAL CALAMITIES

One of the critical roles played by climate change is that it changes weather and environmental conditions all around the world thus affecting the occurrence and intensity of natural disasters. The effects of climate change, especially the long-term phenomenon of global warming, are creating the conditions under which increasingly frequent and intense disasters take place, leaving communities ever so vulnerable. It is our job to see how climate change influences natural disasters and vice versa to build resilience and mitigate future risk (Kelman et al., 2015).

Interaction of Climate Change and Natural Disasters

Many types of natural disasters are happening more frequently, and they are growing more severe due to global warming, which is caused by greater greenhouse gas emissions. Global temperatures have been rising, which means more extreme weather patterns are the result, including prolonged droughts, intense floods, and heat waves. For example, higher temperatures increase the rate of water evaporation, resulting in more acute rainfall and flooding in certain areas and in others, the drought. Floods in South Asia and the 2021 European heatwaves show how climate change is supercharging natural disasters in terms of intensity and their scope. The most visible examples of the effects of climate related disasters are droughts, floods, and heat waves. Droughts are emerging as more frequent and more severe in many parts of the world, especially in arid and semi-arid regions, and they increasingly threaten the security of water and agricultural production. Though the climate change accelerates water cycle causing the larger storms and more intense rainfall, Floods are happening more frequently. Heatwaves, too, are up, as with global warming the world sees temperatures climb to unprecedented highs and raise the chances of heat illness and death. That's just what these examples show: the direct connection between climate change and the threat of natural disasters is growing (Banholzer et al., 2014).

Long Term Effects of Climate Change on Disaster Vulnerability

Climate change is exacerbating vulnerability to natural disasters and further increasing the risks associated with natural disasters in high-risk areas (such as coastal cities and arid regions). Rising sea levels make coastal areas some of the most vulnerable, with greater risk of flooding, storm surge and coastal erosion. New York, Miami, and other coastal cities are seeing their risks of disaster growing with floodwaters poised to eviscerate their infrastructure, swamping their homes and livelihoods. Persistent flooding will make entire islands and coastal communities around the world uninhabitable. Similarly, desertification, caused by climate change, is increasing in the arid regions, including notably in Africa and Asia. It reduces arable land, makes clean water scarce and worsens droughts so that food insecurity and displacement are more likely. Desertification, which has already spread to the areas, has begun so that these communities face daunting and mounting disasters like dust storms and heat waves (Hill and Martinez-Diaz 2020).

Efforts to Combat Climate Change and Enhance Resilience Around the Globe

To mitigate the impacts of climate change and to build our resilience to future disasters, efforts are underway globally to respond to the threat of increasingly disastrous natural calamities related to climate change. The Paris Agreement is one of the international climate agreements that seeks to cut greenhouse gas emissions and maintain global temperature rise below 2 degrees Celsius above pre-industrial levels. These agreements are especially important for reducing the risk of future disaster by persuading countries to use more sustainable methods and move to renewable energy sources. Worldwide, there are also various other climate adaptation strategies being implemented, mostly to build resilience for vulnerable communities. To take but one example, many countries are investing in early warning systems, better infrastructure and planning, and more sustainable land use to lessen the blow from devastating natural disasters. Examples of adaptation measures include mangrove restoration in coastal areas, flood plain management and drought resistant crops to help communities to survive the multiplying effects of the challenges caused by climate change. These strategies, in combination with global efforts to reduce emissions, are critically important for mitigating climate change's risk of disasters and improving resilience of societies worldwide (O'Brien et al., 2006).

THE IMPORTANCE OF DISASTER RESILIENCE: A GLOBAL PERSPECTIVE

In a world in which we are increasingly experiencing weather events more frequently and with greater intensity because of climate change, never has the importance of disaster resilience been greater. Earthquakes, floods, and other catastrophic events pose an increased threat to societies around the world, and thus stronger resilience mechanisms are imperative. Disaster resilience exceeds the response to a disaster and includes the whole through preparedness, risk reduction, and recovery strategies, critical tools to reduce loss and enhance sustainable development. In this section we look at the main concepts, examples from history and how technology has been used to improve global disaster resilience (Paton and Johnston 2017).

Defining Disaster Resilience

Disaster resilience is the ability of communities, organizations, and nations to anticipate, absorb, recover, and adapt, in a timely and efficient manner, to adverse events. The key principles of disaster resilience have been reducing the risks from disasters, engaging, and involving communities, and a sustainable recovery after disasters. The critical elements including robust infrastructure, effective governance, and social capital, provide the foundation for these principles. Resilience building is a continuous and cyclic process, which does not only decrease the effect of disasters but also quickens recovery process and persistence (Neal, 2013).

In the framework of disaster resilience risk assessment and preparedness are key elements. Performing risk assessment involves identifying hazards, vulnerabilities, and potential impacts of risk for an effective risk management and risk mitigation process. On the contrary, there is preparedness, i.e., readiness, discussing equipping a community with the tools and knowledge needed to respond efficiently when rears come. By coming together, these components provide a strong basis for resilience building and avoid damage from natural disasters (Harrison and Williams 2016).

A Brief History of Successful Resilience Works Initiatives

Lots of lessons of successful resilience initiatives exist in many countries and communities around the world for learning by disaster prone regions. For example, Japan is the world's leading country for disaster preparedness in general, and for earthquakes and tsunamis in particular. The country has strategically covered areas of early warning systems, disaster education and infrastructure that is able to withstand seismic activities, which has saved many lives for years. Bangladesh has also made dramatic improvements in saving lives from cyclones. Building of cyclone shelters, upgrading of early warning systems, and including local communities in disaster preparedness have cut fatalities from devastating cyclones in Bangladesh to a tiny fraction of what it was in past decades. The Caribbean islands have also taken steps to strengthen their resilience through strategies including community-based approaches and underpinning their disaster response capabilities. It is important to note that in some of these regions' local knowledge, sustainable development, and international collaboration have been identified as fundamental elements for building resilience (National Research Council, 2011).

Technology and Innovation in Enhancing Resilience

Technology and innovation are key enablers of building global resilience to disasters. In decreasing the impact of natural disasters, modern early warning systems have become indispensable. Using advances in such areas as satellite technology, meteorology, and seismic data monitoring, these systems provide real time data and warning to allow communities enough time to evacuate and prepare before a disaster strikes (Mendelson 2011).

Disaster response technologies like drones, artificial systems intelligence (AI) and geographic information systems (GIS) are reimagining how government agencies and organizations handle crises. These innovations speed up the rate of damage assessment, resource allocation, and more effective rescue action, all essential in post-disaster situations. Additionally, innovations in infrastructure such as floodproof and earthquake resistant construction, and energy grids that are less likely to be damaged are critical to reducing long term damage and speeding recovery. Resilient infrastructure investment enables communities to be better able to withstand disasters with lesser interruptions in essential services (Chang et al., 2014).

Real time disaster management is also ensured by digital tools and platforms that coordinate response efforts between local governments, emergency services and international relief organizations. They all provide real-time communication, resource tracking and data sharing to guarantee that they will help reach the affected area as fast and as efficiently as possible. In a world where information is swiftly transmitted, timely and well-coordinated responses can save lives and minimize economic losses, and these innovations are necessary. Therefore, disaster resilience based on evidenced principles, historical authorities' references, and emerging technology is something that every community must have if they want to protect communities and ensure long-term sustainability in a world increasingly prone to disaster (Jones et al., 2010).

ENHANCING RESILIENCE THROUGH A MULTI STAKEHOLDER APPROACH

Natural disasters need to be worked on collaboratively and by someone who is willing to integrate and be involved with multiple stakeholders across several sectors. Taking a multi stakeholder approach improves the chances of disaster risks being mitigated, and effective response when such things occur. The development and maintenance of a broad disaster resilience strategy is possible through the coordination of efforts among government, NGOs, the private sector, and local communities. We analyze how these different stakeholders participate in preparing for disaster, respond to disaster, and recover from disaster (Mojtahedi and Oo 2017).

Policies and Laws of Government for Enhancing the Disaster Resiliency

Disaster risk reduction and resilience building governance depends fundamentally on governmental policy and associated legislative framework. Guidelines for creating disaster resilient societies are available from both national and international policies like the Sendai Framework for Disaster Risk Reduction. These frameworks urge countries to have a complete plan in place for preparing for and mitigating disaster risks, including prevention, preparedness, and mitigation. As national governments set laws and policies to international standards, disaster resilience must be a priority in development plans. Disaster preparedness and response are also organized by governmental organizations. National disaster management agencies are set up, early warning systems are developed, and disaster relief operations are facilitated. In addition, these organizations are critically important in facilitating multi government collaborations, ensuring that disaster response is as fast as possible and as coordinated as possible to avoid any undue impact on affected populations. They also fund and support local governments for community level resilience initiatives (Dzigbede et al., 2020).

The NGOs and the Private Sector Role

At the community level however, disaster preparedness work is incomplete without non-governmental organizations (NGOs). Grassroots initiatives are often led by them; fostering community-based disaster preparedness programs that increase the local population's capacity to take proactive steps towards risk reduction. NGOs are involved in capacity building activities, carrying out important training and encouraging investments in disaster preparedness in vulnerable areas. With close contact with communities, NGOs can identify individual needs and vulnerabilities to tailor disaster risk reduction efforts to the local context. Private sector is being increasingly recognized as an important partner in disaster resilient efforts. Public-private partnerships (PPPs) can foster sharing of expertise, resources, and technology to significantly enhance disaster preparedness and response capacity. Having private sector partner with national and local governments in building resilient infrastructure, in having business continuity plans, and even establishing disaster risk reduction practices within industries makes sense. Governments and private sectors collaborate in investments that create resilient infrastructure or manage supply chains and are essential in increasing overall resilience (Chen et al., 2013).

Awareness and Preparedness Building within Community

Disaster resilience depends on community awareness and involvement. Communities are helped to understand the risks they face and steps they can take to reduce the risks when education and awareness programmes about disaster preparedness are undertaken. Communities are enabled to respond effectively by training through public campaigns, workshops, and training sessions. These measures create a culture of preparedness so that people and communities can respond better when a disaster happens. Local resilience has been especially enhanced by grassroots initiatives. For example, programs in Community Based Flood Management have been seen effective in countries prone to floods. These programs teach local volunteers how to operate early warning systems, prepare for floods and evacuate. Some other examples are earthquake preparedness drills in areas experiencing seismic activity, in which the community is shown how they ought to react to an earthquake to reduce fatalities and harms. These grassroots initiatives do much to help to build community resilience, and to strengthen ties of cooperation and solidarity among residents when crises arise. Finally, it is concluded that building disaster resilience requires a multi-stakeholder approach that involves the government, NGOs, the private sector, and local communities. With multiple resources and unique strengths, each stakeholder has a major role in working together to create new capacities, close existing gaps, reduce vulnerabilities, and increase preparedness at all levels. By coordinating efforts and sharing responsibility, much resilience to natural disasters can be achieved and the devastating effects of these natural phenomena on the world's societies can be avoided (National Research Council, 2011).

CONCLUSION

Building resilience to natural disasters is a global imperative, especially in an era of ever-increasing environmental instability. As events like earthquakes, floods, wildfires, and hurricanes occur with greater frequency and increasing intensity, due in part to climate change, there is a greater imperative to prepare effectively and to mitigate disaster impact. In short, this chapter argues that resilience extends beyond emergency response to risk reduction, preparedness, recovery, and long-term sustainability. Renowned to be universal in nature and binding on the Parties thereto, the Paris Agreement and Sendai Framework for Disaster Risk Reduction are crucial to shaping international and national disaster resilience policies and hence are important global frameworks. Absolute must haves for disaster mitigation are technological advancements that include early warning systems, artificial intelligence and resilient infrastructure. For instance, real time disaster management tools, drone technology and GIS systems make disaster response and resources allotment easier for the government and organizations. Japan and Bangladesh present important lessons in disaster preparedness: Early warning systems can reduce loss of life and economic damage; communities are key; and being built to last matters. But solo government is not enough to tackle the challenge of how to build disaster resilience is not enough. A multi stakeholder, collaborative approach from NGOs, private enterprises and local communities is needed. Development of resilient infrastructure, disaster risk financing and business continuity planning can be achieved through public private partnerships. The NGO's efforts are crucial to raising community awareness, to offer training and to strengthen local capacity building to prevent disaster risks at the grassroots level. Education and local initiatives like flood management or earthquake drills help communities prepare for them, giving residents the ability to properly respond when an emergency strikes, thus significantly reducing the effect of disasters. Finally, building a globally resilient society to natural disasters needs collaboration among all sectors and all levels of society. Technology innovation, community engagement, sustainable development and worldwide collaboration are combined in such a way that it can minimize the devastating effects of natural disasters and bring societies toward a more secure and sustainable future.

REFERENCES

Arcaya, M. C., Raker, E. J., & Waters, M. C. (2020). The social consequences of disasters: Individual and community change. *Annual Review of Sociology*, *46*(1), 671–691. DOI: 10.1146/annurev-soc-121919-054827

Banholzer, S., Kossin, J., & Donner, S. (2014). The impact of climate change on natural disasters. In Singh, A., & Zommers, M. (Eds.), *Reducing disaster: Early warning systems for climate change* (pp. 21–49). Springer., DOI: 10.1007/978-94-017-8598-3_2

Chang, S. E., McDaniels, T., Fox, J., Dhariwal, R., & Longstaff, H. (2014). Toward disaster-resilient cities: Characterizing resilience of infrastructure systems with expert judgments. *Risk Analysis*, *34*(3), 416–434. DOI: 10.1111/risa.12133 PMID: 24152135

Chen, J., Chen, T. H. Y., Vertinsky, I., Yumagulova, L., & Park, C. (2013). Public–private partnerships for the development of disaster resilient communities. *Journal of Contingencies and Crisis Management*, *21*(3), 130–143. DOI: 10.1111/1468-5973.12021

Dzigbede, K. D., Gehl, S. B., & Willoughby, K. (2020). Disaster resiliency of U.S. local governments: Insights to strengthen local response and recovery from the COVID-19 pandemic. *Public Administration Review*, *80*(4), 634–643. DOI: 10.1111/puar.13249 PMID: 32836460

Harrison, C. G., & Williams, P. R. (2016). A systems approach to natural disaster resilience. *Simulation Modelling Practice and Theory*, *65*, 11–31. DOI: 10.1016/j.simpat.2016.02.008 PMID: 32288694

Hill, A. C., & Martinez-Diaz, L. (2020). *Building a resilient tomorrow: How to prepare for the coming climate disruption*. Oxford University Press. DOI: 10.1093/oso/9780190909345.001.0001

Jakhar, R., Samek, L., & Styszko, K. (2023). A comprehensive study of the impact of waste fires on the environment and health. *Sustainability (Basel)*, *15*(19), 14241. Advance online publication. DOI: 10.3390/su151914241

Jones, L., Jaspars, S., Pavanello, S., Ludi, E., Slater, R., Grist, N., & Mtisi, S. (2010). Responding to a changing climate: Exploring how disaster risk reduction, social protection and livelihoods approaches promote features of adaptive capacity. Overseas Development Institute. https://odi.org/en/publications/responding-to-a-changing-climate-exploring-how-disaster-risk-reduction-social-protection-and-livelihoods-approaches-promote-features-of-adaptive-capacity/

Kelman, I., Gaillard, J.-C., & Mercer, J. (2015). Climate change's role in disaster risk reduction's future: Beyond vulnerability and resilience. *International Journal of Disaster Risk Science*, *6*(1), 21–27. DOI: 10.1007/s13753-015-0038-5

McDaniels, T., Chang, S., Cole, D., Mikawoz, J., & Longstaff, H. (2008). Fostering resilience to extreme events within infrastructure systems: Characterizing decision contexts for mitigation and adaptation. *Global Environmental Change*, *18*(2), 310–318. DOI: 10.1016/j.gloenvcha.2008.03.001

Mendeldon, E. (2011). System and method for providing alarming notification and real-time, critical emergency information to occupants in a building or emergency designed area and evacuation guidance system to and in the emergency exit route (U.S. Patent No. US20110241877A1). Google Patents. https://patents.google.com/patent/US20110241877A1/en

Mojtahedi, M., & Oo, B. L. (2017). Critical attributes for proactive engagement of stakeholders in disaster risk management. *International Journal of Disaster Risk Reduction*, *21*, 35–43. DOI: 10.1016/j.ijdrr.2016.10.017

National Research Council. (2011). *Building community disaster resilience through private-public collaboration*. National Academies Press., DOI: 10.17226/13028

Neal, D. M. (2013). Social time and disaster. *International Journal of Mass Emergencies and Disasters*, *31*(2), 247–270. DOI: 10.1177/028072701303100207

O'Brien, G., O'Keefe, P., Rose, J., & Wisner, B. (2006). Climate change and disaster management. *Disasters*, *30*(1), 64–80. DOI: 10.1111/j.1467-9523.2006.00307.x PMID: 16512862

Paton, D., & Johnston, D. (2017). *Disaster resilience: An integrated approach* (2nd ed.). Charles C Thomas Publisher.

Smith, K. (2013). *Environmental hazards: Assessing risk and reducing disaster* (6th ed.). Routledge. DOI: 10.4324/9780203805305

Chapter 10
Psychological and Social Impact of Natural Disasters:
A Case Study of Wayanad District of India

Anitha Parveen J.
https://orcid.org/0009-0006-3073-7558
Lovely Professional University, Punjab, India

Guddo
https://orcid.org/0000-0002-0322-5606
Lovely Professional University, Punjab, India

ABSTRACT

Kerala's geographic vulnerability to natural disasters, particularly in Wayanad, has been identified by the Centre for Earth Science Studies as one of the most landslide-prone areas in the state, significantly impacting residents' mental health. This study explores the psychological issues arising from such disasters, including PTSD, depression, and anxiety, which often persist for years due to displacement, loss of social support, emotional instability, and chronic distress. Landslides, in particular, have devastating consequences, orphaning children, fracturing families, and disrupting communities. While immediate rescue efforts prioritize physical safety, addressing psychological needs is crucial for long-term recovery. The study highlights the urgent need for increased psychological support, awareness, and policy initiatives to address the mental health challenges faced by disaster survivors. By

DOI: 10.4018/979-8-3693-9745-9.ch010

integrating mental health into disaster management strategies, stakeholders can promote sustainable recovery and well-being for affected individuals and communities in Wayanad.

INTRODUCTION

Natural disasters happen as a result of natural processes and often lead to social distress and loss of lives and property with deep physical and psychological impacts on the survivors. Natural calamities elicit a wide range of emotions, such as fear, helplessness, anxiety, worry, and tension. Serious psychosocial problems will emerge post-disaster, which will affect the health and productivity of the community at large. Further, co-morbid conditions can arise as a result of natural disasters and mental disorders. Studies show that psychological disorders can be two times more prevalent among affected populations when compared to others. Floods and landslides are very common natural disasters globally. There have been nearly 53,000 deaths due to floods in previous decades (Doocy et al., 2013). Similarly, in India the India Space Research Organization (ISRO) updated its "*Land Slide Atlas*" on July 31, 2024; it shows that Wayanad district stands for 13th position among 147 districts in India in terms of socioeconomic indicators linked to landslide vulnerability, and within Kerala, Wayanad stands for the first position. Official records related to landslide deaths show that Kerala witnessed the highest number of casualties in 2018, 2019, and 2020.

The various reports highlighted that the mental health effects among these communities face elevated rates of psychological distress, anxiety, somatization, depression, and post-traumatic stress disorder (Stake et al., 2012; Fernandez et al., 2015). These types of mental health issues may continue for years, with higher enduring being reported 2-5 years after the disaster (Kessleretal., 1995; Goldmann and Galea., 2014). In a small minority, exposure to disaster may cause a high range of mental illnesses. People experience vulnerability because of their greater need for support networks and supply chains (WHO, 2005). Reducing mental health issues is important to address the overall well-being of the Individuals.

The natural disaster affected all age groups; in the case of children, they become orphans, which affects their total mental health for a long time; adults lose their social networking, which affects their social life and leads them to trauma and isolation. Old age people lose their support system, which affects the rest of their lives. So, these previous data revealed that it is very important to address the mental health of disaster survivors. They need immediate psychological first aid (PFA) to get emotional support to reduce distress for their future life. Early interventions, such as psychological support and counselling, can reduce more severe mental health issues. Altogether, it is important to address these psychological issues and their consequences.

Vulnerability of Wayanad District

The Wayanad district of Kerala is naturally vulnerable to disaster and it falls under zone-III, a moderate risk of earth quick zone. The district is particularly susceptible to landslides which have caused significant damage to its infrastructure and livelihoods. The landslide, which occurred in the state of Wayanad, is prone to natural disasters such as landslides and floods. Kerala has a tropical climate with a distinct monsoon season, with heavy rainfall from June to September, and a cooler climate in the highlands. The landslide has prompted the Government and stakeholders to respond to the situation and provide necessary assistance. The state's population is 34,630,192, with a population density of 860 persons per sq. km.

The previous reports pointed out that landslides repeatedly impacted Wayanad, causing damage. The notable previous disasters include The Mundakkai flow of 1984, which resulted in 14 deaths and the loss of substantial agricultural land. In 1992, the Kappikkalam landslide was one of the significant landslides that killed 11 lives, and the Valamthod landslide, which happened on 2007 June 23, also witnessed four deaths. Other significant landslides are Ambalavayal-Kumbaleri, Puthumala, and Mundakai in sequential years. It also causes left families, homelessness, and loss of agricultural land.

The landslide is not only a natural disaster in Wayanad district; the flooding is another recurrent disaster, especially in the monsoon season. Due to torrential rainfall in 2024, a severe landslide invaded eight Indian states, claiming 304 lives. In an unprecedented calamity, a single incidence from Kerala's Wayanad district claimed 285 lives on July 30, which accounts for nearly 94% of total deaths. It indicates the elevated risk of natural disasters and as a result, psychological issues are also increasing simultaneously. The disaster situation can have long-term trauma. Many of them experience long-term mental health struggles because of the loss of houses, families, and agricultural land (Hameed et al., 2019). The survivors experiencing mental health issues include depression, post-traumatic stress disorder, and anxiety, which can exist maybe for years (Neria et al., 2018). The survivors also experienced displacement and loss of social support networks that resulted in emotional instability, grief, and chronic stress (WHO-2019).

Along with this, drought is a major challenge to Wayanad, and the situation became even more alarming with less rainfall in 2016-17. Along with implementing long-term measures for the prevention and mitigation of the drought, urgent measures are called for to address such crises. In July 2024, torrential monsoon rains triggered a series of landslides in eight Indian states, claiming at least 304 lives.

The unparalleled nature of the disaster was brought into sharper relief in one particular instance: 285 people died on July 30 in Wayanad district, nearly 94 percent of the total deaths. Among India's 147 districts, Wayanad ranked 13th in terms of socioeconomic indicators relating to landslide vulnerability, ranging across 17 states and two Union Territories. Within Kerala, Wayanad was among 14 districts rated 'very vulnerable' in ISRO's updated Landslide Atlas published on July 31, 2024. As of August 1, 2024, in Wayanad, the number of missing persons was about 240, raising fears that the death toll could exceed the 1999 toll in Maharashtra. Official landslide death tolls reveal that Kerala recorded the highest losses in 2018, 2019, and 2020 (Government of Kerala, 2024).

This chapter is more relevant to the present scenario as it investigates the psychological, emotional, and social consequences of disaster on affected individuals and communities. Further, it conceptualizes the Government and non-government initiatives aimed at reducing the psychological impact of natural disasters and highlights the suggested policies and programs measures to strengthen mental health services and community resilience in Wayanad district.

METHODOLOGY

The present study employs a qualitative literature review approach to examine the psychological impact of natural disasters in the Wayanad district of Kerala. By synthesizing existing research, this study aims to establish a conceptual framework that elucidates the intricate relationship between natural disasters and their psychological consequences on disaster survivors. To gather relevant literature, a systematic search was conducted using keywords like '*natural hazards*,' '*disaster management*,' '*psychological impact*,' and '*disaster policy and program*.' These keywords were strategically selected to ensure a comprehensive exploration of the topic, encompassing both theoretical and empirical perspectives.

The study adopts a narrative review methodology, which allows for a holistic understanding of the psychosocial factors influencing disaster response and recovery. This approach facilitates the integration of diverse scholarly insights, highlighting the multifaceted nature of psychological trauma resulting from natural disasters among communities in Wayanad. Specifically, the review underscores the prevalence of mental health issues such as post-traumatic stress disorder (PTSD), depression, and anxiety among disaster survivors. It also emphasizes the long-term effects of such trauma, including emotional instability, grief, and chronic distress, which are often exacerbated by displacement and the disruption of social support networks in Wayanad district Kerala. It contributes to a deeper understanding of the psychosocial dimensions of disaster management and underscores the importance of holistic approaches in mitigating the enduring impact of natural disasters.

LITERATURE REVIEW

Despite the growing recognition of the profound mental health consequences of natural disasters, the available literature in the Kerala context remains notably limited. This gap underscores the need for a focused exploration of the psychological issues arising from such events. The review encompasses several key areas, including the prevalence and nature of natural disasters in Wayanad, the immediate and long-term psychological impacts on survivors, the coping mechanisms adopted by affected individuals, and the policies and programs implemented by the government to provide psychological support. By synthesizing these diverse strands of research, this review aims to provide a comprehensive understanding of the psychosocial dimensions of natural disasters in Wayanad, while also identifying critical gaps and areas for future inquiry. This foundational exploration sets the stage for a deeper investigation into the interplay between disaster vulnerability and mental health, emphasizing the need for targeted interventions to address the psychological well-being of disaster-affected communities.

Natural Disaster and Wayanad District

The Wayanad landslide disaster in Kerala's Wayanad district on July 30, 2024, caused extensive destruction and over 250 deaths (Jinoy, 2024). The disaster prompted a coordinated rescue effort involving 700 personnel from various agencies, including the Army, State Police, Fire and Rescue, and the National Disaster Response Force. The disaster also sparked debates about climate change and the ignored 2011 Gadgil Committee report, which recommended declaring 64% of the Western Ghats as an Ecologically Sensitive Area. Experts emphasize the need for engineering solutions, early warning systems, and sustainable land-use policies to address climate change and disaster preparedness in Kerala. Further, Shaji (2024) explored that it was India's worst-ever landslide, affecting an eight-kilometer stretch, crushing homes and infrastructure. The economic impact was severe, with extensive damage to plantations. The disaster is the largest in India's history, displacing six million cubic meters of debris. The Hume Centre for Ecology and Wildlife Biology warned 16 hours before the landslide, but officials deny receiving the alert. Recovery efforts continue, but the scale of devastation highlights the need for better disaster preparedness and sustainable environmental policies.

According to the National Crime Records Bureau's annual reports on accidental deaths and suicides from 2003 to 2022, the recent Wayanad landslide has recorded the most casualty figures in Kerala in the last 20 years. This deadly spell of nature has made history with the highest death tolls from landslides after 2018, when devastating floods took away 135 lives and 285 deaths in 2024, considered the worst of the 21st century.

Azad et al., (2024) stated that Wayanad landslides of 2024 have already caused much destruction, halted the plantation work of workers, and disrupted the life cycle, with loss in material and financial terms and further environmental degradation. While the initial rescue and relief efforts of the state, by NGOs, volunteers, and well-wishers, are good, the permanent rehabilitation and mitigation challenges in the face of the state hence, it is essential to provide access to suitable livelihoods as well as diversify local economies along with an effective post-event response to be able to permanently rehabilitate hundreds of survivors of Kerala's deadliest disaster. Improving preparedness for disaster risks requires Kerala to incorporate strong mechanisms of institutions, hazard maps, and early warning system upgradation systems, all integrated into a comprehensive pre-event strategy. Learning from successful international frameworks and integrating advanced technologies can help Kerala build up its resilience against climate change-induced disasters. Bayback in 2018, major flooding occurred due to excessive rainfall throughout the Monsoon period, predominantly in Kerala. According to Manoj (2019) it was the worst flooding in Kerala for nearly a century and resulted in terms of giant landslides and floods all over the State, and the intensity of havoc was very severe in Wayanad, one of the most sensitive biodiversity areas in the Western Ghats due to corporate interests over governmental and legal measures for environmental sustainability.

Psychological Impacts of Natural Disaster

The psychological impacts of natural disasters have garnered increasing attention in academic discourse, particularly in understanding how communities navigate the aftermath of environmental crises. Koshy and Smith (2023) explore the concept of community resilience in their study, emphasizing the role of state and non-state organizations in either supporting or obstructing community-led recovery efforts. Their research highlights the gestures of resourcefulness, solidarity, and mutual trust that emerge during and after disasters, yet notes the lack of systematic documentation on how these dynamics influence spatial planners, humanitarians, and decision-makers operating under uncertainty. Employing an inductive epistemological approach, the authors draw on exploratory fieldwork and interviews from two case studies: Wayanad, a hill district in India devastated by monsoon floods in 2018 and 2019, and marketplaces in Port-au-Prince, Haiti, following the 2010 earthquake. These cases serve as critical examples of how community meeting spaces can foster resilience and inform institutional responses in the face of uncertainty. However, the study could benefit from a deeper analysis of the long-term psychological effects on individuals within these communities, as well as the role of cultural and social capital in shaping resilience trajectories.

In a similar vein, George et al. (2022) address the gendered dimensions of disaster vulnerability, focusing on the disproportionate impact of extreme weather events on women. Their study, centered on the 2018 Kerala floods, utilizes primary data to examine the socioeconomic determinants of women's vulnerability and proposes policy interventions to mitigate these factors. While the study provides valuable insights into the intersection of gender and disaster resilience, it could further explore the psychological toll of recurrent disasters on women, particularly in terms of mental health and coping mechanisms.

Similarly, Joy et al. (2021) contribute to this discourse by examining the psychological consequences of repeated natural disasters, specifically floods in Kerala. Their findings reveal that continuous victimization can lead to learned helplessness and a decline in psychological well-being, with statistically significant differences observed between individuals affected once, twice, or not at all. The study underscores the urgent need for psychological support and community-based interventions to address the mental health challenges faced by disaster-affected populations. While the recommendations for integrating psychological support into disaster management frameworks are commendable, the study could expand its scope by exploring the role of social networks and community cohesion in mitigating psychological distress. Furthermore, the authors could delve into the potential of pro-environmental behaviors as a coping mechanism, linking psychological resilience to broader environmental sustainability goals.

Collectively, these studies underscore the multifaceted nature of disaster impacts, highlighting the interplay between community resilience, gender vulnerability, and psychological well-being. However, there remains a need for more comprehensive frameworks that integrate these dimensions, particularly in addressing the long-term mental health consequences of recurrent disasters.

Long-term Psychological Problems of Survivors

The long-term psychological effects of disasters on survivors have been a critical area of study, yet significant gaps remain in understanding the mechanisms of long-term adjustment and the role of health-protective factors. Siri Thoresen et al. (2019) highlight that while the mental health impacts of trauma and traumatic bereavement are well-documented, there is limited research on how survivors adapt over extended periods. Their study reveals that individuals who survive catastrophic events or experience bereavement often exhibit elevated levels of anxiety and depression symptoms even 26 years post-disaster. This enduring psychological distress is compounded by a notable decline in social support, suggesting that trauma can inflict irreversible damage on interpersonal relationships. These findings underscore the necessity for sustained mental health interventions and the need to explore protective factors that could mitigate long-term psychological harm.

Similarly, Tanaka et al. (2016) emphasize the scarcity of research on the psychological well-being of survivors beyond the immediate aftermath of disasters. Their study, which examined secondary school students in Sichuan province six years after the Wenchuan earthquake, found that severely traumatized adolescents continued to experience psychological symptoms long after the event. Utilizing secondary data from a five-year mental health and psychosocial support project, the study highlights the critical need for long-term psychological support for survivors, particularly adolescents, who may remain vulnerable to mental health challenges years after the disaster. This research underscores the importance of extending mental health interventions beyond the initial recovery phase to address the persistent psychological needs of survivors.

The variability in psychological outcomes following disasters is further explored by Mohammad Asim et al. (2019), who focus on the mental health consequences of flooding in the Indian subcontinent. Floods, as recurrent natural disasters in this region, not only cause high mortality but also lead to significant mental health disturbances, particularly post-traumatic stress disorder (PTSD). The study notes that the prevalence of PTSD in India varies widely, ranging up to 70%, depending on factors such as the severity of the disaster, socioeconomic status, pre-existing mental health conditions, and the availability of social support. Vulnerable groups, including children, the elderly, and those from lower socioeconomic backgrounds, are disproportionately affected. The study also highlights the ethnic diversity of India as a contributing factor to the variability in psychiatric outcomes, calling for more nuanced research to understand the interplay between cultural, social, and disaster-related factors. It underscores the need for targeted mental health interventions that consider the unique vulnerabilities of affected populations.

Building on these findings, Eamin Z. Heanoy and Norman R. Brown (2024) provide a comprehensive review of the mental health consequences of natural disasters, emphasizing their increasing frequency and intensity. While most individuals exposed to disaster-related trauma do not develop mental disorders, a significant subset experiences prolonged psychological effects. They propose a transitional impact framework to understand the mental health consequences of disasters, highlighting the importance of assessing both direct and indirect impacts, as well as short- and long-term outcomes. It would provide a more holistic understanding of disaster-related mental health outcomes and inform the development of culturally sensitive interventions. Therefore, future research should prioritize longitudinal studies, cross-cultural comparisons, and the development of innovative assessment tools to better understand and mitigate the long-term psychological impacts of disasters.

Government Initiatives and Disaster Recovery

The Government of India has implemented various programs and policies aimed at providing psychological support to disaster survivors, recognizing the profound emotional and mental health challenges they face. However, the effectiveness of these initiatives often hinges on the broader governance framework and the integration of risk management strategies. Turkey and Saud (2024) emphasize that landslides, triggered by natural causes such as prolonged rainfall or earthquakes, have devastating consequences in India, particularly in regions like Wayanad District, Kerala. The destruction of agriculture, infrastructure, and human life underscores the urgent need for effective risk management. While natural disasters are inevitable, they identify significant governance gaps during the recovery and post-recovery phases of landslide hazards over three decades. The key stakeholders—including the District Collector, State Disaster Management Authority, and various departmental agencies lacks at the level of cohesive strategies and inter-departmental coordination. Therefore, risk reduction strategies need stakeholder involvement and institutional reforms to enhance future disaster preparedness and mitigation efforts.

Complementing this perspective, Koshy (2022) explores Decision-Making Under Uncertainty (DMUU) in Wayanad, focusing on both short-term and long-term uncertainties. The finding reveals that while decentralized governance enabled effective short-term responses to floods during the 2018 and 2019 monsoons, long-term decision-making remains hampered by institutional weaknesses and unsustainable practices such as unregulated quarrying. It highlights the need of institutional reforms, cross-sectoral collaboration, and adaptive decision-making frameworks to address these challenges, offering valuable insights into the complexities of disaster governance in ecologically vulnerable regions.

The 2018 Kerala floods, which severely impacted Mananthavady town in Wayanad, serve as a case study for evaluating disaster management strategies. Bigith V.B. et al. (2024) highlights the role of government initiatives and local organizations in supporting recovery efforts. However, it also points to the need for more robust disaster management frameworks to mitigate the long-term socio-economic impacts of such events, through deeper understanding of disaster resilience and community-centric approaches in disaster recovery.

Further insights into community resilience are provided by Mohan et al. (2024), who investigates the impact of the 2018 Kerala floods on local businesses. They found that businesses with strong organizational structures and adaptive capacities were better equipped to withstand the disaster. It highlights the importance of sustainable development practices to support small and medium enterprises (SMEs) in building resilience, highlighting the broader implications for economic stability in disaster-prone regions.

Importantly, addressing these gaps requires a multi-faceted approach that integrates stakeholder collaboration, institutional reforms, and adaptive decision-making frameworks. By fostering resilience at both the institutional and community levels, regions like Wayanad can better navigate the complexities of disaster recovery and build a more sustainable future.

PSYCHOLOGICAL IMPACT OF NATURAL DISASTERS

This section delves into the psychological impact of natural disasters in Wayanad District, expanding upon their immediate consequences, long-term effects, and affected regions. How natural disasters like floods, landslides, and droughts affect the psyche can, in fact, threaten people's and the community's lives. The cooler climes of Wayanad, situated in the State of Kerala, India, suffer the same calamities, affecting the mental well-being of the people inside.

Short-Term or Immediate Effects

The immediate effects of natural disasters on affected populations, particularly for vulnerable people, are often severe and debilitating, manifesting in a range of acute mental health reactions. Survivors frequently experience shock and denial in the initial phase, characterized by emotional numbness and disbelief, which serve as psychological defense mechanisms to cope with the overwhelming reality of the event. The aftermath is also marked by heightened anxiety, pervasive fear, and hypervigilance, as individuals remain in a state of constant alertness, anticipating further threats (Jones & Patel, 2019). Additionally, the loss of property, livelihoods, and loved ones often triggers profound grief and despair, compounding emotional distress (Brown et al., 2021). Furthermore, exposure to traumatic events during disasters can lead to the development of post-traumatic stress disorder (PTSD) in vulnerable individuals, with symptoms such as intrusive memories and flashbacks emerging shortly after the event. These immediate mental health impacts highlight the urgent need for timely psychological interventions and support systems to mitigate long-term consequences.

Long-Term Effects

The long-term effects of natural disasters on mental health are profound and multifaceted, often leading to severe psychological consequences, including post-traumatic stress disorder (PTSD), anxiety, depression, substance abuse, and relationship strain. PTSD, characterized by symptoms like flashbacks, nightmares, and avoidance of trauma-related triggers, is a common outcome of such events, significantly impairing individuals' daily functioning (American Psychiatric Association, 2013). Additionally, the stress and trauma associated with disasters exacerbate the development of anxiety and depression, which can become chronic if unaddressed. Some survivors may turn to substance abuse as a maladaptive coping mechanism, further compounding their mental health challenges (Jones & Patel, 2019). Furthermore, the disruption caused by natural disasters often strains interpersonal relationships, leading to increased conflicts, reduced communication, and a breakdown in social cohesion (Smith et al., 2020). These interconnected issues highlight the need for comprehensive and sustained mental health interventions to address the enduring impact of natural disasters on affected populations.

Vulnerable Groups and Disasters

Natural disasters disproportionately affect populations due to their heightened psychological vulnerability, including children and adolescents, women, older adults, and individuals with pre-existing mental health conditions. Children and adolescents are particularly susceptible due to their limited access to resources, dependence on adults, and reduced capacity to exert control during disasters, which can exacerbate feelings of helplessness and anxiety (Peek, 2008). Women face unique psychological challenges, often linked to their caregiving roles and increased risk of domestic violence in post-disaster settings, intensifying their stress levels (Enarson, 2012). Older adults experience compounded psychological pressure due to age-related health deterioration and social isolation, which hinder their ability to cope effectively with disaster-related stressors (Fernandez et al., 2015). Additionally, individuals with pre-existing mental health conditions, such as anxiety disorders or severe mental illnesses, are at heightened risk of psychological instability, as their susceptibility to stress and trauma is often amplified, leading to exacerbated symptoms or relapse (North & Pfefferbaum, 2013). Addressing the needs of these vulnerable groups requires targeted interventions that consider their unique challenges and vulnerabilities.

REHABILITATION STRATEGIES

Wayanad District is frequently afflicted by natural disasters like floods, landslides, and droughts. Despite these challenges, the local people has developed robust coping mechanisms to mitigate the impacts of such calamities. This section examines the rehabilitation strategies employed in Wayanad District, focusing on individual coping mechanisms and the resilience of the community. These strategies highlight the interplay between individual agency and collective action in disaster management.

Individual Strategies

The residents of Wayanad District face recurrent natural disasters, necessitating the adoption of diverse individual strategies to enhance their preparedness and adaptability. These strategies include risk assessment, economic diversification, social support networks, and stress management techniques.

Risk Assessment and Preparedness: A significant proportion of the population in Wayanad District engages in proactive risk assessment and preparedness measures. These include assembling emergency kits, formulating evacuation plans, and staying informed through weather forecasts (Smith et al., 2020). Such practices enable individuals to respond swiftly and effectively during disasters, minimizing potential losses.

Economic Diversification: To reduce vulnerability, many residents have diversified their economic activities. Instead of relying solely on agriculture, which is highly susceptible to climatic fluctuations, individuals engage in mixed livelihoods such as livestock rearing, small-scale businesses, and agro-based enterprises (Jones & Kumar, 2019). This diversification not only stabilizes household incomes but also enhances resilience against economic shocks caused by disasters.

Social Support Networks: Social support networks play a critical role in disaster recovery, particularly in resource-constrained settings. In Wayanad, families, friends, and community members provide emotional, psychological, and material support to those affected by disasters (Gupta & Nair, 2021). These networks are instrumental in fostering solidarity and collective recovery, particularly in the absence of formal institutional support.

Coping Skills and Stress Management: The psychological toll of natural disasters is significant, and residents of Wayanad employ various stress management techniques to cope with trauma. Practices such as meditation, yoga, and breathing exercises are widely adopted to alleviate stress and promote mental well-being (Patel et al., 2018). These techniques not only aid in individual recovery but also contribute to the overall resilience of the community.

Building Community Resilience

Community resilience is a cornerstone of disaster management in Wayanad District. It is manifested through community-based disaster risk reduction initiatives, strong social capital, traditional knowledge, and collaborations with external agencies.

Community-Based Disaster Risk Reduction and Management: Wayanad District has established community-based disaster risk reduction (CBDRR) committees that play a pivotal role in disaster preparedness and response. These committees conduct training sessions, raise awareness, and coordinate emergency responses, ensuring that communities are well-equipped to handle disasters (Shaw & Sharma, 2020). Such grassroots initiatives empower local populations and enhance their capacity to manage risks effectively.

Social Capital and Community Networks: The strong social fabric of Wayanad District facilitates collective action during crises. Social capital, characterized by trust, reciprocity, and shared norms, enables communities to mobilize resources and support one another in times of need (Putnam, 2000). This cohesion is particularly evident in the formation of self-help groups and community networks that provide immediate assistance during disasters.

Traditional Knowledge and Practices: Indigenous communities in Wayanad possess traditional knowledge that has been honed over generations to adapt to environmental challenges. For instance, their agricultural practices are designed to withstand both floods and droughts, ensuring food security even in adverse conditions (Fernandes & Thampi, 2017). This knowledge, often overlooked in modern disaster management frameworks, offers valuable insights into sustainable resilience-building.

Collaboration with Government Agencies and NGOs: Local communities in Wayanad actively collaborate with government agencies and non-governmental organizations (NGOs) to access resources, expertise, and logistical support. These partnerships enhance the effectiveness of disaster response and recovery efforts, bridging the gap between local initiatives and institutional frameworks (Kumar & Singh, 2022).

The rehabilitation strategies employed in Wayanad District underscore the importance of integrating individual and community-level approaches to disaster management. While individual strategies focus on preparedness and adaptability, community resilience is built through collective action, traditional knowledge, and external collaborations.

RESPONSE STRATIES

The Government of Kerala has demonstrated a proactive approach to addressing mental health challenges in disaster-affected communities by initiating a comprehensive mental health and psychosocial project, *PARIRAKSHA*. This initiative aims to achieve three primary objectives: early identification of individuals with mental health issues, provision of appropriate care within local communities, and the implementation of a structured stepped-care approach (Kerala Government, n.d.). These objectives align with global best practices in disaster mental health, emphasizing the importance of community-based interventions and early intervention to mitigate long-term psychological trauma (World Health Organization, 2013). To operationalize these goals, the Kerala government has implemented a multi-faceted framework of policies and programs. For instance, the *District Mental Health Programme* (DMHP) provides psychological first aid, counselling, and long-term mental health support to survivors, while the *DISHA Helpline* (1056) offers 24/7 post-traumatic psychological support during crises (Kerala State Disaster Management Authority, n.d.). These initiatives reflect a commitment to integrating mental health care into disaster response, a critical step often overlooked in traditional disaster management frameworks (Math et al., 2015).

Furthermore, the Kerala State Disaster Management Authority (KSDMA) has introduced specialized initiatives such as *Psychosocial Support Teams* (PSS), which deploy social workers, psychologists, and trained volunteers to disaster-affected areas for immediate trauma care and counselling. The *School Mental Health Program* prioritizes psychological support for children, recognizing their heightened vulnerability to post-disaster trauma (KSDMA, n.d.). Additionally, training and capacity-building programs for first responders, ASHA workers, and community health workers ensure that mental health support is accessible at the grassroots level. These efforts are complemented by community-based interventions led by the *Kudumbashree Women's Network*, which engages women in disaster-affected areas to promote mental well-being and resilience through self-help groups and disaster resilience training (Kudumbashree, n.d.). Such community-driven approaches are critical for fostering long-term psychological recovery and social cohesion (Norris et al., 2008).

Rehabilitation and long-term psychological support are also central to Kerala's disaster mental health strategy. Programs focusing on livelihood support aim to address economic distress, a significant contributor to post-disaster trauma, while specialized interventions, such as those following the 2018 Kerala floods, highlight the government's collaboration with NGOs and international organizations to provide large-scale, sustained counselling support (KSDMA, n.d.). These initiatives underscore the importance of integrating mental health care into broader disaster recovery and rehabilitation efforts, ensuring that psychological well-being is prioritized alongside physical and economic recovery (WHO, 2013). Overall, Kerala's policy initiatives serve as a model for other regions, demonstrating the critical role of structured, community-centered mental health interventions in disaster management.

CONCLUSION

Natural disasters greatly affect the lives of individuals, communities, and societies, not only concerning the physical environment but also considering the psychological, social, and emotional wellbeing of the affected. This chapter highlights the necessity of addressing such complex consequences of disasters, including psychological trauma, social disintegration, and environmental degradation. It highlights the need for a comprehensive, multifaceted approach that integrates immediate disaster response, short-term restoration, long-term reconstruction, psychological support for victims, and environmental amelioration. Following these recommendations can help avert the consequent disasters, enhance resilience, and encourage sustainable development, which together act as a buffer against the occurrence of future disasters and promote a culture of disaster preparedness and risk reduction. In addition, disaster management would be more inclusive and effective, as well as sustainable, by giving priority to the needs of vulnerable groups such as children, women, and the elderly while promoting community-based initiatives and community participation. It is a priori and multifaceted that disaster management is essential to mitigate disaster impact and achieve a safe and sustainable world for everyone.

REFERENCES

American Psychiatric Association. (2013). *Diagnostic and Statistical Manual of Mental Disorders* (5th ed.). American Psychiatric Publishing.

Asim, M., Mekkodathil, A., Sathian, B., Elayedath, R., Kumar, R. N., Simkhada, P., & van Teijlingen, E. (2019). Post-traumatic stress disorder among the flood-affected population in the Indian subcontinent. *Nepal Journal of Epidemiology*, *9*(1), 755–758. DOI: 10.3126/nje.v9i1.24003 PMID: 31210997

Azad, P., Muhammed Salim, A. P., & Anoop Das, K. S. (2024). *Socioeconomic and ecological impact of Wayanad landslides 2024: Lessons for future resilience in Kerala.*

Bigith, G., Sooraj, Rajat Nainwal, & Kumara, H. S. (2024). *Assessing the impact and response to the disaster in Mananthavady town: A case study. IPO Conference Series: Earth and Environment Science*. DOI: 10.1088/1755-1315/1409/1/012023

Brown, L., Green, R., & Taylor, S. (2021). Mental Health and Natural Disasters: A Comprehensive Review. *Journal of Traumatic Stress*, *34*(2), 123–135.

Doocy, S., Daniels, A., Packer, C., Dick, A., & Kirsch, T. D. (2013). The human impact of floods: A historical review of events and systemic response. *Prehospital and Disaster Medicine*, *28*(1), 1–9. PMID: 23158909

Enarson, E. (2012). *Women and disasters: A gender analysis*. International Federation of Red Cross and Red Crescent Societies.

Fernandes, S., & Thampi, K. (2017). *Traditional Knowledge and Disaster Resilience: Lessons from Kerala*. Sage Publications.

Fernandez, J., Black, M., Jones, L., Wilson, L., Salvador-Carulla, T., Astell-Burt, D., & Black, D. (2015). Flooding and mental health: A systematic mapping review. *PLoS One*, *10*(4), e0119929. DOI: 10.1371/journal.pone.0119929 PMID: 25860572

Fernandez, L. S., Byard, D., Lin, C. C., Benson, S., & Barbera, J. A. (2015). Frail elderly as disaster victims: Emergency management strategies. *Prehospital and Disaster Medicine*, *17*(2), 67–74. DOI: 10.1017/S1049023X00000200 PMID: 12500729

George, B., Kumar, R., & Banerjee, S. (2022). Vulnerability of women in the face of climate change: A study of Wayanad District of Kerala, India. *Journal of Chinese Economic and Foreign Trade Studies*, *15*(3), 279–297. DOI: 10.1108/JCEFTS-09-2021-0058

Goldmann, S., & Galea, S. (2014). Mental health consequences of disasters. *Annual Review of Public Health, 35*(1), 169–183. DOI: 10.1146/annurev-publhealth-032013-182435 PMID: 24159920

Government of Kerala. (2024). *Report on drought and landslide vulnerability in Wayanad district.*

Gupta, R., & Nair, S. (2021). Social Support Networks in Disaster Recovery: Evidence from India. *Journal of Disaster Studies, 15*(3), 45–60.

Hameed, S., Sadiq, A., & Din, A. (2019). The Increased Vulnerability of Refugee Population to Mental Health Disorders. *Kansas Journal of Medicine, 11*(1), 1–12. DOI: 10.17161/kjm.v11i1.8680 PMID: 29844851

Heanoy, E. Z., & Brown, N. R. (2024). Impact of natural disasters on mental health: Evidence and implications. *Health Care, 12*(18), 1812. PMID: 39337153

Jones, A., & Patel, K. (2019). Anxiety and Fear in Disaster Survivors: A Meta-Analysis. *Psychological Trauma: Theory, Research, Practice, and Policy, 11*(4), 456–464.

Jones, P., & Kumar, R. (2019). Economic Diversification as a Strategy for Disaster Resilience. *The International Journal of Environmental Studies, 76*(4), 512–528.

Jose, P. J. (2024, August 17). *Wayanad landslides: When nature rewrote the map. Frontline Magazine.* Retrieved from https://frontline.thehindu.com/environment/wayanad-mundakkai-chooralmala-landslides-army-bodies-kerala-climate-change-impact-gadgil-western ghats/article68528102.ece

Joy, J. L. K., Ramachandran, M., & George, S. (2021). Learned helplessness, psychological wellbeing, and pro-environment care behaviour among victims of frequent floods in Kerala. *Journal of Neurosciences in Rural Practice, 12*(1), 137–144. DOI: 10.1055/s-0040-1721566 PMID: 33531773

Kerala Government. (n.d.). *Pariraksha: Comprehensive mental health and psychosocial project.* Retrieved from https://www.kerala.gov.in/pariraksha

Kerala State Disaster Management Authority. (n.d.). *DISHA Helpline (1056): 24/7 post-traumatic psychological support.* Retrieved from https://disha1056.com/

Kerala State Disaster Management Authority. (n.d.). *Psychosocial Support Teams (PSS) and School Mental Health Program.* https://sdma.kerala.gov.in/psychosocial-support/

Kessler, R. C., Sonnega, A., Bromet, E., Hughes, M., & Nelson, C. B. (2013). Posttraumatic stress disorder in the National Comorbidity Survey. In *Fear and anxiety* (pp. 22–34). Routledge.

Koshy, M., Aranya, R., & Refstie, H. (2022). Handling compounded uncertainty in spatial planning and humanitarian action in unexpected floods in Wayanad, Kerala: Towards a contextualized contingency planning approach. *Planning Theory & Practice*, *23*(1), 33–51. DOI: 10.1080/14649357.2022.2143548

Koshy, M., & Smith, D. (2024). Community resilience implications for institutional response under uncertainty. *Sustainable Development (Bradford)*, *32*(2), 1412–1427. DOI: 10.1002/sd.2678

Kudumbashree. (n.d.). *Community-based interventions for mental well-being and resilience*. Retrieved from https://www.kudumbashree.org/

Kumar, A., & Singh, V. (2022). Collaborative Disaster Management: Role of NGOs and Government Agencies. *Disaster Prevention and Management*, *31*(2), 89–104.

Math, S. B., Nirmala, M. C., Moirangthem, S., & Kumar, N. C. (2015). Disaster management: Mental health perspective. *Indian Journal of Psychological Medicine*, *37*(3), 261–271. DOI: 10.4103/0253-7176.162915 PMID: 26664073

Mathew, M. (2019). *Environmental degradation and disaster in Wayanad: A study on the role of corporate interests and governmental measures.*

Mohan, C., & Joy, R. (n.d.). Entrepreneurs' resilience to natural disasters: Emerging entrepreneurs from flood-affected areas in Kerala. *The Empirical Economic Letters*, *20*(3), 33–41.

Neria, Y., Nandi, A., & Galea, S. (2008). Post-traumatic stress disorder following disasters: A systematic review. *Psychological Medicine*, *38*(4), 467–480. DOI: 10.1017/S0033291707001353 PMID: 17803838

Norris, F. H., Stevens, S. P., Pfefferbaum, B., Wyche, K. F., & Pfefferbaum, R. L. (2008). Community resilience as a metaphor, theory, set of capacities, and strategy for disaster readiness. *American Journal of Community Psychology*, *41*(1-2), 127–150. DOI: 10.1007/s10464-007-9156-6 PMID: 18157631

North, C. S., & Pfefferbaum, B. (2013). Mental health response to community disasters: A systematic review. *Journal of the American Medical Association*, *310*(5), 507–518. DOI: 10.1001/jama.2013.107799 PMID: 23925621

Patel, S. (2018). Stress Management Techniques in Disaster-Affected Populations. *Journal of Mental Health and Resilience*, *10*(1), 22–35.

Peek, L. (2008). Children and disasters: Understanding vulnerability, developing capacities, and promoting resilience. *Children, Youth and Environments*, *18*(1), 1–29. DOI: 10.1353/cye.2008.0052

Putnam, R. (2000). *Bowling Alone: The Collapse and Revival of American Community*. Simon & Schuster.

Shaji, K. A. (2024, September 6). *A month after India's deadliest landslide ever, Wayanad villages begin to recover. Down To Earth*. Retrieved from https://www.downtoearth.org.in/natural-disasters/a-month-after-indias-deadliest-landslide-ever-wayanad-villages-begin-to-recover

Shaw, R., & Sharma, A. (2020). *Community-Based Disaster Risk Reduction*. Springer.

Smith, J. (2020). Risk Assessment and Preparedness in Disaster-Prone Regions. *Disaster Resilience Journal*, *8*(2), 112–130.

Smith, J., White, T., & Lee, M. (2020). Psychological Responses to Natural Disasters: A Cross-Cultural Study. *International Journal of Disaster Risk Reduction*, *45*, 101–112.

Sphere India. (2021). *Preliminary Assessment Report – Wayanad Landslides, Kerala*. Retrieved from https://www.sphereindia.org.in/sites/default/files/2024-08/Wayanad%20Preliminary%20Assessment%20Report%20Aug%202024.pdf?utm_source=chatgpt.com

Stanke, V. Murray, R. Amlot, J. Nurse, R. Williams (2012). *The effects of flooding on mental health: outcomes and recommendations from a review of the literature*, PLoS Curr 4 e4f9f1fa9c3cae.

Tanaka, E., Shibata, H., Sugiura, K., Hoshino, N., & Nishikawa, J. (2016). Long-term psychological effects of the Wenchuan earthquake on adolescent survivors: A six-year follow-up study. *Journal of Affective Disorders*, *204*, 342–348002E. DOI: 10.1016/j.jad.2016.08.001

Thoresen, S., Birkeland, M. S., Arnberg, F. K., Wentzel-Larsen, T., & Blix, I. (2019). Long-term mental health and social support in victims of disaster: Comparison with a general population sample. *BJPsych Open*, *5*(1), e2. DOI: 10.1192/bjo.2018.74 PMID: 30762498

Turkey, N. A., & Jawahar Saud, S. (2024, June). Assessment of Governance Gaps in Landslide Recovery and Post Recovery Phase Using Analytical Hierarchy Process (AHP)—A Case Study from Wayanad District, Kerala, India. *In International Conference on Structural Engineering and Construction Management* (pp. 483-499). Cham: Springer Nature Switzerland. DOI: 10.1007/978-3-031-70431-4_36

World Health Organization. (2013). *Building back better: Sustainable mental health care after emergencies*. Retrieved from https://www.who.int/publications/i/item/9789241564571

World Health Organization. (2019). *Mental Health in Emergencies*. World Health Organization.

Chapter 11
Social, Psychological, Motivational, and Educational Effects of 6 February Earthquakes:
A Qualitative Study on Teachers

Ceyda Akilli
https://orcid.org/0000-0003-1928-1124
Fırat University, Turkey

ABSTRACT

As a result of two earthquakes with magnitudes of 7.7 Mw and 7.6 Mw in Kahramanmaraş on 6 February 2023, 11 cities in Turkey were severely damaged, and over 50,000 citizens lost their lives. This study aims to explore the social, psychological, and motivational effects of these earthquakes on teachers and their impact on educational activities. A qualitative phenomenological design was used, with 46 teachers from Elazığ, Malatya, and Diyarbakır, areas declared disaster zones, participating. Data was collected using a semi-structured interview form and analyzed through descriptive and content analysis. Results revealed that the earthquakes negatively impacted teachers socially, psychologically, and motivationally, leading to trauma-related disorders, phobias, and reduced professional motivation. Some female teachers, however, noted increased solidarity and empathy. The findings also showed disruptions in educational quality, decreased academic performance, increased absenteeism, and a negative shift in school climate.

DOI: 10.4018/979-8-3693-9745-9.ch011

1. INTRODUCTION

Natural disasters are unpredictable events that occur suddenly, have multidimensional effects, and cannot be prevented by humans (Hutton et al., 2011). Disasters have varying effects based on their type and strength. In addition to physical damages, they can cause psychological and social traumas as well as physical damages since their timing cannot be predicted in advance (Raschky, 2008). Earthquake, as a type of natural disaster, can be physically very destructive in proportion to its severity and its effects can spread over a long period of time although they last for a short time. Earthquakes, one of the most destructive types of natural disasters, are characterized by their sudden onset and brief duration, yet their physical destruction can be immense, particularly in areas with high seismic activity (Bilham, 2004). The degree of destruction caused by an earthquake is closely related to its magnitude, depth, and the population density of the affected region (Kanamori & Brodsky, 2004). In addition to immediate physical damage to infrastructure, homes, and roads, earthquakes can lead to long-term disruptions, including displacement of communities, loss of life, and economic instability (Olshansky et al., 2012). The aftershocks that often follow a major earthquake can further exacerbate the damage and prolong recovery efforts, making earthquakes particularly challenging in terms of both immediate disaster response and long-term recovery and resilience planning (Reid, 1911).

Earthquakes cause negative effects on both living and non-living beings. These effects may be in the form of damage to the lives, belongings, as well as short and long term psychological problems as a result of the earthquake being placed in people's subconscious (Köknel, 1987; Watt, 2002; Öztürk, 2013). Earthquake, which is a commonly experienced natural disaster, brings out feelings of shock, confusion, anger, helplessness, powerlessness, loss of confidence, loss of control and fear of death. The homes of people, which mean the safest place for everyone, no longer have this meaning and start to carry a meaning based on fear, anxiety and insecurity (Sivri, 2023). Therefore, earthquakes have short, medium and long-term effects on people and the environment.

Since Turkey is prone to earthquakes, it has experienced this disaster many times in the historical process. As a result of two earthquakes with a magnitude of 7.7 Mw and 7.6 Mw in Kahramanmaraş on 6 February 2023, 11 cities were severely damaged and more than 50 thousand citizens died. As a result of the earthquakes, many families experienced loss of life and property (Abay & Abay Çelik, 2023), forced internal migration was carried out, and psychological traumas were faced (Aydın & Özüdoğru, 2023). As a result of the earthquakes, the education sector was also damaged to a great extent, and face-to-face education was temporarily terminated and the distance education process was initiated. After the Kahramanmaraş earthquakes, 1,842 institutions were completely destroyed, 637 institutions were moderately damaged and 17,951 institutions were slightly damaged (Ateşgöz, 2023). Education was suspended for 3 to 7 weeks in 11 provinces that were declared as disaster zones.

It is seen that the earthquake has multidimensional effects that will be reflected on the educational processes in the affected regions both during and after the earthquake. It is inevitable that the educational structures damaged during the earthquake, as well as the educational staff and students, will be psychologically affected by the effects of the earthquake and its aftermath. The interruption of face-to-face education due to the damage caused after the earthquake, the housing problems experienced by teachers and students, the traumas experienced by those who lost their relatives in the earthquake and the ongoing effects are reflected in the functioning of educational processes. Educators and students had to continue their educational activities with the feelings of fear and sadness caused by the earthquake. In a study conducted by Arslan (2023) on school administrators, it was concluded that after the earthquake, absenteeism problems increased significantly in schools. This rise in absenteeism was attributed to multiple factors, including the displacement of families and the psychological trauma experienced by students and staff. The study found that students' motivation to engage in learning also decreased sharply due to the disruption of normal routines, emotional distress, and the overall uncertainty caused by the disaster (Arslan, 2023). This aligns with previous research showing that natural disasters often lead to a decline in student performance and school attendance, as psychological factors such as trauma and anxiety can hinder concentration and motivation (Peek, 2008; Kousky, 2016). Furthermore, Arslan's (2023) study indicated that students' fear and anxiety increased substantially following the earthquake. This finding is supported by other research that demonstrates how exposure to natural disasters can heighten anxiety levels and contribute to post-traumatic stress disorder (PTSD) symptoms in children and adolescents (La Greca et al., 2010). Teachers, too, were deeply affected by the earthquake, with many reporting difficulty in returning to their professional responsibilities while managing their own personal losses and trauma. Previous studies corroborate this, emphasizing that educators in disaster-stricken areas often experience burnout, emotional exhaustion, and difficulty in providing emotional support to students when they themselves are grappling with the aftermath of a disaster (Kataoka et al., 2009). Another consequence identified in the study was the movement of students to other cities as families sought safer environments or relocated due to housing damage. This out-migration of students, combined with an influx of new students transferring from other schools, caused additional challenges for educational continuity. Schools had to quickly adapt to fluctuating student numbers, with administrative burdens increasing and the need for psychological support services intensifying (Arslan, 2023). Such population shifts are consistent with disaster studies that highlight the long-term impacts of displacement and migration on school systems, often resulting in overcrowding, teacher shortages, and strained resources (Masten & Narayan, 2012).

In order to evaluate the impact of the earthquake on teachers and educational activities, it is essential to collect data directly from those affected. First-hand data from teachers who experienced the disaster can provide valuable insights into the psychological, social, and educational challenges that emerged in the aftermath. However, despite the significant time that has passed since the 6 February earthquakes, research on this topic remains limited. While over a year has elapsed, most of the existing studies focus primarily on the economic and financial impacts of the disaster (Tetik & Akbulut, 2023; Dündar, 2023), its social and cultural dimensions (Erdoğan, 2023), and the media's portrayal of the event (Maden, 2023; Ayan & Keten, 2023). This emphasis on economic and social aspects has left a gap in understanding the profound impact of the earthquake on the education sector, particularly on teachers and the day-to-day functioning of schools. Research into how educational organizations and teachers were affected is crucial for developing comprehensive recovery strategies. Past studies on disasters emphasize that schools play a vital role in community recovery and in providing a sense of normalcy for children, but this role is often compromised when teachers themselves are struggling with the psychological and emotional aftermath of the event (Masten & Osofsky, 2010; Peek, 2008). There is a pressing need for more research that specifically addresses the effects of earthquakes on educational organizations, including the psychological well-being of teachers and the disruptions to educational activities. Collecting data from teachers working in the earthquake region would offer essential insights into how the disaster affected both teaching staff and students. Teachers' perspectives are particularly valuable, as they are at the frontline of the education system and can provide detailed accounts of how the earthquake disrupted learning environments, affected student performance and attendance, and increased their own stress and burnout levels (Kataoka et al., 2009).

In this study, the aim is to determine the psychological, social, motivational, and educational effects of the earthquake from the perspective of teachers. This approach will not only contribute to filling the research gap but will also inform future policy decisions aimed at supporting teachers and educational institutions in disaster-prone areas. Understanding teachers' experiences will shed light on both the immediate and long-term needs of the education sector following such disasters, providing a foundation for more effective disaster management and resilience strategies in schools (Masten & Narayan, 2012).

1.1 Purpose of the Study

The aim of this study is to determine the effects of the 6 February earthquakes on teachers and educational activities. In line with this general purpose, answers to the following questions will be sought.

1. What are the social and psychological effects of the 6 February earthquakes and their aftermath on teachers?
2. How did the 6 February earthquakes and its aftermath affect teachers' work motivation?
3. According to teachers' views, what are the reflections of the 6 February earthquakes and their aftermath on education activities?

2. METHOD

Qualitative research method was used in the study. According to Marrow & Smith (2000), qualitative research is used to explain and understand the meanings created by the participants of the activity or content. The aim of qualitative research is to understand how people make sense of the world they live in and to identify the main aspects of this meaning-making process (Merriam, 2013). In this study, it is aimed to examine the impact of the earthquakes of 6 February 2023 on the educational activities in Elazığ, Malatya and Diyarbakır, which are the cities declared as disaster zones after the earthquakes of 6 February 2023, from the perspective of teachers. Therefore, qualitative research method was adopted in the study.

2.1. Research Design

This research is a study conducted in the phenomenological design, which is included in qualitative research methods. Phenomenology aims to collect data by centring on phenomena that we are aware of but do not have an in-depth and detailed understanding and focuses on the awareness of all aspects of phenomena (Yıldırım & Şimşek, 2016). In addition, although phenomenology has many application forms, its common point is that it prioritises subjective experience (Kocabıyık, 2016). In the study, the phenomenological design was preferred since it was aimed to determine the effects of earthquakes on teachers and educational activities based on the subjective experiences of teachers in the 6 February earthquakes.

2.2. Study Group

In this study, criterion sampling, one of the purposive sampling methods, was preferred. In criterion sampling, attention is paid to the fact that the sampled people have certain qualifications (Büyüköztürk, et al., 2014). Two criteria were taken into consideration when determining the teachers to be included in the study group. The first one is that the teachers in the study group work in one of the cities declared as the disaster area after the earthquakes of 6 February, and the second one is that the teachers who experienced 6 February earthquakes. According to these criteria, data were collected from a total of 46 teachers working in Elazığ (f=21), Malatya (f=12) and Diyarbakır (f=13). Of the teachers in the study group, 24 were female and 22 were male. 7 of them work in kindergarten, 22 in primary school, 9 in secondary school and 8 in high school. Their professional experience ranged between 1 and 27 years (16 of them 1-5 years, 13 of them 6-10 years, 17 of them 11 years or more).

2.3. Data Collection Tool

In the study, a semi-structured interview form was used to collect data from teachers. The interview form includes questions to determine the teachers' experiences during and after the 6 February earthquakes. In the process of preparing the semi-structured interview form, firstly, a literature review was conducted. Then, a question list consisting of the questions that were thought to be included in the interview form was prepared. The candidate questions in this list were submitted to the opinions of three field experts working in the Department of Educational Sciences, Faculty of Education, Fırat University in terms of the scope of the research and the clarity and comprehensibility of the questions. In line with the expert opinions, a semi-structured interview form consisting of 3 questions was formed. The questions in the semi-structured interview form are given below.

1. How did the earthquakes of 6 February 2023 and its aftermath affect you socially and psychologically? Please explain.
2. How did the 6 February 2023 earthquakes and its aftermath affect your motivation to work? Please explain.
3. What are your feelings and thoughts about the reflections of the 6 February 2023 earthquakes and their aftermath on the education and training process in schools? Please explain.

2.4. Data Collection

During the data collection process, the purpose of the research was shared with the teachers in the study group and the "Informed Voluntary Consent Form" was shared with the participants to ensure that the research data would not be used for any purpose contrary to scientific publication ethics, that the personal information of the participants would never be shared, that participation was voluntary and that they could withdraw from the interview if they wished. The semi-structured interview form was shared online with the participants who agreed to participate and they were asked to answer the questions via the online form. In this context, a total of 46 teachers responded. In order to ensure the confidentiality of the participants, each participant was given a code number to be used in the analysis and reporting process.

2.5. Data Analysis

Both descriptive analysis and content analysis methods were used to analyse the data. According to Yıldırım and Şimşek (2016), descriptive analysis is used to process data that do not require in-depth analysis, while content analysis requires closer and in-depth examination of the data obtained and reaching concepts and themes that explain these data. In descriptive analysis, data can be organised according to predetermined themes, and a general framework for data analysis can be created based on the research interview questions. In content analysis, the main purpose is to reach concepts that can explain the data obtained. In order to ensure the confidentiality of the participants during the data analysis process, they were assigned pseudonyms as P-1, P-2 etc.

2.6. Validity and Reliability

The best way to ensure validity and reliability in qualitative research is to provide diversity of data. In addition to ensuring validity and reliability, data triangulation also supports the generalisability of research results (Creswell & Miller, 2000; Yıldırım & Şimşek, 2016). In addition to data triangulation, method triangulation, diversification of data sources, analysing data by different researchers and analysing data based on different theories are also approaches used to ensure validity and reliability (Patton, 2014). In addition, including detailed descriptions and direct quotations frequently in the analysis and reporting process contributes to reliability (Creswell & Miller, 2000). In this study, data diversification was provided by including teachers working in cities affected by the 6 February earthquakes at different levels in the study group. In addition, care was taken to ensure the validity and reliability criteria by analysing the data obtained from the study completely and supporting them with direct quotations. In addition, the data were coded by both researchers and the consistency of the coding was checked.

3. FINDINGS

3.1. Effects of the 6 February Earthquakes on Teacher Motivation

The first question asked to the teachers within the scope of the research was "How did the earthquakes of 6 February 2023 and its aftermath affect your motivation to work? Please explain.". When the responses of the teachers in the study group were analysed, it was found that most of the teachers' motivation was negatively affected by the earthquake; however, some teachers stated that the earthquake did not affect their motivation or their motivation increased. The effect of the earthquake on teachers' motivation is given in Table 1.

Table 1. The effect of 6 February earthquakes on teacher motivation

Category	f	%
Negatively Affected	39	84,78
Not impressed	5	10,87
Positively Impressed	2	4,35
Total	46	100

As seen in Table 1, a great majority of the teachers stated that the earthquake affected their motivation negatively (f=39). However, some teachers stated that the earthquake did not affect their motivation (f=5) and some teachers stated that they were positively affected (f=2). One of the striking results of the study is that almost all of the participants who stated that they were not affected or positively affected by the earthquake were female teachers (f=6).

> *"Since education was suspended in schools after the earthquake, my motivation to work was not affected since I started school after the break. I acted with the awareness that I had a duty to increase student motivation and achievement, which decreased after the earthquake and the break in education." (P-13)*

After the 6 February earthquakes, education was suspended for at least 1 month in 11 cities declared as disaster areas. In cities with high destruction rates, schools were closed for nearly 3 months. The teacher with the pseudonym P-13 stated that since the motivation of the students decreased as a result of the interruption of education after the earthquake, her motivation was not affected in order to adapt the students to the education process and to be a role-model for the students with his strong stance.

"I had to act more logically because I was working with students. I have to motivate the students, so my motivation to work was not affected." (P-28)

Similarly, the teacher with the pseudonym P-28 stated that she did not allow the earthquake to affect her own motivation in order to keep the motivation and psychology of the students strong.

"It affected my motivation positively. I wanted to help people." (P-14)

"My motivation to work increased more. I wanted to help people, each human salvation was a high motivation for us." (P-4)

The teachers with the pseudonyms P-4 and P-14 stated that helping people after the earthquake positively affected their motivation for their job. When motivation theories are analysed, it is known that the factors that increase or decrease the motivation of each individual are different from each other (Bogg & Roberts, 2004). After the 6 February earthquakes, teachers who stated that their motivation increased or did not change generally stated that they experienced this situation due to their sense of duty or the responsibility of being a model for students. In addition, the participants stated that the feeling of helping earthquake victims or students provided motivation for their profession. It is thought that the fact that the majority of the participants who responded in this category are women may be due to the fact that activities such as helping and being a role model are generally associated with womanhood and motherhood.

"After the earthquakes; I tried to minimize the psychological dimension of the situation my students experienced with activities appropriate to their level. Of course, while doing this, I tried not to reflect my own fears to my students. I tried to take my students away from the earthquake stories and bring them back to their social life at school. At the same time, I was interested in the orientation and psychological support process of the earthquake survivor student who came to my class. I organized activities such as acquaintance games and picnics so that my students could mingle with their new friend and support them. I organized an aid campaign with my students for earthquake victims in need and tried to relieve their anxiety by doing something and transferring them to the right place." (P-17)

The teacher with the pseudonym P-17 stated that he was actually in some fears and anxieties, but he suppressed these fears and anxieties for the psychological health of the students and did not reflect them to the students. After the earthquake, citizens living in cities with a high rate of destruction moved to cities with a lower rate of destruction, although they were located in the disaster zone. For this reason, the number of earthquake survivor students in the schools in the city where the research was conducted is quite high. The teacher with the pseudonym P-17 stated that she focused on this issue and carried out studies to ensure the adaptation of earthquake victims, organised aid campaigns to the earthquake region and developed a positive perspective towards her profession with these studies.

Almost 85% of the teachers in the study group stated that the earthquake on 6 February negatively affected their motivation. The results of the analyses on the reasons for the negative impact of the earthquake on their motivation are given in Table 2.

Table 2. Reasons for the negative effects of the earthquake on motivation

Theme	Category	f
Health	Psychological problems	7
	High pulse rate	1
	Heart rhythm disturbance	1
	Stress disorder	1
Negative Emotions	Fear	12
	Anxiety	5
	Lack of concentration	3
	Grief	2
	Questioning life	2
Aftershocks	Thinking that the school building is not earthquake resistant	4
	Don't think there will be new earthquakes	3
Management	Failures in the crisis management process	4

When Table 2 is examined, the reasons for the negative effects of the 6 February earthquakes on teachers' motivation towards work were divided into themes in the dimensions of health (f=10), negative emotions (f=24), aftershocks (f=7) and management (f=4).

The majority of the loss of motivation experienced in the health theme was determined as psychological problems (f=7). When the distribution of the responses in the category of psychological problems is examined, the frequencies of female (f=4) and male (f=3) teachers are quite close to each other.

> *"We had difficulty recovering psychologically for a long time. We felt like an earthquake was happening during the lessons." (P-1)*
>
> *"The earthquake feeling never went away. My pulse is constantly high. I have a heart rhythm disorder." (P-2)*
>
> *"The February 6 earthquake affected us deeply. It affected me and my family psychologically by this disaster that happened right next to us. Although more than a year has passed, even the word earthquake has affected our mood a lot. The mobilization of all citizens as a country after the earthquake and the mea-*

sures taken against reconstruction and psychological traumas made me a little happy....." (P-7)

As can be seen in the responses of the participants with the pseudonyms P-1, P-2 and P-7, the earthquakes negatively affected the psychological health of the teachers and this situation decreased their motivation towards their profession. Teachers faced problems such as heart rhythm disorder, high pulse rate and stress disorder. After the earthquakes, psycho-technical aids were provided to the individuals residing in the disaster zone. As emphasised by the participant with the pseudonym P-7, these supports were effective in reducing the psychological problems of individuals after the earthquake and distracting them from the negative emotions caused by the earthquake.

The most frequently cited category in the theme of negative emotions was fear (f=12). Teachers stated that they were in fear and anxiety due to the earthquakes.

"Both the students and us teachers could not be motivated to work for a long time because of this terrible process. We did not want to go to school, we did not want to do any work, our concentration was broken. Unfortunately, both the students and us teachers had the thoughts of "what does it matter if I work anymore?" in their lives for a long time." (P-19)

When the responses of the participant with the pseudonym P-19 are analysed, it is seen that earthquakes cause concentration disorder along with fear and individuals develop a critical perspective on the meaning of life. This situation emotionally distanced teachers and students from school for a while.

"It had a negative impact on work motivation. Because the destruction was so great, our minds were on our citizens who lost their lives in the earthquake. The university students, primary/secondary/secondary school students/colleagues etc. who lost their lives affected us deeply and we had difficulty in recovering our motivation for a while." (P-25)

After the 6 February earthquakes, around 53 thousand citizens lost their lives. The participant with the pseudonym P-25 stated that he was in mourning due to the deaths after the earthquake.

Teachers who experienced loss of motivation due to the earthquake stated that they generally reconnected to their profession with the happiness of being together with their students. While the teachers helped the students overcome this difficult process, they also healed their own wounds. The participant with the pseudonym P-44 expressed this situation in a striking way:

"After the earthquake, I had a hard time adapting to school, I didn't even want to get out of bed. But after I came to school and saw my students, I realized that I had to recover immediately and heal my children's wounds. They were going through great trauma inside. They were scared because they had experienced the earthquake first hand. Then I entertained them in the classroom and had fun

and we healed each other's wounds to some extent. May God never let it happen to our country again." (P-44)

After the 6 February earthquakes, more than a thousand aftershocks were experienced within a month. This situation is among the negative factors affecting the motivation of the participants. Participants who stated that they did not trust the school building after the aftershocks experienced loss of motivation (f=4). Similarly, there were participants who thought that there would be a new big earthquake after the aftershocks and therefore had anxiety (f=3).

"My motivation to work decreased. I feel uneasy because the school building is old. It is always in the back of my mind." (P-10)

All of the participant comments in the management theme are related to the lack of crisis management (f=4).

"The earthquake's devastating impact was caused not only by the natural disaster but also by many negligences. It was also caused by the negligence of those responsible from the bottom to the top who did not do their part well and their negligence. All these people were once educated in the same classrooms where we used to teach and have passed through these paths. But we realized that they were not sufficiently informed. It made us focus more effort and more sensitively on informing students about this issue in order not to experience the same things again." (P-11)

In Turkey, especially in recent years, compulsory trainings have been given to all institutions in the fields of disaster management, crisis management and first aid. However, as a result of the 6 February earthquakes, it is thought that these trainings are not sufficient for a catastrophic natural disaster.

3.2. Social and Psychological Effects of the 6 February Earthquakes

The second question of the research was "How did the earthquakes of 6 February 2023 and its aftermath affect you socially and psychologically?". With this question, the social and psychological effects experienced by the teachers after the earthquakes were tried to be examined in depth.

The social effects of the 6 February earthquakes on the participant teachers are as shown in Table 3:

Table 3. Social effects of the 6 February earthquakes

Theme	Category	f
Negative Effects	Anti-sociality	10
	Not wanting to be alone	8
	Avoiding crowds	2
Positive Effects	Solidarity	5
	Empathy	1

As seen in Table 3, the social effects of the 6 February earthquakes on teachers were analysed under two themes: negative (f=20) and positive (f=6) effects. Among the negative effects, the most frequently cited category was being anti-social, not wanting to be together with people and wanting to be alone (f=10).

"My socialization has decreased and there is always a state of uneasiness. The psychological low of the children affected me negatively." (P-10)

"Socially, I was more isolated. Psychologically, the thought of helplessness was dominant, albeit temporarily." (P-33)

"After February 6, there was a period in which there was inevitably a breakdown in communication between people in terms of social aspects, a weakening of human relations, and as a result, a psychological withdrawal and psychological loneliness emerged." (P-39)

As can be understood from the statements of the participants with the pseudonyms P-10, P-33 and P-39, the social lives of the teachers were damaged after the earthquakes of 6 February. When the statements of the participants are analysed, it is understood that the psychological process they experienced after the earthquakes pushed them to isolation and isolation from social life.

While there were participants who said that they isolated themselves from social life after the earthquakes, the number of participants who avoided being alone was also quite high (f=8).

".....socially, there was a panic situation. Not being able to stay in places for too long, not being able to be alone, being afraid to leave the children alone at home ... etc." (P-12)

"....I struggled for a long time with my fear of not being able to be on my own while continuing my profession. It took me a long time to return to my personal and social routines." (P-17)

The participants with the pseudonyms P-12 and P-17 stated that they were afraid of being alone after the earthquakes and that they were afraid of being alone.

The participants who stated that the 6 February earthquakes had positive effects on their social lives stated that their feelings of solidarity increased and their empathy skills strengthened in the post-earthquake process.

"We were in solidarity socially. It affected us positively. But our psychology deteriorated because of our loss.... I always sat at the table with sadness. I felt like I was under rubble while sleeping. It took me a long time to get out of this psychology. Although our psychology seems to be fine now, it will leave a mark on me for the rest of my life." (P-25)

"I learned that we can cope with situations where we need to act together socially. I think it also taught me to empathize." (P-42)

As can be understood from the statements of the participants above, after a major natural disaster such as the 6 February earthquakes, some individuals distance themselves from social life and live an introverted life, while others strengthen their feelings of unity and solidarity and mobilise. When the demographic variables of the participants are analysed, male teachers are the ones who stated that they became lonely after the earthquakes, while female teachers are the ones who stated that their feelings of solidarity and empathy strengthened.

The psychological effects of the 6 February earthquakes on the participants are given in Table 4.

Table 4. Psychological effects of the 6 February earthquakes

Category	f
Fear	12
Anxiety	7
Monophobia	7
Despair	5
Anxiety	5
Sadness	4
Malnutrition	4
Sleep disturbance	3
Claustrophobia	2
Grief	2
Reluctance to live	2
Panic attack	2

As seen in Table 4, the greatest psychological effect of the 6 February earthquakes on the participants was the feeling of fear (f=12). The participants stated that they experienced intense fear of a new earthquake, fear of losing their loved ones, and fear of being trapped under rubble. Participants also experienced anxiety along with fear (f=7).

"It affected us very badly and even after more than a year, we are very scared even in aftershocks." (P-29)

As the participant with the pseudonym P-29 stated, the aftershocks experienced after the 6 February earthquakes kept the feeling of fear alive.

Most of the participants reported experiencing monophobia (f=7) after the 6 February earthquakes, while some participants reported claustrophobia (f=2). Some of the statements of the participants who experienced fear of being alone and fear of enclosed spaces are as follows:

"Psychologically, I was worried about my children and loved ones. It affected our social life and caused us to enter closed environments and be on our toes." (P-14)

"There was fear that an earthquake could happen at any moment and fear of separation from my loved ones." (P-38)

"Psychologically, being in closed spaces made us more cautious and careful." (P-11)

After the 6 February earthquakes, the participants also faced the fear of death and were afraid of losing their loved ones. In addition, participants who lost their families or loved ones described the psychological difficulties of this situation with the following statements:

"I was very affected psychologically. Especially the fear of losing my loved ones and the fear of death affected me a lot." (P-18)

"I lost many of my friends, I had to move from the house I was in during the earthquake, which exhausted me and my family, and we had very difficult days psychologically." (P-30)

After the 6 February earthquakes, participants also stated that they experienced nutritional (f=4) and sleep disorders (f=3).

"I woke up frequently at night or could not sleep at all, I was afraid to leave my loved ones for a long time, I could not make daily plans, my eating and drinking balance was disturbed." (P-18)

"Of course, it affected my psychology badly. I didn't want to see anyone, I didn't want to eat, when I put food in my mouth, I would think of the people in the earthquake zone and I would stop eating. But of course, as time passed and I saw that people's wounds were healed a little more, I started to continue my old life." (P-43)

As can be seen from the statements above, the 6 February earthquakes left many different psychological effects on the participants. Although the psychological expressions of the participants were similar to each other, fear, anxiety and fear of enclosed spaces came to the fore. However, while talking about these effects, the participants also stated that they started to recover psychologically over time.

3.3. Reflections of the 6 February Earthquakes on Education

The last question posed to the participants within the scope of the research was "What are your feelings and thoughts about the reflections of the 6 February earthquakes and their aftermath on the education and training process in schools?". With this question, the effects of earthquakes on the education and training process were tried to be addressed in depth. By analyzing the participant responses, the themes of "education process" and "student" were reached. The categories created after the participant responses are given in Table 5.

Table 5. Reflections of the 6 February earthquakes on education

Theme	Category	f
Education process	Education and training process disrupted	13
	School climate deteriorated	5
	Activities to raise disaster awareness increased	5
	Teacher productivity down	3
	Overcrowded classrooms	3
Students	Anxiety disorder in students has started	8
	Academic achievement drops	7
	Students started to lose motivation	6
	Absenteeism increased	6
	Students experiencing loss	6

When the reflections of the 6 February earthquakes on education are analyzed under the theme of education, it is seen that the most cited category is the disruption of the education process (f=13). There are basically two reasons for this disruption. The first one is that face-to-face education was suspended for 1 month to 3 months in 11 cities in the disaster zone from the day after the earthquake and distance education was started. The second reason is that thousands of schools in the disaster zone were destroyed or damaged due to the earthquake.

> *"First of all, the repercussions have been negative. We have suffered. On top of that, schools were canceled, and not only that, our students were allowed to be absent. Our students did not attend school and this negatively affected education." (P-25)*
>
> *"Education and training was disconnected like during the pandemic. Unfortunately, it seems that it will take a long time for students to recover academically and psychologically " (P-1)*

As seen in the participant statements above, the long break in the education and training process caused disconnections and disruptions. Participants also stated that this situation caused the school climate to deteriorate (f=5).

"In our school, the trauma experienced by the guest students coming from the city center and the housing problems were reflected on all our students. Our school was not damaged, but we acted with the thought that there would be an aftershock at any moment." (P-23)

As mentioned by the participant with the pseudonym P-23, after the break in education, the students of the schools whose buildings were demolished started to receive education in the intact school buildings. This led to overcrowded classrooms (f=3). The mourning process, fear of earthquake, forced relocation and anxiety disorder caused the school climate to deteriorate and created a negative atmosphere in the school.

"Children have become more sensitive to noise and shaking, so even at the slightest noise, they act with the thought of "earthquake"." (P-32)

Participants stated that activities for disaster awareness increased after the 6 February earthquakes (f=5). However, it was also stated that teachers could not perform their profession efficiently in this process (f=3).

"Psychologically, since our minds were always on the earthquake zone, from time to time I felt reluctance towards education and training and the restlessness of not being able to do anything." (P-40)

6 February earthquakes had much deeper and more negative repercussions on students. Participants generally stated that students started to suffer from anxiety disorders (f=8).

"Negative attitudes have developed among students. The belief that life is too short was reflected in students' behavior. Students experienced psychological negativity. And for a few weeks, focusing on lessons had a negative impact on the normal course of life." (P-16)

"I saw deep scars on the students. They constantly talked about the earthquake and felt like an earthquake was going to happen at any moment. Students' motivation decreased and they were psychologically damaged" (P-36)

As can be understood from the statements of the participants above, the anxiety disorder experienced by the students even affected their attitudes and behaviors and decreased their motivation. The reason for the anxiety experienced by the students is fear and the sadness of losing their relatives or families.

"Children's psychology has been very damaged by the earthquake. Some of them lost their relatives and friends. Some of them have lost their motivation and excitement towards life and naturally towards school." (P-27)

"It has been very difficult for bereaved children to continue school. Facing death or the fear of death completely disconnected them from their studies. It was not

possible for learning to take place because there was an earthquake in every moment. This state of theirs affected me a lot too, every moment sadness, every moment anxiety, etc. We learned once again that all this hustle and bustle is empty and that human life is more valuable than anything else." (P-9)

As stated by the participants with the pseudonyms P-27 and P-9, the loss of relatives, relatives and loved ones, and the fact that the students were exposed to this pain at a young age were reflected in their educational lives and a decrease in their academic achievement was observed (f=7). The majority of the participants who stated that students' academic achievement decreased are working at the high school level.

The fact that education was suspended after the earthquakes and high levels of anxiety and fear increased the tendency of students to be absent (f=6).

"Education was suspended after the 6 February earthquake. We made up for the break, but since attendance was not compulsory, some students had absenteeism problems." (P-13)

"Distance education was introduced, but it caused serious disruptions in education due to the fact that not every student attended the class." (P-18)

4. RESULTS AND DISCUSSION

In this study, the motivational, social and psychological effects of two major earthquakes with the magnitudes of 7.7 Mw and 7.6 Mw, which occurred 9 hours apart on February 6, 2023, centered in Kahramanmaraş, on teachers and the reflections of earthquakes on education from the perspective of teachers were investigated. As a result of this research, it was tried to reveal the measures to be taken and policies to be implemented for teachers in similar situations that may occur in the future by investigating how natural disasters that can be described at the level of disaster affect the professional lives of teachers.

According to the results of the research, a large majority of the participants were negatively affected by the 6 February earthquakes in terms of motivational dimension. Participants who were negatively affected in terms of motivation stated that the earthquake also had an impact on their health. After the earthquakes, the participants stated that they experienced psychological problems, high pulse rate, heart rhythm disturbances and stress disorders. This situation has similar results with the study conducted by Bhavnani (2006). The majority of teachers are psychologically vulnerable after earthquakes (Bhavnani, 2006). In addition, teachers stated that they experienced negative emotions such as fear, anxiety and lack of concentration after the earthquakes and that their motivation towards work was negatively affected. Negative emotions such as fear and anxiety are frequently observed in individuals after earthquakes (Kruhl et al., 2018). Lack of concentration is a common post-disaster symptom, especially in social professions that require attention, such as teaching (Basnet, 2020).

The occurrence of aftershocks is also among the factors that negatively affect teachers' motivation. Participants who thought that school buildings were not earthquake-resistant or who believed that aftershocks would trigger a new major earthquake were negatively affected in the motivation dimension. Aftershocks are one of the factors that increase the stress level of teachers by creating a sense of uncertainty and risk in the post-disaster process (Kennedy et al., 2008). After the disaster, making school buildings earthquake-resistant and taking necessary security measures help teachers and students feel safer in coping with disasters (Basnet, 2020).

Disruptions in the crisis management process are among the factors that negatively affect teachers' motivation. In a study conducted by Norris et al. in 2022, it was found that teachers' feelings of trust and motivation increased in schools where crisis management was carried out effectively in the post-disaster process. When teachers think that crisis management is not carried out effectively or not coordinated properly, they experience uncertainty about what they should do in the post-disaster process; this situation may decrease teachers' motivation and make it difficult for them to focus on their work (Coppola, 2015). School administrators and senior management mechanisms have an important role in the crisis management process. Administrators' making fast and effective decisions, acting in coordination, inspiring confidence and activating assistance mechanisms prevent uncertainties that may be experienced in schools (Bendimerad, 2003).

As an interesting finding, some of the female participants stated that the 6 February earthquakes positively affected their motivation. These participants stated that the earthquakes strengthened their feelings of motivation, solidarity and empathy. This is in line with the theory of natural disaster vulnerability and natural disaster resilience established by Zakour and Gillespie (2013). Zakour and Gillespie (2013) state that vulnerability is experienced after major disasters, but resilience increases after the disaster. The fact that the participants mentioned positive motivational effects after the disaster supports the theory of disaster resilience. This finding shows that disasters can increase people's resilience and positively affect their motivation. The results obtained from the research show that disasters can have positive psychological effects especially on women. This finding shows that women's resilience and adaptation skills are strengthened and they experience positive motivational changes after disasters (Ariyabandu, 2006). Women's resistance and resilience in coping with disasters can be associated with their different roles in society (Juran, 2012). In particular, due to gender roles, women are often seen as having the responsibility of providing care and support within the family. In disaster situations, these roles become more prominent and women often assume an important role in meeting the basic needs of family members such as safety, nutrition, health and emotional support. Women's resilience in coping with disasters, their assumption of these responsibilities and their different roles in society are influencing factors (Jenkins & Phillips, 2008).

The negative effects of the 6 February earthquakes on teachers in the socialization dimension are asociality or not wanting to be alone. At this point, according to the personality traits or stress coping characteristics of the participants, the negative social effects are differentiated into two extreme poles of loneliness or reluctance to be alone. Some teachers may develop asocial tendencies due to the post-earthquake stress and uncertainty. These individuals may be affected by the fear of the earthquake and may avoid establishing relationships with others, stay away from social activities, and tend to isolate themselves (Potter et. al., 2015). On the other hand, some teachers may desire more social interaction after the earthquake. These individuals may seek support to cope with stress and may avoid being alone and turn to more social settings (Kemp et. al., 2011).

Some female participants stated that the earthquake affected them positively in the social dimension and that they fulfilled their social responsibility duties by being in solidarity. Women generally have strong social ties and these social support networks help them cope with and endure the stress of disasters (Arora-Jonsson, 2011). Therefore, women's social roles and participation in disasters represent an important element in the post-disaster recovery process of the society. Women's strong social ties and role in solidarity enable the community to come together, support and recover after a disaster (Alston, 2009). Women's focus on their social responsibilities contributes to strengthening social solidarity and making the post-disaster reconstruction process more effective and sustainable. The findings of Kun et al. (2013) following the 7.9 Mw earthquake in China in 2008 revealed that women were more negatively affected than men in the social dimension. However, this study was conducted 5 years after the earthquake. This may indicate that women's tendency to be exposed to the negative effects of the earthquake in the social dimension increases as time passes. In addition, Kun et al. (2013) did not conduct this study on teachers. Therefore, it is thought that the level of resilience of female teachers against earthquakes in the social dimension is higher than women working in other professions or housewives. This may be explained by the fact that the teaching profession is a social service field that touches the lives of individuals.

The 6 February earthquakes' psychological effects on the participants were fear of earthquakes, anxiety, fear of being alone, and fear of confined spaces. In a study conducted by Zhang et al. in 2016, 3 months after the 7.0 magnitude Chinese earthquake, it was revealed that teachers had to struggle with psychological problems and negative emotions. Earthquake fear is associated with individuals' anxiety about future earthquakes or similar natural disasters (Kuntz, 2015). Anxiety reduces the ability to cope with uncertainty and prevents teachers from focusing on their profession by making it difficult to focus on normal activities (Allen et. al., 2011). Fear of being alone causes individuals to avoid social relationships or feel lonely. Fear of confined spaces causes anxiety about being indoors during or after an earthquake. These psychological effects are closely related to the stress and trauma experienced by individuals after the earthquake. These emotional difficulties can be long-lasting and may prevent teachers from continuing their daily activities and fulfilling their duties (Alarcon, 2011).

When the reflections of the 6 February earthquakes on education are examined from the perspective of the participants, it can be said that there were disruptions in the education process and the school climate deteriorated. After the earthquakes, there were disruptions in the teaching process due to the damage to schools and security concerns. Students' and teachers' motivation decreased, which led to a negative impact on the learning environment. The environment of stress and uncertainty after the earthquakes led to an increase in student absenteeism, which led to disruptions in the educational process. After major earthquakes around the world, the education process is negatively affected, teaching activities are interrupted, and this causes disruptions in the process (Alcocer et al., 2020; Mutch, 2021). These findings are contrary to the findings of Seyle et al.'s 2013 study. Seyle et al. found that psychological processes experienced after natural disasters do not have an impact on education. However, this may be due to differences in variables such as the magnitude of the disaster, social structure, pre-disaster preparations, disaster management, cultural and social characteristics. Seyle et al. conducted their research 4 years after the 6.3 magnitude Indonesian earthquake. Since the earthquake in question was of lower magnitude compared to the February 6 earthquake and the study was conducted 4 years after the earthquake, it is thought that the finding that the earthquake did not have negative reflections on education was reached.

A positive development for the education and training process was the increase in activities to raise disaster awareness after the 6 February earthquakes. Similarly, the study conducted by Dixit et al. in 2013 proved that schools play a major role in raising awareness after major natural disasters. Disaster awareness-raising and awareness-raising activities and practical drills are critical to increase the community's ability to cope with disasters (Paton & Johnston, 2017). Increasing disaster awareness and resilience in post-disaster societies makes individuals and society more resilient to disasters. Schools taking a leadership role in this regard helps students, parents and other community members to develop disaster awareness and be prepared for disasters (Tuladhar et al., 2015). Thus, it helps the society to become more resilient and conscious against disasters.

From the participants' perspective, the effects of the 6 February earthquakes on students include anxiety disorder, decline in academic achievement, loss of motivation, increased absenteeism, and negative emotions experienced by students who lost family members or relatives. These findings are similar to the results of the study conducted by Di Pietro (2018) after the L'Aquila earthquake. The stress and difficulties experienced by students distract them and reduce their motivation, which leads to a decrease in their academic performance (Rodgers et al., 2021). Similarly, in the structural equation model created by Yeon et al. (2020) after the 5.4 magnitude earthquake in South Korea, it was concluded that students had negative emotions such as anger and fear as a result of the earthquake and that these emotions had a high impact on their academic achievement. Students who lose their relatives as a result of a traumatic event such as an earthquake are introduced to death at a young age and this situation reduces students' desire to learn and sense of achievement (Shaheen, 2008).

In conclusion, the 6 February earthquakes negatively affected teachers socially, psychologically and motivationally. Teachers had to cope with negative emotions due to the earthquakes, and psychological traumas turned into physical illnesses in some teachers. Teachers developed various phobias and lost their motivation towards their profession. It was concluded that some female teachers, unlike other teachers, experienced positive emotions such as solidarity and empathy. After the earthquake, disruptions occurred in the education process. The school climate deteriorated. Students' academic achievement decreased and absenteeism rates increased.

Based on the findings of the study, it is concluded that the measures to be taken before, during and after the disaster and disaster management are important in reducing the negative effects of the earthquake. It can be said that earthquake-resistant buildings of educational institutions and informing teachers and students about this issue will reduce the negative emotions experienced by both teachers and students during and after the earthquake. In addition, it can be said that psychological support after the earthquake reduces the negative effects of the process on teachers and students. After the 6 February earthquakes, psychological support was mainly given to students. The findings of the study showed that teachers need psychological support as much as students. Especially stress and anxiety management should be emphasized.

Providing disaster and crisis management trainings to school administrators and teachers will enable better management of the process in educational institutions during earthquakes and similar disasters. This may reduce uncertainty and fear. Because the uncertainty experienced by administrators and teachers feeds students' negative emotions such as stress, fear and anxiety. By professionally managing the crisis caused by the disaster, student absenteeism can be reduced and the effects of the earthquake on academic achievement can be minimized. Thus, a safe, supportive and motivating school climate can be established during disasters.

5. LIMITATIONS AND RECOMMENDATIONS

In this study, the psychological, social, motivational and educational effects of the 6 February earthquakes were examined from the perspective of teachers. Of course, the effects of the earthquake covering these dimensions are not limited to those discussed in this study. There is a need for much more research examining the reflections of the 6 February earthquakes that affected 11 provinces and their districts on both educational stakeholders and educational activities. First of all, the results of the current study are limited to teachers working in Malatya, Elazığ and Diyarbakır among the 11 provinces that were declared disaster areas during the earthquake. Research can be conducted on teachers, administrators and students working in the other 8 cities, especially in Kahramanmaraş, Adıyaman and Hatay provinces and districts, which were most affected by the earthquake. In this study, the study group consisted of teachers. Our knowledge on the effects of the earthquake on school administrators and school management can be increased through qualitative and quantitative research on school administrators. In addition, by triangulating the data, research can be conducted to include the internal (teachers, students, administrators, non-educators) and external stakeholders (parents, inspectors, local administrations, NGOs, trade unions, other stakeholders) of the school in the study group.

REFERENCES

Alarcon, G. M. (2011). A meta-analysis of burnout with job demands, resources, and attitudes. *Journal of Vocational Behavior, 79*(2), 549–562. DOI: 10.1016/j.jvb.2011.03.007

Alcocer, S. M., Murià-Vila, D., Fernández-Sola, L. R., Ordaz, M., & Arce, J. C. (2020). Observed damage in public school buildings during the 2017 Mexico earthquakes. *Earthquake Spectra, 36*(2), 110–129. DOI: 10.1177/8755293020926183

Allen, J., Jimmieson, N. L., Bordia, P., & Irmer, B. E. (2007). Uncertainty during organizational change: Managing perceptions through communication. *Journal of Change Management, 7*(2), 187–210. DOI: 10.1080/14697010701563379

Alston, M. (2009). Drought policy in Australia: Gender mainstreaming or gender blindness. *Gender, Place and Culture, 16*(2), 139–154. DOI: 10.1080/09663690902795738

Ariyabandu, M. (2006). Gender issues in recovery from the December 2004 Indian Ocean tsunami: The case of Sri Lanka. *Earthquake Spectra, 22*(3), 759–775. DOI: 10.1193/1.2204838

Arora-Jonsson, S. (2011). Virtue and vulnerability: Discourses on women, gender and climate change. *Global Environmental Change, 21*(2), 744–751. DOI: 10.1016/j.gloenvcha.2011.01.005

Arslan, M. (2023). Okul müdürlerinin 6 Şubat Kahramanmaraş Depremi sonrası okullarda karşılaştıkları sorunlar ve bu sorunlara ilişkin çözüm önerileri. *International Journal of Social Humanities Sciences Research, 10*(97), 1550–1559.

Ayan, V. M., & Keten, E. T. (2023). Medyanın yapısal sorunları bağlamında afet haberciliği: 6 Şubat 2023 Kahramanmaraş deprem haberlerinin gazetelerde çerçevelenmesi. *Niğde Ömer Halisdemir Üniversitesi İletişim Fakültesi Akademik Dergisi, 2*(2), 116–132.

Basnet, B. K. (2020). Earthquake and its impacts on education: Aftemath Nepal quake 2015. *The European Educational Researcher, 3*(3), 101–118. DOI: 10.31757/euer.332

Bendimerad, F. (2003). Disaster Risk Reduction and Sustainable Development. World Bank Seminar on The Role of Local Governments in Reducing the Risk of Disasters, Held in Istanbul, Turkey, 28 April–2 May 2003, 57–75.

Bhavnani, R. (2006). *Natural disaster conflicts*. Harvard University.

Bilham, R. (2004). Earthquakes in India and the Himalaya: Tectonics, geodesy and history. *Annals of Geophysics*, *47*(2/3), 839–858.

Bogg, T., & Roberts, B. W. (2004). Conscientiousness and health-related behaviors: A meta-analysis of the leading behavioral contributors to mortality. *Psychological Bulletin*, *130*(6), 887–919. DOI: 10.1037/0033-2909.130.6.887 PMID: 15535742

Coppola, D. P. (2015). *Introduction to international disaster management*. Kindle edition, B&H Publishing, UK.

Creswell, J. W., & Miller, D. L. (2000). Determining validity in qualitative inquiry. *Theory into Practice*, *39*(3), 124–130. DOI: 10.1207/s15430421tip3903_2

Di Pietro, G. (2018). The academic impact of natural disasters: Evidence from L'Aquila earthquake. *Education Economics*, *26*(1), 62–77. DOI: 10.1080/09645292.2017.1394984

Dixit, A. M., Shrestha, S., Guragain, R., Pandey, B., Sen Oli, K., Adhikari, S., Acharya, S., Jimee, G., Upadhyay, B., Sangachhe, S., Shrestha, N., Pradhan, S., Dhungel, R., Khatiwada, P., Baskota, A., Poudel, A., Tandingan, M., Mathema, N., Parajuli, B., & Marasini, N. (2018). *Risk Management*. Response, Relief, Recovery, Reconstruction, and Future Disaster Risk Reduction., DOI: 10.1016/B978-0-12-812808-4.00005-5

Dündar, H. Ç. (2023). 6 Şubat Kahramanmaraş depremlerinin iş dünyasına etkileri. *Çevre Şehir ve İklim Dergisi, 2(*4), 262-281.

Erdoğan, B. (2023). Depremin sosyolojisi: 6 Şubat felaketinin toplumsal ve kültürel boyutları. *TRT Akademi*, *8*(18), 718–725. DOI: 10.37679/trta.1306900

Hutton, C., Kienberger, S., Amoako Johnson, F., Allan, A., Giannini, V., Allen, R. (2011). Vulnerability to climate change: People, place and exposure to hazard. Advances in Science and Research, 7, 37.45.

Jenkins, P., & Phillips, B. (2008). Battered women, catastrophe, and the context of safety after Hurricane Katrina. *NWSA Journal*, *20*(3), 49–68. DOI: 10.1353/ff.2008.a256898

Juran, L. (2012). The Gendered Nature of Disasters: Women Survivors in Post-Tsunami Tamil Nadu. *Indian Journal of Gender Studies*, *19*(1), 1–29. DOI: 10.1177/097152151101900101

Kanamori, H., & Brodsky, E. (2004). The physics of earthquakes. *Reports on Progress in Physics*, *54*(8), 1429–1496. Advance online publication. DOI: 10.1088/0034-4885/67/8/R03

Kataoka, S. H., Langley, A. K., Wong, M., Baweja, S., & Stein, B. D. (2009). Responding to students with PTSD in schools. *Child and Adolescent Psychiatric Clinics of North America, 18*(3), 563–579.

Kemp, S., Helton, W. S., Richardson, J. J., Blampied, N. M., & Grimshaw, M. (2011). Sleeplessness, stress, cognitive disruption and academic performance following the September 4, 2010, Christchurch earthquake Australas. *Disaster Trauma Studies, 2*, 11–18.

Kennedy, J., Ashmore, J., Babister, E., & Kelman, I. (2008). The meaning of 'build back better': Evidence from posttsunami Aceh and Sri Lanka. *Journal of Contingencies and Crisis Management, 16*(1), 24–36. DOI: 10.1111/j.1468-5973.2008.00529.x

Kocabıyık, O. O. (2016). Olgubilim ve gömülü kuram: Bazı özellikler açısından karşılaştırma. *Trakya Üniversitesi Eğitim Fakültesi Dergisi, 6*(1), 55–66.

Köknel, Ö. (1987). *Zorlanan insan*. Altın kitaplar Yayınevi, İstanbul.

Kousky, C. (2016). Impacts of natural disasters on children. *The Future of Children, 26*(1), 73–92. DOI: 10.1353/foc.2016.0004

Kun, P., Tong, X., Liu, Y., Pei, X., & Luo, H. (2013). What are the determinants of post-traumatic stress disorder: Age, gender, ethnicity or other? Evidence from 2008 Wenchuan earthquake. *Public Health, 127*(7), 644–652. DOI: 10.1016/j.puhe.2013.04.018 PMID: 23806187

Kuntz, J. R. C. (2015). Protracted effect: Surveying teachers experiences in the aftermath of the christchurch earthquakes. *Natural Hazards Review, 16*(1), 401–414. DOI: 10.1061/(ASCE)NH.1527-6996.0000147

La Greca, A. M., Silverman, W. K., Lai, B., & Jaccard, J. (2010). Hurricane-related exposure experiences and stressors, other life events, and social support: Concurrent and prospective impact on children's persistent posttraumatic stress symptoms. *Journal of Consulting and Clinical Psychology, 78*(6), 794–805. DOI: 10.1037/a0020775 PMID: 20939624

Maden, S. (2023). 6 Şubat 2023'te Kahramanmaraş'ta Yaşanan Depremler Ekseninde Türkiye'de Deprem Haberciliğine Bakış: Prof. Dr. Süleyman İrvan ile Söyleşi. *Etkileşim, 6*(11), 406–420. DOI: 10.32739/etkilesim.2023.6.11.202

Masten, A. S., & Narayan, A. J. (2012). Child development in the context of disaster, war, and terrorism: Pathways of risk and resilience. *Annual Review of Psychology, 63*(1), 227–257. DOI: 10.1146/annurev-psych-120710-100356 PMID: 21943168

Masten, A. S., & Osofsky, J. D. (2010). Disasters and their impact on child development: Introduction to the special section. *Child Development*, *81*(4), 1029–1039. DOI: 10.1111/j.1467-8624.2010.01452.x PMID: 20636680

Mutch, C. (2021). The role of schools in helping communities copes with earthquake disasters: The case of the 2010–2011 New Zealand earthquakes. In Xu, J., Lu, Y., & Penning-Rowsell, E. C. (Eds.), *Earthquake disasters: Prevention and reconstruction*. Routledge. DOI: 10.4324/9781003173328-5

Norris, F., Friedman, M., Watson, P., Byrne, C., Diaz, E., & Kaniasty, K. (2002). 60,000 disaster victims speak: Part I: An empirical review of the empirical literature, 1981–2001. [Crossref. PubMed. ISI.]. *Psychiatry*, *65*(3), 207–239. DOI: 10.1521/psyc.65.3.207.20173 PMID: 12405079

Olshansky, R. B., Hopkins, L. D., & Johnson, L. A. (2012). Disaster and recovery: Processes compressed in time. *Natural Hazards Review*, *13*(3), 173–178. DOI: 10.1061/(ASCE)NH.1527-6996.0000077

Öztürk, M. K. (2013). A research on earthquake experiences of primary school teacher candidates. [Hacettepe Unıversity Journal of Education]. *Hacettepe Üniversitesi Eğitim Fakültesi Dergisi*, *28*(1), 308–319.

Paton, D., & Johnston, D. (2017). *Disaster resilience: an integrated approach*. Charles C Thomas Publisher.

Patton, M. Q. (2014). Nitel araştırma ve değerlendirme yöntemleri. M. Bütün ve S. B. Demir (Çev. Ed.). Ankara: Pegem Akademi.

Peek, L. (2008). Children and disasters: Understanding vulnerability, developing capacities, and promoting resilience—An introduction. *Children, Youth and Environments*, *18*(1), 1–29. DOI: 10.1353/cye.2008.0052

Potter, S. H., Becker, J. S., Johnston, D. M., & Rossiter, K. P. (2015). An overview of the impacts of the 2010-2011 Canterbury earthquakes. *International Journal of Disaster Risk Reduction*, *14*(1), 6–14. DOI: 10.1016/j.ijdrr.2015.01.014

Raschky, P. A. (2008). Institutions and the losses from natural disasters. *Natural Hazards and Earth System Sciences*, *8*(4), 627–634. DOI: 10.5194/nhess-8-627-2008

Reid, H. F. (1911). *The elastic-rebound theory of earthquakes*. University of California Press.

Rodgers, J., Hassan, W., Motter, C., & Thornley, J. (2021). Impacts of the 2018 M7.1 Anchorage earthquake on schools. *Earthquake Spectra*, *37*(3), 1849–1874. DOI: 10.1177/8755293020988022

Seyle, D. C., Widyatmoko, C. S., & Silver, R. C. (2013). Coping with natural disasters in Yogyakarta, Indonesia: A study of elementary school teachers. *School Psychology International*, *34*(4), 387–404. DOI: 10.1177/0143034312446889

Shaheen, M. A. (2008). Earthquake effects on educational institutions and libraries of Azad Kashmir: An appraisal. *Library Review*, *57*(6), 449–456. DOI: 10.1108/00242530810886724

Sivri, S. (2023). Depremin Psikolojik Etkileri. https://www.acibadem.com.tr/hayat/depremin-psikolojiketkileri/#:~:text=Do%C4%9Fal%20bir%20felaket%20olan%20deprem,temelli%20bir%20anlam%20ta%C5%9F%C4%B1maya%20ba%C5%9Flar. (Erişim Tarihi: 13.04.2024).

Tetik, N., & Akbulut, İ. İ. (2023). *6 şubat 2023'te yaşanan depremin ekonomik ve finansal etkileri: İhracat üzerinden bir inceleme. Kahramanmaraş merkezli depremler sonrası için akademik öneriler*. Özgür Yayınları.

Tuladhar, G., Yatabe, R., Dahal, R. K., & Bhandary, N. P. (2015). Disaster risk reduction knowledge of local people in Nepal. *Geoenvironmental Disasters*, *2*(1), 2–12. DOI: 10.1186/s40677-014-0011-4

Watt, F. (2002). *Depremler ve yanardağlar*. (Çev. Deniz Yurtören), 18. Basım. Tübitak popüler bilim kitapları: 71, Ankara.

Yeon, D.-H., Chung, J.-B., & Im, D.-H. (2020). The effects of earthquake experience on disaster education for children and teens. *International Journal of Environmental Research and Public Health*, *17*(15), 1–14. DOI: 10.3390/ijerph17155347 PMID: 32722236

Yıldırım, A. ve Şimşek, H. (2016). Sosyal bilimlerde nitel araştırma yöntemleri. Ankara: Seçkin Yayıncılık.

Zhang, Jun MDa; Zhang, Ye MDa; Du, Changhui MSc; Zhu, Shenyue MAb; Huang, Yalin MSa; Tian, Yulian MAd; Chen, Decao MAa; Li, Haimin MSa; Gong, Yao MSa; Zhang, Mengmeng MSa; Gu, Bo MSe. (2016). Prevalence and risk factors of posttraumatic stress disorder among teachers 3 months after the Lushan earthquake: A cross-sectional study. *Medicine 95*(29), 1-6.

Chapter 12
Preparedness on Seismic Risk:
A Case Study of a Community in Oaxaca, Mexico

Tatiana Gouzeva
Instituto Politecnico Nacional, Mexico

ABSTRACT

The book chapter presents some results on children's seismic risk knowledge and perception for a sample size of n=522. The study has been conducted in Huajuapan de Leon, Oaxaca, Mexico. The municipality is located in a highly seismic region. The data was collected by employing a questionnaire-based survey. Some of the main findings are the following: a) earthquakes (72.4%) and floods (34.1%) were recognized as a threat, but not drought, however, the lack of water is one of the major concerns of the community, b) around 40% perceived that they lack the knowledge on seismic risk; 57% recognized they live in a highly seismic region, around 95% responded that they know what to do in case of an earthquake while in a classroom, c) most of the students perceived that earthquakes would happen sometime in the future, e.g., 48.3% considered an earthquake would occur "next year", and d) around 68% of the participants "sometimes" feel scared or upset when thinking of earthquakes.

INTRODUCTION

Children and youth are one of the most vulnerable groups to disasters triggered by natural hazards. It is believed that disasters triggered more than 60% of the internal displacements in 2021 worldwide; moreover, more than half of the displaced people were children and youth individuals (IDMC, 2023). Some of the consequences of

DOI: 10.4018/979-8-3693-9745-9.ch012

internal displacement are educational disruptions which can have long term impacts such as disengagement from school, fewer educational and professional opportunities, among others (Kousky, 2016). Given the great importance of the subject, children and youth's vulnerabilities have been taken into consideration in the 2030 UN agenda for Sustainable Development Goals (SDGs). The UN-SDGs aims at building an inclusive, sustainable, and resilient future for people and planet (UN-SDG, 2023). The 2030 agenda proposed 17 SDGs of which at least four are related to children, i.e.: Goals 1 ("End poverty in all its forms everywhere"), 4 ("Quality in education"), 11 ("Make cities inclusive, safe, resilient and sustainable"), and 13 ("Take urgent action to combat climate change and its impacts") (UN-SDG, 2023).

For example, regarding Goal 1, and target 1.5.,

"By 2030, build the resilience of the poor and those in vulnerable situations and reduce their exposure and vulnerability to climate-related extreme events and other economic, social and environmental shocks and disasters."

Similarly, Goal 4, target 4.1, which states that,

"By 2030, ensure that all girls and boys complete free, equitable and quality primary and secondary education leading to relevant and Goal-4 effective learning outcomes."

Finally, Goal 11, target 11.B, states the following:

"By 2020, substantially increase the number of cities and human settlements adopting and implementing integrated policies and plans towards inclusion, resource efficiency, mitigation and adaptation to climate change, resilience to disasters, and develop and implement, in line with the Sendai Framework for Disaster Risk Reduction 2015-2030, holistic disaster risk management at all levels."

However, given the latest published data on children and youth (IDMS, 2023), these targets are difficult to achieve, and more actions need to be done. For example, the earthquakes that occurred on September 7 (Mw 8.1) and 19 (Mw 7.1), 2017 in Mexico, hit the states of Chiapas, Mexico City, Morelos, Oaxaca and Puebla (Santos-Reyes and Gouzeva, 2023). It is believed that 320 municipalities were declared in a state of emergency: 118 in Chiapas, 16 in Mexico City, 33 in Morelos, 41 in Oaxaca and 112 in Puebla. Moreover, it is estimated that 7 million children and adolescents live in the affected areas. The earthquakes caused the death of 369 people and 32 were children, and 155,674 homes were damaged. In Mexico City, 14,908 schools were damaged, of which 288 require complete reconstruction. It was estimated that 4 million students would be temporarily relocated to other schools (UNICEF, 2018).

The more recent strong earthquakes that occurred on 6 February 2023 hit Türkiye and Syria killing over 50,000 people and displaced millions with about 6.2 million children were affected (Reliefweb, 2024).

A great deal of effort has been made by scholars to investigate and develop approaches to better understand human behaviour to disasters triggered by natural hazards. Overall, there is a consensus among researchers that well prepared communities to natural hazards make them more resilient to disasters (Zhang & Zheng, 2023; Lv, et al., 2024; Bodas, et al., 2022). It has been argued that a factor that influences preparedness behavior and consequently resilience, is risk perception (Paton, 2022; Lindell and Hwang, 2008). Risk perception (RP) is defined as people's judgements and evaluations of hazards they are exposed to; RPs are interpretations of the world based on past experiences or beliefs, and immersed in values, norms, and cultural issues of communities (Finucane & Holup 2006, Rohrmann, 2003; Slovic 2000). Kellen, et al. (2011) defines flood RP as a study on human consciousness, emotions, and behaviors. It has also been found that knowledge influences risk perception (Wachinger et al., 2013). For example, having little knowledge about the causes of floods have lower RP (Działek et al. 2013b); conversely, a knowledgeable community, for example, on flood risk, usually increases its RP (Raaijmakers et al. 2008). Similar findings have been reported regarding seismic risk (Santos-Reyes & Gouzeva, 2023; Finis, et al., 2010; Ronan, et al., 2001).

Regarding children and adolescents in the context of earthquakes, several studies have been published addressing seismic hazard and risk perception, preparedness, among others (Santos-Reyes and Gouzeva, 2023; Baytiyeh and Ocal, 2016; Antronico, et al., 2023; Finis, et al., 2010; Ronan, et al., 2001). However, research on RP and seismic risk knowledge and perception is lacking in the context of a local community in Mexico. The chapter addresses these issues for a convenience sample of $n = 600$ of children in the municipality of Huajuapan de Leon, Oaxaca, Mexico. The aims of the case study are the following:

1. Awareness of the following hazards: earthquakes, floods, hurricanes, landslides, extreme weather (drought).
2. Earthquake knowledge on issues such as knowing what to do in case of an earthquake in the classroom, whether the participants know that they live in a highly seismic region.
3. Earthquake risk perception.
4. Their perception of fear of earthquakes.

The chapter is organized as follows: first, a very brief description of some evidence of seismic risk of the area of the case study is presented. Second, a description of the employed materials and methods are briefly described; third, the most relevant results are presented. Fourth, a discussion of the importance of water and related work is presented in the section; also, some limitations and future directions are given within the discussion section. Finally, some relevant conclusions are reported.

Seismic Risk in Oaxaca, Mexico

Mexico is a highly seismic country and located between five tectonic plates: Pacific, Caribbean, Cocos Rivera, and North American (Santos-Reyes, 2024; Ramos-Cruz and Ruiz-Garcia, 2024;) (Fig. 1). Similarly, Oaxaca is also one of the most seismic states of the country. The Oaxaca region is located between 15.00° and 18.50° North Latitude and 94.50° and 98.75° West Longitude. The seismic activity is associated with the subduction process of the Cocos Plate beneath the North American Plate. This is known as the subduction zone and is the area where approximately 85-90% of the earthquakes that occur in the country are located (SISPLADE, 2000).

Figure 1. Earthquakes in Mexico during 2023

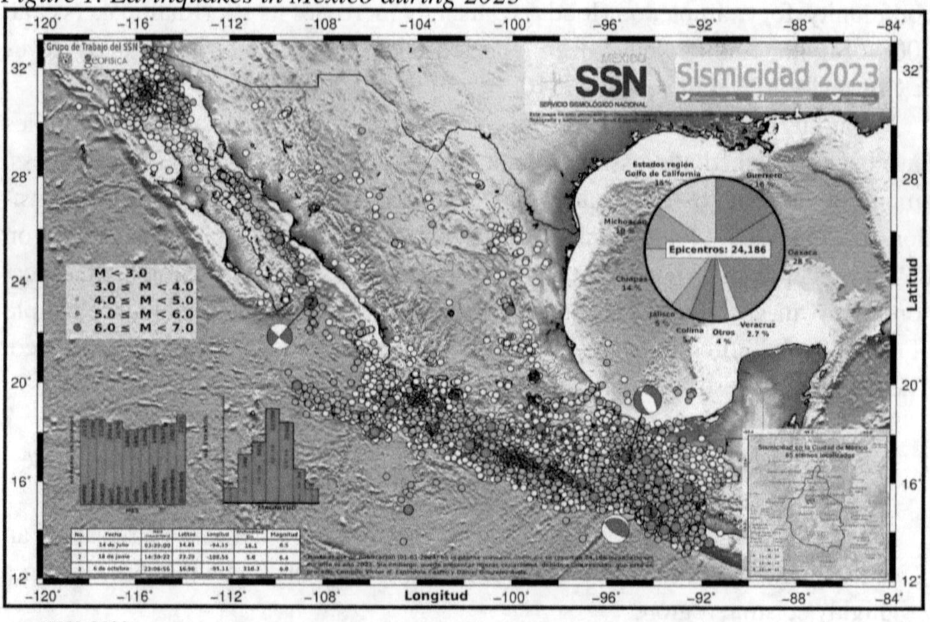

(SSN, 2023).

The Oaxaca state is divided into eight seismic zones (Fig. 2), one of them is the Huajuapan zone (zone 2), where the present case study took place. Huajuapan de León city's geographical coordinates are 17° 43' - 18° 03' north latitude and 97° 42' - 97° 55' west longitude and its altitude fluctuate between 1,400 and 2,300 meters above sea level (Fig. 3) (AtlasRiesgo, 2021).

In Huajuapan seismic zone, some of the strongest earthquakes were the one on July 19, 1882, which was felt in a large part of the country, and the earthquake on October 24, 1980 (Mw=7). Other relevant earthquakes with epicentres in this zone are shown in Table 1.

Figure 2. Oaxaca state (left) and seismic zones (right)

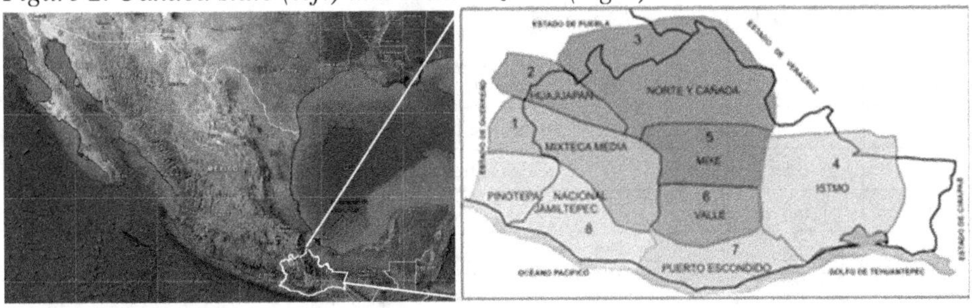

(SISPLADE, 2000).

Figure 3. Seismic zones of Huajuapan de Leon, Oaxaca

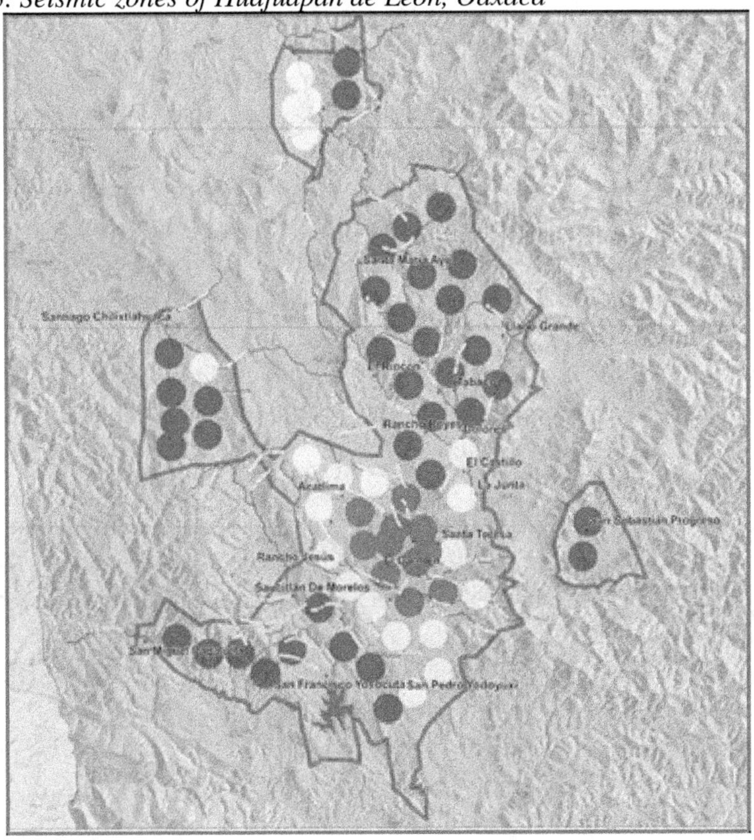

(AtlasRiesgo, 2021). (Grey circle=Low risk; Yellow circle=Medium risk; and Red circle=High risk).

Table 1. Some of the earthquakes that originated in Huajuapan seismic zone.

Year	Magnitude Mw	Epicentre, Region
1882	7.3	Huajuapan de Leon, Oaxaca
1911	7.0	19 km east from Huajuapan de Leon, Oaxaca
1928	6.5	With epicentre 9 km northeast of Acatlan, Puebla
1945	6.5	31 km southeast of Tehuacan, Puebla
1978	7.6	32 km northwest of San Pedro Pochutla, Oaxaca
1980	7.1	19 km west of Acatlan, Puebla
1999	7.0	Tehuacan, Puebla
2018	5.9	10 km southwest of Huajuapan de Leon, Oaxaca

DISASTER RISK MANAGEMENT

Several approaches to disaster management systems have been proposed, e.g., from a systemic perspective (Beard and Santos-Reyes, 2003; Santos-Reyes & Beard, 2005) and the so-called disaster risk management cycle (Shirazi and Nohegar, 2023; Sharma, et al., 2022; Wilhite, et al., 2005), among others. But what is meant by disaster and disaster risk management? There is still not a consensus of their meanings among scholars. For example, the UNIDR (2009) proposes the following definition of a disaster:

> *"A serious disruption of the functioning of a community or a society involving widespread human, material, economic or environmental losses and impacts, which exceeds the ability of the affected community or society to cope using its own resources." (p.9).*

However, what is clear is that disasters may be the result of an interaction among capacity, hazard, and vulnerability, and having serious consequences in terms of {a} loss of life/property/economic, among others, and {b} causing disruption of communities' way of life, among others.

On the other hand, disaster risk management is being defined as (UNISDR, 2009):

> *"The systematic process of using administrative directives, organizations, and operational skills and capacities to implement strategies, policies and improved coping capacities in order to lessen the adverse impacts of hazards and the possibility of disaster." (p.10).*

Figure 4 shows a disaster risk management cycle that is widely used by practitioners and governmental organizations in charge of Civil protection. Overall, the upper half of the cycle refers to Pre-disaster ("A"), the period of time before the earthquake disaster occurrence. For example, preparedness involves planning for a seismic emergency, conduct random earthquake drills, seismic risk education. Hence, the present case study may be part of the activities within preparedness.

The lower half, post-disaster, involves all the activities involved in the assessment of the impact of the disaster, response, recovery, and reconstruction. For example, during response phase, the mobilization of emergency services and first responders for search and rescue of those that may be trapped under the rubble following an earthquake. (For further details of each of the phases of the disaster risk management, see the provided references.)

Figure 4. Disaster risk management cycle.

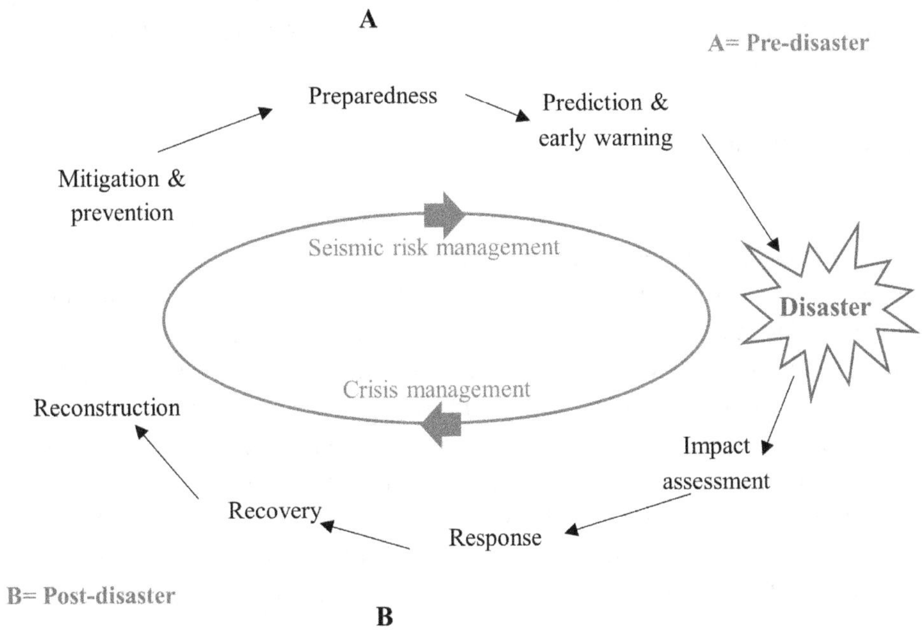

MATERIALS & METHODS

Rationale of the Study

The present study is part of an ongoing research project on multi-hazards being conducted since 2015; in particular, the research project aims at gaining a better understanding of the level of preparedness to disasters triggered by natural hazards of communities living along the Pacific coast of the country. Regarding seismic risk, these communities experience high frequency of earthquakes (Fig. 1). As already mentioned, communities in Oaxaca state and the municipality of Huajuapan are vulnerable to earthquakes (Table 1), and there is a lack of studies on this field. Hence, the aim of this study is to address seismic risk (and other natural hazards) awareness and preparedness of a particular population, i.e., children.

Area of Study

The area of study of the research project is in the state of Oaxaca, Mexico; six schools agreed to take part in the survey in the city of Huajuapan de Leon, Oaxaca (Fig. 1). The data were collected in January 2017 and 30 June 2017. As already mentioned, the city is located in a highly seismic area of the state and country (Fig. 2, and Table 1).

Data Collection

The population of interest has been one of the most vulnerable to seismic risk, i.e., children. For this research project, a convenience sampling was employed to students whose age ranged from 8-14 years old. To administrate the questionnaires, parents of the student participants were asked for their consent by means of the school heads. Students were made aware of the questionnaire's scientific goals and were reassured that their identities would be kept anonymous. Moreover, respondents were informed of their right to withdraw at any time. Moreover, the study was conducted in accordance with the University´s research ethical code, through the informed consent letter, which appeared at the beginning of the questionnaires.

The questionnaire included several sections, however, only those applicable for children and for the present case study are included in the following subsections.

Hazard Awareness And Risk Perception

To assess the level of awareness of the potential hazard in the community, students were asked the following question, "*are you aware of the following hazards?*" Respondents rated their answers according to the following: "Yes" or "No", and by considering the following natural hazards: "floods", "hurricanes", "earthquakes", "landslides", extreme weather (i.e., drought).

Regrading seismic risk perception, the following question was considered in the questionnaire, "*how likely is that an earthquake would occur in the future?*" The possible responses were the following: "Next year", "Within 10 years", "Within my lifetime", and "Never".

General Earthquake Knowledge

To assess the general seismic knowledge of participants of the study, they were asked the following questions:

"Do you know that you live in a highly seismic region?" Respondents rated their answers by selecting "Yes" or "No".

"What degree of knowledge do you think you have about earthquakes?" Students rated their answers according to a Likert type scale: "Very poor", "Poor", "Acceptable", "Good", and "Very good".

Finally, the following question:

"Do you know what to do in case of an earthquake if you are in the classroom?" The possible responses were either "Yes", or "No".

Fear of Earthquakes

Regarding fear of earthquakes, respondents were asked the following question. "Does talking or thinking about earthquakes scare you" The possible answers to the question were "Always", "Sometimes", and "Not at all".

Statistical Methods

To analyze the descriptive information, frequency analysis has been conducted. The relationship between the dependent and independent variables has been assessed by employing Pearson Chi-square tests. The Statistical Package for Social Science (SPSS), ver. 25, has been employed to the analysis of the collected data. All the conducted tests were two-tailed, and the statistically significant results acceptance criteria were yield at $p<.05$.

RESULTS & DISCUSION

Sample Characteristics

The sample size of the case study was $n = 600$, however, 78 respondents were excluded due to missing data, among others; hence, the employed sample was $n = 522$. The sample characteristics shows that the percentage of women were slightly higher than men (52.1% (272/522) vs. 47.9% (250/522)), as shown in Figure 5. The range of age of the participants of the study was from 8-14 years old (Mean = 10.45; Standard deviation = 1.415).

Figure 5. Sex of the participants of the study.

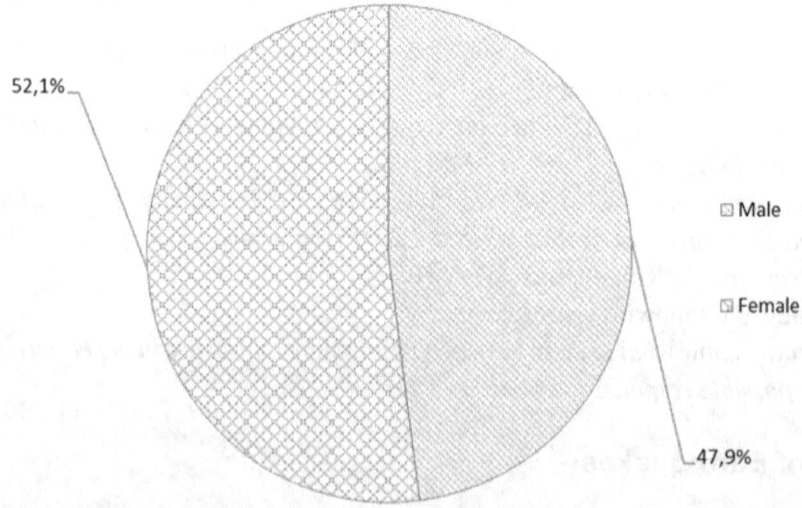

Seismic Risk Awareness

There is an agreement among scholars that hazard and risk awareness are key to motivating communities in adopting mitigation and preparedness behaviors (Lindell, et al., 2021; Lechowska, 2018; Santos-Reyes, 2020, 2024; Yildiz, et al., 2021; Antronico, et al., 2023). In an attempt to investigate the participants' awareness of some the hazards that threaten their community, the following natural hazards were considered in the study: earthquakes, floods, hurricanes, landslides, and extreme weather (drought) (Fig. 6).

As expected, earthquake hazard scored the highest percentage (72.4%, 378/522), followed by floods (34.1%, 178/522), and extreme weather (e.g., drought) was perceived as not representing a big threat to the participants and their community (5.6% (29/522)). However, the lack of water is one of the major concerns of the community and this has not been recognized by the participants of the study. For example, it has been reported that around 100 municipalities in the district of Huajuapan, Nochistlan, Tlaxiaco, Putla, and Juxtlahuaca, registered severe drought (Cruz, 2023), i.e.:

> *"They (the municipalities) somehow maintain the degree of drought D1, and D2, severe and extreme drought (respectively); this means that the precipitation that occurred during the rainy months, May to November, has been practically below 50 percent, in some cases even less than 20 percent; the precipitation deficit has come dragging for more than six months in these municipalities, and this causes secondary consequences such as water shortages in some urban centres, as well*

as loss of crops, loss of livestock, that are typical consequences of the drought phenomenon." (Cruz, 2023).

Figure 6. Earthquake and other natural hazards.

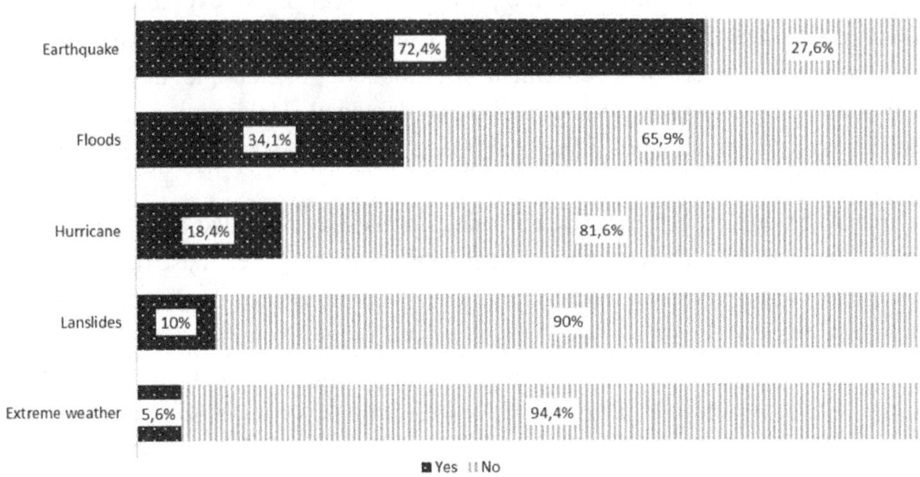

Figure 7. Flood risk perception.

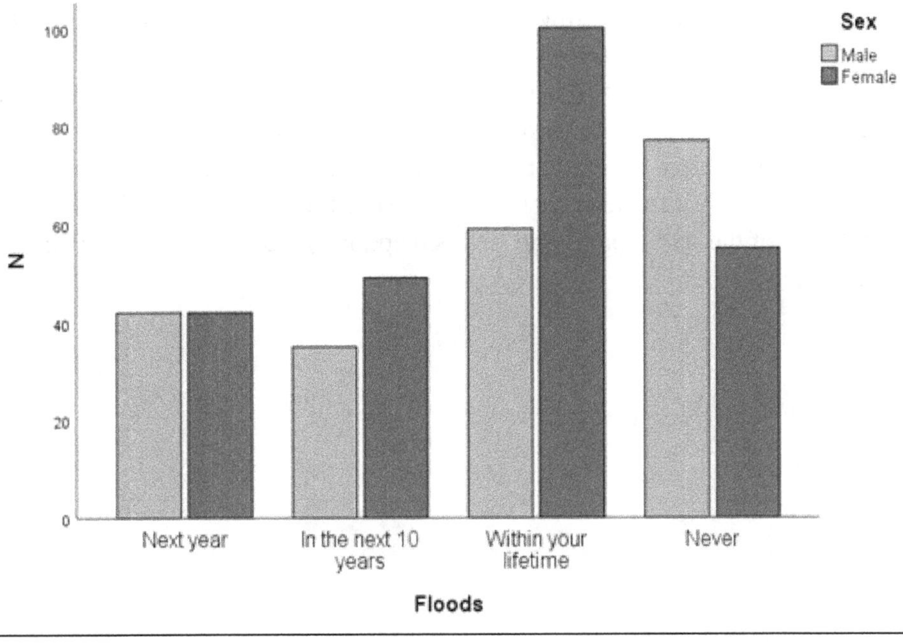

The findings have also highlighted that floods were perceived as a threat (34%) to the participants. (It has been found a significant relationship between flood risk perception and the variable related to the sex of the participants, χ^2 (3, n = 459) = 12.274, p = .003, Cramer's V = .176; Figs. 6&7). The finding may contradict with what have just been mentioned in the previous subsection; an explanation may be due to the fact that during the raining season, and with irregular settlements, when the Mixteco river overflows, settlements are flooded. For example, in 2009, a total of 6 municipalities were affected, 43 houses damaged, 150 people evacuated, 20 transferred to a shelter due to the overflowing of the Mixteco and Salado rivers (Sandoval, 2009).

It may be argued that earthquakes and floods may regarded as High-Impact, Low-Frequency (HILF) events in this region of the country, i.e., these risks have the potential to cause severe damage on the local community but rarely occur (or low frequency events). As already mentioned, lack of water (drought), a high frequency event, is being perceived by respondents as something normal (i.e., a low impact event). These issues may have influenced to their perceptions of these hazards; however, this requires further research in the context of natural hazards and children.

More generally, it should be highlighted that HILF events have been addressed in socio-technical systems (Li et al., 2023), but there are not that many studies in the context of disasters triggered by natural hazards.

Seismic Risk Perception

Regarding the question "*How likely is that an earthquake would occur in the future?*" The frequency data showed that only 6.3% of the participants of the study perceived that an earthquake "never" would occur (Fig. 8). That is, most of the students perceived that earthquakes would happen sometime in the future, i.e., 48.3% considered than an earthquake would occur "next year". However, when conducting a Chi-square test, there was not statistical significance between the variable sex of the participants and risk perception (χ^2 (3, n = 447) = 2.032, p = .56, Cramer's V= .067).

Figure 8. Seismic risk perception.

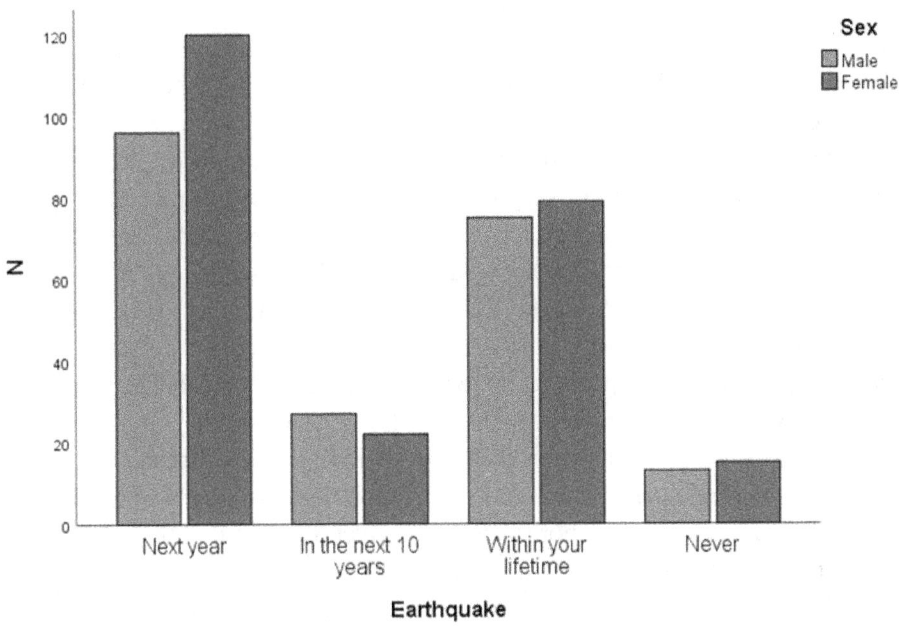

Figure 9. Knowledge on seismic risk.

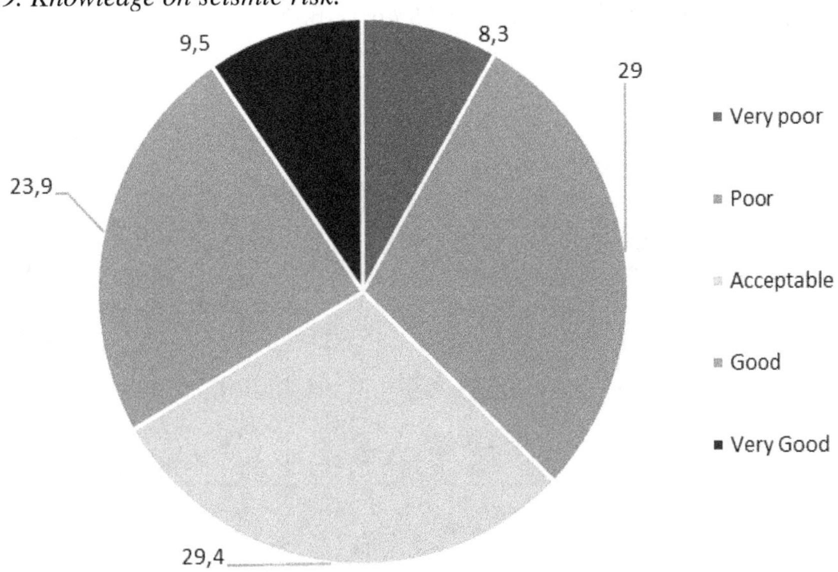

Earthquake Knowledge

Overall, to increase the level of community awareness, knowledge on seismic risk, among other, (and other natural hazards) is essential (Baytiyeh and Ocal, 2016; Paton, et al., 2015; Yildiz, et al., 2021; Andronico et al., 2023; Lechowska, 2018; Lindell, et al., 2021; Bandecchi, et al., 2019).

When students were asked about their level of earthquake knowledge, it was found that around 40% perceived that they lack the knowledge on seismic risk (Fig. 9).

Further questions were asked by the participants in the study, for example, a question on whether they know that they live in a highly seismic region. The results highlighted that around 57% of the students recognized this fact (Fig. 10). It has also been found a significant relationship between the level of knowledge and the variable sex of the participants to the extent that women considered as having a "poor" knowledge when compared with men who considered to having a "very poor" knowledge on seismic risk (χ^2 (4, $n = 507$) = 13.603, $p = .009$, Cramer's $V = .164$).

Overall, the participants of the study self-reported as having poor knowledge on seismic risk. The finding did not surprise us, given the fact that there is a lack of seismic risk education in the educational programs of schools in the state. Several studies have found that those with more earthquake knowledge are more likely to be prepared for earthquakes (Yildiz, et al., 2021, Ronan et al., 2001). Therefore, there is a need to design and implement an educational program not only on seismic risk, but for multi-hazards in all schools in the Oaxaca state.

Figure 10. Knowledge on living in a highly seismic region.

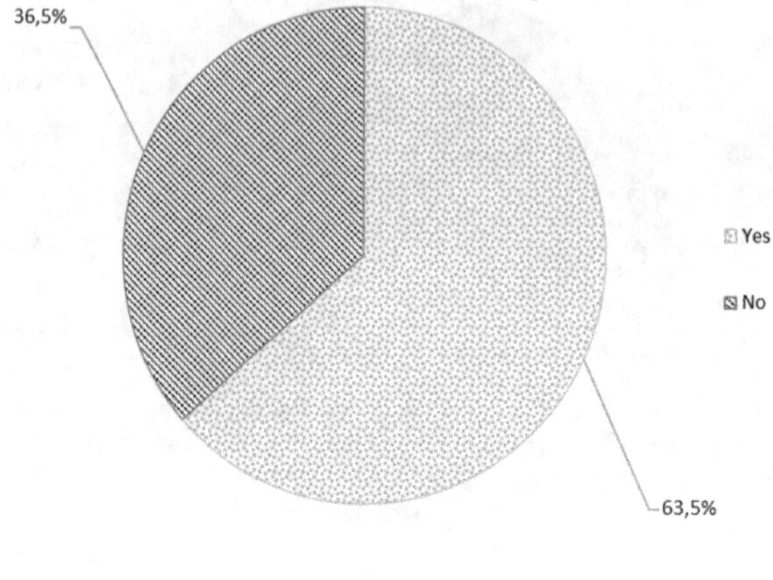

Research has shown that communities' religious belief has a significant influence on preparedness to earthquakes and other natural hazards (Sun, et al., 2018; Yari, et al., 2019; Oven and Bankoff, 2020; Santos-Reyes, 2024). For example, regarding seismic risk perception of a rural community in Kazahastan, it has been reported that "people were also fatalistic, trusting in God to protect them and justify their own inactivity." (Oven and Bankoff, 2020; p. 181).

Regarding the question on why earthquakes occur, most of the participants of the study responded correctly that earthquakes originated due to the slide of tectonic plates (63,8%), Figure 11. It should be mentioned that most of the inhabitants of Huajuapan preach Catholicism, however, the findings have highlighted that a religious belief did not influence on their perception how earthquakes are originated, i.e., around five percent of the participants of the study perceived earthquakes as acts of God.

Figure 11. Earthquake precursors.

Regarding the question on knowing what to do during an earthquake in the classroom, the results have highlighted that most of the students responded "Yes" (94.8%), i.e., they know what protective actions to take. However, their answers were self-reported and future research should be conducted to assess, for example, specific actions during a seismic emergency (see for example, Santos-Reyes & Gouzeva, 2023; Yildiz, et al., 2023; Finnis, et al., 2010).

Fear of Earthquakes

Fear may be regarded as an emotion in anticipation of danger or threat and has been a subject of research by several authors in the context of disasters triggered by natural hazards and related emergencies (Cvetkovic et al., 2019; Ronan et al., 2001; Santos-Reyes and Gouzeva, 2023). Hence, a question related to fear of earthquakes was included in the questionnaire.

The participants of the study were asked the following question "*Does talking or thinking about earthquakes scare you*" The frequency data showed that 10.3% (54/519) "always", 67.8% (354/519) "sometimes", and 21.4% (111/519) "not at all". The results also highlighted that women "always" (14.8%) feel more scared than men (27.8%) when thinking of earthquakes (χ^2 (2, n = 519) = 19.733, $p < .001$, Cramer's $V = .195$).

As with the case of earthquake knowledge, the results highlighted the need of psychological preparedness of the participants of the study. Studies have found that children receiving seismic information at school "reported a significantly stronger belief in their ability to cope with a future hazard and significantly reduced hazard-related fears compared to children who reported not receiving information from a teacher" (Ronan et al., 2001, p.36).

Limitations of the Study & Future Work

As with any study, the presented analysis has limitations and the most relevant of these is the lack of resources to collect data in urban and remote indigenous communities. It is important to have access to such regions to be able to understand their response to earthquakes (and other natural hazards). Future research should address not only children but also adults in these communities. Moreover, a disaster management system should be designed and implemented in such regions where the presented findings should be considered when designing preventive measures (Fig. 4). Last but not least, the present study was for convenience, a random sample of the population vulnerable to multi-hazards should be pursued.

CONCLUSIONS

Children are one of the most vulnerable groups to disasters triggered by earthquakes. The most recent earthquake that occurred in Türkiye and Syria have demonstrated this fact, i.e., around 6.2 million children in both countries were affected and needed humanitarian assistance (UNICEF, 2023; El-Khani, et al., 2023).

The book chapter presented some results of seismic risk perception, knowledge, and preparedness of children in Huajuapan, Oaxaca, Mexico. Some specific findings of the study were the following:

1. Regarding hazard awareness, the two most recognized threats to the participants of the study were earthquakes (72.4%) and floods (34.1%); extreme weather (e.g., drought), on the other hand, was perceived as not representing a big threat to them and their community. However, the lack of water is one of the major concerns of the community.
2. Regarding seismic risk perception, most of the students perceived that earthquakes would happen sometime in the future, e.g., 48.3% considered than an earthquake would occur "next year".
3. Regarding earthquake knowledge, around 40% perceived that they lack the knowledge on seismic risk; 57% recognized that they live in a highly seismic region. Moreover, around 95% responded that they know what to do in case of an earthquake while in a classroom.
4. Finally, around 68% of the participants "sometimes" feel scared or upset when thinking of earthquakes.

More generally, it is hoped that studies such as the presented herein may help to better understand children's perceptions and prepare to risk disaster.

ACKNOWLEDGEMENT

This project was funded under the following grants: SIP-IPN: No-20240864.

REFERENCES

AltlasRiesgo. (2021). Atlas de riesgos-Municipio de Heroica Ciudad de Huajuapan de León, Oaxaca [Risk Atlas-Municipality of heroica City of Huajuapan de león, Oaxaca]. Secretaria de Desarrollo Agrario, Territorial y Urbano (SEDATU).

Antronico, L., Coscarelli, R., Gariano, S. L., & Salvati, P. (2023). Perception of climate change and geo-hydrological risk among high-school students: A local-scale study in Italy. *International Journal of Disaster Risk Reduction*, *90*, 103663. DOI: 10.1016/j.ijdrr.2023.103663

Bandecchi, A. E., Pazzi, V., Morelli, S., Valori, L., & Casagli, N. (2019). Geo-hydrological and seismic risk awareness at school: Emergency preparedness and risk perception evaluation. *International Journal of Disaster Risk Reduction*, *40*, 101280. DOI: 10.1016/j.ijdrr.2019.101280

Baytiyeh, H., & Ocal, A. (2016). High school students' perception of earthquake disaster: A comparative study of Lebanon and Turkey. *International Journal of Disaster Risk Reduction*, *18*, 56–63. DOI: 10.1016/j.ijdrr.2016.06.004

Beard, A. N., & Santos-Reyes, J. (2003). A Safety Management System Model with Application to Fire Safety Offshore. *The Geneva Papers on Risk and Insurance. Issues and Practice*, *28*(3), 413–425. https://link.springer.com/content/pdf/10.1111/1468-0440.00234.pdf. DOI: 10.1111/1468-0440.00234

Bodas, M., Peleg, K., Stolero, N., & Adini, B. (2022). Risk Perception of Natural and Human-Made Disasters—Cross Sectional Study in Eight Countries in Europe and Beyond. *Frontiers in Public Health*, *10*, 825985. DOI: 10.3389/fpubh.2022.825985 PMID: 35252099

Cruz, G. (2023). 100 municipios de la Mixteca con sequía severa y extrema: Conagua [100 municipalities of the Mixteca with severe and extreme Drought: Conagua]. https://www.xeouradio.com/2023/11/10/100-municipios-de-la-mixteca-con-sequia-severa-y-extrema-conagua/

Cvetkovic, V., Ocal, A., & Ivanov, A. (2019). Young adults' fear of disasters: A case study of residents from Turkey, Serbia and Macedonia. *International Journal of Disaster Risk Reduction*, *350*, 101095. DOI: 10.1016/j.ijdrr.2019.101095

Działek, J., Biernacki, W., & Bokwa, A. (2013). Challenges to social capacity building in flood-affected areas of southern Poland. *Natural Hazards and Earth System Sciences, 13*(10), 2555–2566. DOI: 10.5194/nhess-13-2555-2013

El-Khani, A., Calam, R., Sabine-Rakotomalala, L. C., & Maalouf, W. (2023). Urgent help needed for children affected by the earthquake in Syria and Turkiye. *Lancet, 10*(4), 246–247. Advance online publication. DOI: 10.1016/S2215-0366(23)00060-3 PMID: 36870354

Finnis, K. K., Johnston, D. M., Ronan, K. R., & White, J. D. (2010). Hazard perceptions and preparedness of Taranaki youth. *Disaster Prevention and Management, 19*(2), 175–184. DOI: 10.1108/09653561011037986

Finucane, M. L., & Holup, J. L. (2006). Risk as value: Combining affect and analysis in risk judgements. *Journal of Risk Research, 9*(2), 141–164. DOI: 10.1080/13669870500166930

IDMC. (2023). Children and youth in internal displacements. Internal Displacement Monitoring Centre (iDMC). https://www.internal-displacement.org/global-report/grid2022/#download

Kellens, W., Zaalberg, R., Neutens, T., Vanneuville, W., & De Maeyer, P. (2011). An analysis of the public perception of flood risk on the Belgian coast. *Risk Analysis, 31*(7), 1055–1068. DOI: 10.1111/j.1539-6924.2010.01571.x PMID: 21231949

Kousky, C. (2016). Impacts of natural disasters on children. *The Future of Children, 26*(1), 73–92. DOI: 10.1353/foc.2016.0004

Lechowska, E. (2018). What determines flood risk perception? A review of factors of flood risk perception and relations between its basic elements. *Natural Hazards, 94*(3), 1341–1366. DOI: 10.1007/s11069-018-3480-z

Lechowska, E. (2018). What determines flood risk perception? A review of factors of flood risk perception and relations between its basic elements. *Natural Hazards, 94*(3), 1341–1366. DOI: 10.1007/s11069-018-3480-z

Li, J., Meng, L., Zhang, Z., & Yang, K. (2023). Low-frequency, high impact: Discovering important rare events from UGC. *Journal of Retailing and Consumer Services, 70*, 103153. DOI: 10.1016/j.jretconser.2022.103153

Lindell, M., Bostrom, A., Gotz, J. D., & Prater, C. S. (2021). Evaluating hazard awareness brochures: Assessing the textual, graphical, and numerical features of tsunami evacuation products. *International Journal of Disaster Risk Reduction, 61*, 102361. DOI: 10.1016/j.ijdrr.2021.102361

Lindell, M. K., & Hwang, S. N. (2008). Households' perceived personal risk and responses in a multihazard environment. *Risk Analysis*, *28*(2), 539–556. DOI: 10.1111/j.1539-6924.2008.01032.x PMID: 18419668

Lv, Y., Sarker, N. I., & Firdaus, R. B. R. (2024). Disaster resilience in climate-vulnerable community context: Conceptual analysis. *Ecological Indicators*, *158*, 111527. DOI: 10.1016/j.ecolind.2023.111527

Oven, K., & Bankoff, G. (2020). The neglected country(side): Earthquake risk perceptions and disaster risk reduction in post-Soviet rural Kazakhastan. *Journal of Rural Studies*, *80*, 171–184. DOI: 10.1016/j.jrurstud.2020.08.048

Paton, D. (2019). Disaster risk reduction: Psychological perspectives on preparedness. *Australian Journal of Psychology*, *71*(4), 327–341. DOI: 10.1111/ajpy.12237

Paton, D., Anderson, E., Becker, J., & Petersen, J. (2015). Developing a comprehensive model of hazard preparedness: From the Christchurch earthquake. *International Journal of Disaster Risk Reduction*, *14*, 37–45. DOI: 10.1016/j.ijdrr.2014.11.011

Raaijmakers, R., Krywkow, J., & van der Veen, A. (2008). Flood risk perceptions and spatial multi-criteria analysis: An exploratory research for hazard mitigation. *Natural Hazards*, *46*(3), 307–322. DOI: 10.1007/s11069-007-9189-z

Ramos-Cruz, J. M., & Ruiz-Garcia, J. (2024). Mean period prediction model for Mexican interpolate and intermediate-depth intraslab earthquake ground motions. *Soil Dynamics and Earthquake Engineering*, *178*, 108448. DOI: 10.1016/j.soildyn.2023.108448

Reliefweb. (2024). Türkiye/Syria – One in 3 children displaced by Türkiye earthquakes still homeless while needs at record levels in Syria. Türkiye/Syria – One in 3 children displaced by Türkiye earthquakes still homeless while needs at record levels in Syria - Syrian Arab Republic | ReliefWeb

Rohrmann, B. (2008). Risk perception, risk attitude, risk communication, risk management: A conceptual appraisal. https://tiems.info/dmdocuments/events/TIEMS_2008_Bernd_Rohrmann_Keynote.pdf

Ronan, K. R., Johnston, D. M., Daly, M., & Fairley, R. (2001). School Children's Risk Perceptions and Preparedness: A Hazards Education Survey. *Australasian Journal of Disaster and Trauma Studies*, *1*, 1–42.

Sandoval, A. (2009). Se desbordan los ríos Salado y Mixteco; afectados 6 municipios de Huajuapan [The Salado and Mixteco rivers overflow; 6 municipalities of Huajuapan affected]. https://www.xeouradio.com/2009/07/01/se-desbordan-los-ros-salado-y-mixteco-afectados-6-municipios-de-huajuapan/#:~:text=Un%20total%20de%206%20municipios,intensa%20lluvia%20que%20azot%C3%B3%20Huajuapan

Santos-Reyes, J. (2020). Factors motivating Mexico City residents to earthquake mass evacuation drills. *International Journal of Disaster Risk Reduction*, *49*, 101661. DOI: 10.1016/j.ijdrr.2020.101661 PMID: 32455104

Santos-Reyes, J. (2024). Awareness and risk perception of a mult-hazard megacity: The case of adolescent students. *Safety Science*, *171*, 106382. DOI: 10.1016/j.ssci.2023.106382

Santos-Reyes, J., Beard, A. N., & Smith, R. A. (2005). A Systemic Analysis of Railway Accidents, *Journal of Rail and Rapid Transit: Proc. IMechE. Proceedings of the Institution of Mechanical Engineers. Part F, Journal of Rail and Rapid Transit*, *219*(2), 47–65. DOI: 10.1243/095440905X8745

Santos-Reyes, J., & Gouzeva, T. (2023). Adolescents' Responses to the 2017 Puebla Earthquake in Mexico City. *Journal of Disaster Research*, *18*(7), 771–183332. DOI: 10.20965/jdr.2023.p0771

Sharma, S., & Kumar, V. Saruchi. (2022). Community approach toward disaster resilience. In: Siddhartha Bhattacharyya, Kumar Mondal, Bhanu Prakash (eds.), *Cognitive Data Models for Sustainable Environment- A volume in Cognitive Data Science in Sustainable Computing* (pp. 121-161). Elsevier. https://doi.org/DOI: 10.1016/B978-0-12-824038-0.00003-1

Shirazi, N. A., Omidvar, B., & Nohegar, A. (2023). Introduction to decision-making and disaster management systems. In Reza, M., Omidvar, B., & Makarem, M. (Eds.), *Crises in Oil, Gas and Petrochemical Industries-Volume 1: Disasters and Environmental Challengues* (pp. 3–24). Elsevier., DOI: 10.1016/B978-0-323-95154-8.00008-6

SISPLADE. (2000). Macrosismos en el estado de Oaxaca [Macrosisms in the state of Oaxaca]. http://sisplade.oaxaca.gob.mx/sisplade/PeligrosDocs/Tomo2/tomo2%20macrosismos%20en%20el%20estado%20de%20oaxaca.pdf

Slovic, P. (2000). *The perception of risk*. Earthscan.

SSN. (2023). Mapa de sismicidad anual 2023 [Annual sesimicity map 2023]. http://www.ssn.unam.mx/sismicidad/mapas-de-sismicidad-anual/

Sun, L., Deng, Y., & Qi, W. (2018). Two impact pathways from religious belief to public disaster response: Findings from a literature review. *International Journal of Disaster Risk Reduction*, *27*, 588–595. DOI: 10.1016/j.ijdrr.2017.10.004

UN-SDGs (United Nations-Sustainable Development Goals). (2023). Sustainable development goals. https://sdgs.un.org/goals

UNICEF. (2018). El impacto de los terremotos en México en los niños, niñas y adolescentes [The impacto of the earthquakes in Mexico on children and adolescents]. https://www.unicef.org/mexico/informes/impacto-de-los-terremotos-en-m%C3%A9xico-en-los-ni%C3%B1os-ni%C3%B1as-y-adolescentes

UNICEF. (2023). 100 days after earthquakes hit Turkiye and Syria, millions of children continue to face desperate conditions. https://reliefweb.int/report/syrian-arab-republic/100-days-after-earthquakes-hit-turkiye-and-syria-millions-children-continue-face-desperate-conditions-enar

UNISDR. (2009). 2009 UNISDR terminology on disaster risk reduction. United nations International Strategy for Disaster Reduction (UNISDR), Geneva, Switzerland. 2009 UNISDR Terminology on Disaster Risk Reduction - World | ReliefWeb

Wachinger, G., Renn, O., Begg, C., & Kuhlicke, C. (2013). The risk perception paradox— Implications for governance and communication of natural hazards. *Risk Analysis*, *33*(6), 1049–1065. DOI: 10.1111/j.1539-6924.2012.01942.x PMID: 23278120

Wilhite, D. A., Hayes, M. J., & Knutson, C. L. (2005). Drought preparedness planning: building institutional capacity. In: Wilhite, D. (ed.),2005. Drought and water crises: science, technology, and management issues, CRC Press, 93-121.

Yari, A., Zarezadeh, Y., & Ostadtaghizadeh, A. (2019). Prevaence of fatalistic attitudes toward earthquake disaster risk management in citizens of Tehran, Iran. *International Journal of Disaster Risk Reduction*, *38*, 101181. DOI: 10.1016/j.ijdrr.2019.101181

Yildiz, A., Teeuw, R., Dickinson, J., & Roberts, J. (2021). Children's perceptions of flood risk and preparedness: A study after the May 2018 flooding in Golcuk, Turkey. *Progress in Disaster Science*, *9*, 100143. DOI: 10.1016/j.pdisas.2021.100143

Zhang, M., & Zheng, W. (2024). Building disaster resilience through emergency plan updates: A case study of Ya'an, China. *International Journal of Disaster Risk Reduction*, *100*, 104173. DOI: 10.1016/j.ijdrr.2023.104173

Chapter 13
Resilience and Renewal: The Literary Reflections and Societal Impact of Disasters in England

Gülten Akgül
https://orcid.org/0000-0002-0476-6826
The Ministry of National Education, Turkey

ABSTRACT

This chapter explores the historical and literary significance of the 1665 Great Plague and the 1666 Great Fire of London, their profound impact on social, economic, and cultural transformations in England. Through a qualitative approach, the research combines the historical analysis of primary sources such as Samuel Pepys' diaries and Daniel Defoe's "A Journal of the Plague Year" with literary analysis. Findings highlight how the Great Plague disrupted social structures, while also sparking acts of solidarity within communities. The aftermath of the Great Fire facilitated urban rebuilding, leading to modern urban planning and improved sanitation systems. The chapter also examines the psychological resilience of individuals and communities during these crises, particularly through religious and community responses. The research reveals the long-term impact of these events on cultural memory and resilience-building. By examining the historical strategies like quarantine measures and fire brigades, the study offers valuable insights for modern disaster preparedness and management.

DOI: 10.4018/979-8-3693-9745-9.ch013

INTRODUCTION

Throughout history, humanity has faced numerous disasters, both natural and man-made, that have left indelible marks on societies, cultures, and individual lives. Disasters, by their very nature, bring about destruction, suffering, and a loss of normality, but they also serve as defining moments for human endurance, adaptation, and renewal. The term "disaster" refers to any event that disrupts the ordinary course of life, whether through natural phenomena like earthquakes, floods, and storms, or through human-induced catastrophes such as wars, economic crashes, or industrial accidents. These events have shaped not only the physical landscapes of nations but also the cultural and psychological landscapes of societies. Throughout history, major natural disasters such as earthquakes, volcanic eruptions, droughts, and floods have prompted large-scale relief efforts, often with significant international involvement. Recognizing the importance of disaster response, the United Nations designated the 1990s as the International Decade for Natural Disaster Reduction (Alexander, 2018).

The significance of disasters is not merely in their capacity for destruction but in the ways in which human beings respond to them. Disasters force individuals and communities to confront their vulnerabilities, reassess their values, and rebuild their lives from the rubble of what has been lost. The resilience exhibited by individuals and communities in the wake of such events is often nothing short of extraordinary, demonstrating the capacity for recovery, adaptation, and even transformation. In many ways, disasters can be viewed as the moments of crisis that catalyze societal reflection and growth, challenging existing structures and beliefs while giving rise to new forms of solidarity, resilience, and social renewal.

A disaster occurs when an extreme natural event interacts with a human population, and its severity depends on the population's vulnerability and society's ability to provide relief and recovery. Both archaeological and historical evidence suggest that some villages were abandoned due to climatic changes. For example, it was once thought that settlements like Hound Tor in Devon struggled to cultivate land at high altitudes due to the climate fluctuations of the early 14th century. In times of crisis, self-reliance has played an essential role in land use and settlement patterns. Historians previously linked ecological challenges to famines and poor harvests, attributing them to expansion into infertile lands, such as moorlands and heaths, which resulted in lower agricultural yields (Dyer, 2020). Disasters can be categorized based on their causes as either natural or technological. Natural hazards stem from environmental processes, while technological disasters arise from human activities, such as industrial accidents or infrastructure failures. Over the past century, the frequency of hydro-meteorological disasters—those related to water and weather—has increased at a much higher rate than geological or biological disasters.

Between 1964 and 2013, natural disasters affected 6.9 trillion people globally and caused $2.6 trillion in damages. Hydro-meteorological disasters accounted for 74% of all disaster events, 97% of people affected, and 71% of total economic losses during this period (Keim, 2015).

A natural disaster is defined as a catastrophic event resulting from Earth's natural processes, often causing significant environmental, economic, and human consequences. These disasters are closely linked to the workings of the planet, arising from geological, meteorological, and oceanic activities. Their direct effects include destruction of ecosystems, loss of life, and financial hardship, while their indirect impacts manifest as long-term environmental changes, economic instability, and shifts in societal structures. Throughout history, humans have sought explanations for these catastrophic events, shaping cultural narratives and scientific understandings over time. Since ancient times, humans have attempted to interpret natural phenomena beyond their comprehension. Various creation myths, often referred to as cosmic architecture, have played a crucial role in forming early explanations of natural disasters. The Greek philosopher Aristotle (384-322 BCE) theorized that the physical world was composed of four primary elements—earth, water, fire, and air—each with inherent properties of heat, cold, moisture, and dryness. These elements were not only foundational to early scientific thought but were also embedded in myths and legends that explained their influence on human life (Mendia-Landa, 2013). Effective development can help mitigate the impact of disasters by reducing societal vulnerabilities. However, in many instances, development itself has contributed to increased exposure to hazards. Disasters are not extraordinary events but rather part of the everyday fabric of human existence. Consequently, the term "natural disaster" is somewhat misleading, as these events arise from societal vulnerabilities rather than nature alone. Vulnerability is shaped by historical and contemporary processes of change, including development practices and socio-economic conditions (Lewis & Kelman, 2025).

Literature, as a reflection of society, plays a crucial role in capturing the emotional, psychological, and moral dimensions of disasters. Through the written word, authors have chronicled the experiences of individuals and communities who have faced calamities, offering insights into how people make sense of suffering, loss, and survival. Literary works about disasters often explore the human condition in its most vulnerable state, examining the themes of survival, guilt, morality, and hope. They provide a means of processing trauma, offering both a space for catharsis and a medium for reflection on the broader implications of disaster on societal values and human relationships.

The literary treatment of disasters serves several purposes. First, it preserves the collective memory of catastrophic events, ensuring that future generations understand the impact of these tragedies. Second, literature allows readers to vicariously experience the emotional and psychological aftermath of disasters, offering them a deeper understanding of the human condition. Third, it provides a space for cultural commentary, allowing writers to critique societal responses to disasters and explore the moral and ethical questions that arise in their wake. Through the lens of disaster, literature grapples with the complexities of human resilience, survival, and the potential for renewal.

Disasters, both large and small, have been central to the development of cultural narratives throughout history. From ancient mythologies to modern-day fiction, stories of destruction and recovery have been used to explore the fundamental human experiences. The themes of loss, survival, and rebirth are universal, resonating across cultures and generations. While the forms and causes of disasters may vary, the emotional and social consequences are often strikingly similar. In the aftermath of a disaster, communities must come together to rebuild, to mourn, and to reassess their priorities. Literature, through its portrayal of these processes, offers valuable insights into how societies navigate the path from devastation to renewal.

One of the most enduring themes in literature about disasters is the tension between destruction and creation. Disasters bring about profound loss, but they also create space for change, innovation, and reinvention. The destruction of the old order often gives rise to the potential for new beginnings, as communities and individuals seek to rebuild their lives and their world. This duality—destruction leading to creation—is a central theme in many disaster narratives, both in literature and in real-life responses to calamities. The tension between these forces is often reflected in the psychological and emotional journeys of the characters who experience the disaster firsthand. These characters may undergo a process of mourning, despair, and hopelessness, but ultimately, they must find a way to move forward, to rebuild, and to create new meaning in the face of loss.

Moreover, disasters frequently challenge established social structures and norms, revealing the fragility of institutions and systems that people have come to rely on. This disruption often exposes underlying inequalities and injustices, prompting calls for social and political change. In the aftermath of a disaster, societies may be forced to reconsider their priorities, values, and approaches to governance. Literature serves as a powerful vehicle for exploring these societal shifts, offering a critique of the systems that failed and envisioning new ways of structuring society in the wake of disaster. Through the lens of disaster, writers have the freedom to interrogate the moral and ethical dimensions of social order, questioning the assumptions that underpin the established ways of living.

Another key aspect of the literary representation of disaster is its ability to evoke a range of emotional responses. The portrayal of suffering and loss in literature can elicit deep empathy from readers, allowing them to connect with the characters and communities who experience the disaster. This emotional engagement is a crucial part of the healing process, both for those who have lived through the disaster and for those who read about it. By engaging with the emotional aftermath of disaster, literature fosters a sense of shared humanity and solidarity. It reminds readers that the impact of catastrophe is not confined to a single individual or community but is felt across the broader social fabric.

Disasters also expose the limits of human agency and control, highlighting the ways in which individuals and societies are often at the mercy of forces beyond their control. Natural disasters, in particular, underscore the unpredictable and uncontrollable nature of the world, forcing human beings to confront their own vulnerability. This sense of powerlessness is often a central theme in disaster literature, as characters struggle to make sense of their experiences and find meaning in the face of overwhelming forces. Yet, even in the face of such adversity, the human spirit often finds a way to persevere, demonstrating an enduring capacity for resilience and renewal. In the early modern period, people were highly vulnerable to environmental forces, with floods, storms, fires, and earthquakes impacting all levels of society across Europe. However, destruction often gave rise to new developments. Reports of these disasters were published as pamphlets and translated into different languages, indicating a shared sense of calamity that transcended religious divisions. Yet, a closer examination reveals that these publications also reflected complex ideas of national identity and collective experience (Barker, 2020). Natural disasters have always shaped life on Earth, but only in recent years have economists begun analyzing their effects on economic development. Interest in this field has grown due to concerns that global warming may increase the frequency or intensity of certain extreme weather events and natural hazards (Berlemann et al., 2020).

In particular, this chapter will examine how disasters have been portrayed in English literature and how these depictions reflect the societal responses to calamities. England, with its rich literary tradition, offers numerous examples of how writers have engaged with the theme of disaster and its aftermath. From the destruction of natural disasters to the societal upheaval caused by war and industrialization, English literature has long grappled with the implications of catastrophe. Through the study of these works, a deeper understanding can be gained of how literature reflects and shapes the societal understanding of resilience, survival, and renewal. The exploration of disasters in literature not only serves to preserve historical memory but also offers a platform for considering the long-term impact of these events on societal values, collective identity, and cultural imagination. This chapter will explore how the literary representations of disaster and resilience in England have contributed to shaping the ways in which society perceives and responds to catastrophe.

METHODOLOGY

The methodology for this study is designed to explore the societal impact of natural disasters in England, focusing on the themes of resilience and renewal. The approach combines qualitative analysis, historical research, and case study methods to gain a comprehensive understanding of how England has responded to and recovered from natural disasters throughout history. The following steps outline the methodology used in this study:

1. Literature Review

A thorough literature review was conducted to gather existing research and analyses on the societal impact of natural disasters in England. This review included academic articles, books, government reports, and historical accounts to provide a broad understanding of how natural disasters have shaped societal responses in the UK. The literature review covered various types of natural disasters, including flooding, storms, coastal erosion, and extreme weather events. Key topics in the literature included resilience, community recovery, government policies, and the role of social solidarity during post-disaster recovery.

2. Historical Case Studies

The study utilized historical case studies to provide concrete examples of natural disasters and their long-term impact on society. These case studies focused on the significant natural disasters in England. The case studies were used to examine how affected communities exhibited resilience in their immediate response, as well as how renewal efforts contributed to rebuilding and long-term recovery.

3. Government and Policy Analysis

The study also analyzed relevant government policies, regulations, and initiatives aimed at improving resilience and promoting renewal after natural disasters. The role of emergency response agencies, local government involvement, and national strategies for disaster management were explored to assess the effectiveness of these policies in fostering resilience.

4. Content Analysis of Literature

The study analyzed how natural disasters and their aftermath have been represented in literature. A particular focus was given to literature, exploring how English authors have depicted post-disaster recovery and the societal impact of such events. This content analysis helped to understand how cultural narratives surrounding resilience and renewal have shaped public perceptions of natural disasters.

5. Limitations

While the study aims to provide a comprehensive analysis of resilience and renewal after natural disasters in England, there are some limitations. The scope of the research is confined to the analysis of natural disasters in England, which may not capture the full global diversity of disaster responses. Additionally, the study relies on historical case studies, which may limit the ability to capture the evolving nature of disaster resilience and recovery efforts in real-time.

THE NATURAL DISASTERS: EARTHQUAKES, FLOODS, AND THE FORCES OF NATURE

Historically, England has been relatively fortunate in terms of natural disasters, particularly when compared to other regions of the world. However, when disasters did strike, they were often chronicled in ways that magnified their emotional and psychological impact on the population. Earthquakes, floods, and storms, while less frequent, still featured prominently in English literary history as symbols of both external chaos and internal moral or spiritual turmoil. One of the most notable representations of natural disaster in English literature comes from the Great Flood of 1607. Different interpretations of the biblical flood provide insight into how natural disasters were perceived during the 17th and 18th centuries. By examining these interpretations, we gain a deeper understanding of early modern European perspectives on nature, humanity, and their interrelationship (Kempe, 2003). The flooding of the Bristol Channel, which caused widespread devastation in the southwest of England, has been immortalized in works like *The Tempest* by William Shakespeare. The play, which is centered around the magical events of a shipwreck and the subsequent recovery of a lost kingdom, draws on the motif of natural catastrophe as a reflection of both personal and political upheaval. Shakespeare's portrayal of the tempest speaks to the fear and awe that natural disasters inspire while also pointing to their potential as agents of transformation. Much like the tempest in the play, the flood can be seen as an external manifestation of an internal crisis, offering characters the opportunity for redemption and self-realization.

Similarly, the floods that devastated much of England in the 19th century were reflected in the works of writers like Charles Dickens. In *Our Mutual Friend* (1864-65), Dickens uses the River Thames as both a literal and symbolic force. The river, which floods and inundated parts of London, mirrors the social and moral decay of the city's population. Dickens' use of the natural disaster as a metaphor for societal decline provides an insightful commentary on the ways in which humans are at the mercy of both natural forces and social corruption. The Industrial Revolution, which began in the late 18th century and continued into the 19th century, brought about significant technological and economic change in England. However, it also led to severe social and environmental consequences. The rise of factory work, urbanization, and industrialization led to numerous accidents, diseases, and environmental damage. These disasters were often depicted in literature as a reflection of the moral and social costs of unchecked progress. Natural disasters have long been a crucial component of global environmental dynamics. While natural catastrophes have always coexisted with the Earth's evolving natural patterns, human activities have increasingly influenced their frequency and severity, particularly since the 19th century. Industrialization and large-scale engineering projects of the 20th century, alongside the complexities of modern global socio-economic structures, have not only intensified human-made disasters but have also altered environmental conditions, further degrading natural habitats and living conditions (Kondratyev et al., 2006).

In works like *Hard Times* (1854) by Charles Dickens, the dehumanizing effects of industrialization are explored, showing how technological advancement can lead to human suffering and moral bankruptcy. The story is set in the fictional industrial town of Coketown, where the relentless pursuit of efficiency and profit results in a bleak, lifeless environment. The factory system, with its harsh working conditions and lack of concern for the well-being of the workers, is depicted as a disaster in itself—a kind of social and moral catastrophe that erodes the humanity of both the workers and the employers. The industrial disasters of the 19th century also found their way into the works of authors like Elizabeth Gaskell. In her novel *Mary Barton* (1848), Gaskell portrays the struggles of working-class families in Manchester during the early days of industrialization. The devastating effects of economic inequality, child labor, and unsafe working conditions are depicted as a kind of disaster, one that leads to the destruction of both the physical and social fabric of society. Gaskell's treatment of industrial disaster is notable not only for its social critique but also for its exploration of human resilience. Despite the suffering they endure, the characters in Mary Barton continue to struggle for justice and survival, exemplifying the resilience of the working class in the face of systemic disaster.

While natural disasters and industrial accidents have been devastating, it is the human-made catastrophes—wars, political upheavals, and societal conflict—that have often inspired the most profound literary reflections. England has been involved in numerous wars throughout its history, from the Wars of the Roses to the Napoleonic Wars, and these conflicts have shaped the nation's identity and literature in lasting ways. Wars, with their widespread destruction and loss of life, have often been portrayed in literature as tragedies that reveal the darker aspects of human nature.

In *The Charge of the Light Brigade* (1854), Alfred Lord Tennyson immortalizes the ill-fated military charge during the Battle of Balaclava in the Crimean War. The poem serves as a reflection on the futility of war and the sacrifice of the soldiers who, despite being sent on a doomed mission, carried out their orders with courage and honor. The disaster of the Light Brigade's charge, while a symbol of heroism, also highlights the human cost of political and military decisions. Tennyson's work is a poignant example of how wars and military disasters are not only political events but are deeply personal tragedies, affecting the lives of individuals and communities.

The devastation of the First and Second World Wars in the 20th century also found expression in English literature, with works like *Mrs. Dalloway* (1925) by Virginia Woolf and *All Quiet on the Western Front* (1929) by Erich Maria Remarque providing stark portrayals of the psychological trauma and disillusionment caused by war. These works explore the personal toll of disaster on soldiers and civilians alike, with Woolf's *Mrs. Dalloway* focusing on the post-war struggles of individuals trying to rebuild their lives in the face of lingering psychological scars. Remarque's novel, though written by a German author, has had a profound impact on English literature, with its depiction of the brutality and futility of war resonating with English readers. Both works reflect the moral and psychological dimensions of war as a disaster, showing how human lives are irrevocably altered by the ravages of conflict. The literary treatment of disasters in England is not limited to the portrayal of physical destruction or societal collapse. Many works also focus on the psychological and emotional aftermath of calamities, delving into the trauma and suffering experienced by individuals and communities. The impact of disaster on the human psyche has been a recurring theme in English literature, with writers often exploring how people cope with loss, grief, and the challenge of rebuilding their lives.

In *Frankenstein* (1818), Mary Shelley explores the psychological effects of a disaster that is both personal and societal. The novel's protagonist, Victor Frankenstein, creates a being that he cannot control, leading to a series of tragic events that result in the destruction of his family and his own moral decay. The disaster of his scientific ambition is both a personal and societal catastrophe, one that reflects the dangers of unchecked progress and the psychological consequences of failure and guilt. Shelley's work raises important questions about the responsibility of individuals for the consequences of their actions, particularly when those actions have the potential to cause widespread harm. Similarly, the works of authors like Thomas Hardy, particularly in novels such as *Tess of the d'Urbervilles* (1891), explore the emotional toll of disaster on individuals. Tess, a tragic heroine, faces a series of personal disasters that lead to her downfall. Hardy's portrayal of her life as a series of unfortunate events highlights the theme of fate and the psychological consequences of social and moral misfortune. The emotional aftermath of Tess' suffering is deeply tragic, reflecting the harshness of a society that does little to protect its most vulnerable members.

Disasters, whether they are natural, industrial, political, or psychological, have played a central role in shaping the history and literature of England. From the devastating floods and earthquakes of the past to the industrial accidents and wars that have marked the nation's development, English literature has been a means of processing and reflecting on these traumatic events. Through the works of authors like Shakespeare, Dickens, Gaskell, and Hardy, literature has served as both a mirror to the past and a means of exploring the complex emotional, psychological, and social consequences of disaster. The literary representations of disaster offer not only a record of historical events but also a deeper understanding of the human condition, highlighting themes of resilience, survival, and renewal. As England continues to face new challenges, the lessons learned from its literary reflections on disaster remain relevant, offering insights into the enduring power of the human spirit in the face of catastrophe.

THE LITERARY REFLECTIONS OF THE NATURAL DISASTERS IN ENGLAND

The natural disasters, though relatively rare in England compared to other parts of the world, have nevertheless had profound impacts on the country's history, culture, and literary output. The unpredictability of nature and the devastation that such events bring often serve as fertile ground for exploration in literature, where writers have used natural catastrophes to examine the human response to adversity, survival, and renewal. This section focuses on how natural disasters—earthquakes, floods, storms, and other catastrophic events—have been depicted in English literature, reflecting both the immediate emotional response to such events and the deeper philosophical and social implications of living in the face of nature's uncontrollable forces.

One of the most significant natural disasters that affected England was the Great Flood of 1607. This catastrophic event, which struck the southwest of England and Wales, caused widespread devastation along the Bristol Channel, claiming many lives and altering the landscape. The flood, though not as widely recognized as some other natural disasters, had a lasting impact on the local communities and the collective memory of the time. In literature, natural disasters such as floods have been used as metaphors for larger social and political upheavals. One of the most notable works drawing from the themes of flood and storm is William Shakespeare's *The Tempest* (1611). Although *The Tempest* is not directly about a historical flood, Shakespeare uses the imagery of a storm at sea, followed by a shipwreck, as a metaphor for chaos, loss, and renewal. The storm that wreaks havoc on the lives of the characters in the play symbolizes not only nature's power but also the personal and political turmoil faced by the protagonist, Prospero. In this way, the tempest becomes a reflection of the unpredictable forces that can sweep through human life, causing disaster but also offering a chance for reconciliation and growth.

Charles Dickens' famous opening lines from *A Tale of Two Cities*—"It was the best of times, it was the worst of times..."—aptly illustrate the duality of human behavior during periods of extreme social stress. Communities facing disasters often exhibit both admirable and troubling responses. This theme has been explored in literature by writers such as Boccaccio, who described the Black Death in 14th-century Florence, Defoe, who chronicled England's plagues, and Chaucer, who depicted similar crises. Historians from ancient Greece, including Thucydides and Herodotus, as well as modern scholars like Teggart and Toynbee, have also examined how disasters reveal both positive and negative aspects of human nature (Dynes & Quarantelli, 1975). Similarly, in the 19th century, floods in England, such as those caused by the rising waters of the Thames, became a common motif in literature. Writers like Charles Dickens often used water and floods to symbolize societal decay or to highlight the emotional turbulence of characters. In *Our Mutual Friend* (1864-65), Dickens uses the river as a prominent symbol of both destruction and potential renewal. The flooded city of London, portrayed as an industrial and moral wasteland, contrasts sharply with the hope for redemption offered through human connection and reform. While earthquakes are rare in England, their infrequent occurrences still made their mark on both historical and literary consciousness. The most significant earthquake in England's history occurred in 1580, when a tremor shook the country and caused widespread fear and confusion. The event, though not as devastating in terms of loss of life as other global earthquakes, had a psychological impact on the population. In the aftermath, writers began to use earthquakes as symbols of divine wrath, societal instability, and the fragility of human existence.

The earthquake of 1580, along with others in the following centuries, found its way into the literary works of the time, often appearing in the works of poets and playwrights who were concerned with themes of fate and divine judgment. One such example is John Milton's *Paradise Lost* (1667), where Milton uses natural disasters, including earthquakes, as a means of illustrating the chaotic effects of sin on the world. In the epic poem, the fall of Adam and Eve leads to the disordering of the natural world, and Milton's depiction of the natural disasters that follow serves as a symbolic representation of the broken harmony between humanity and the divine.

The frequent storms and coastal disasters in England have long been subjects of fascination for writers. England's geographical location, surrounded by seas, means that storms have historically been a constant threat to its coastal areas. These storms have not only caused damage but have also inspired numerous literary works that engage with the symbolism of nature's power. From the violent storms at sea in works like Shakespeare's *The Tempest* to the destruction of homes and livelihoods in coastal towns, storms are frequently used to represent chaos, danger, and change. In works such as *The Rime of the Ancient Mariner* (1798) by Samuel Taylor Coleridge, the storm at sea serves as both a literal and metaphorical event. In Coleridge's poem, the mariner's ship is caught in a fierce storm, which ultimately leads to a supernatural encounter. The storm represents both the unpredictable forces of nature and the mariner's moral and spiritual crisis. This interplay between nature's fury and human fate is a common thread in literary depictions of storms, particularly in the context of England's maritime history.

In the 19th century, coastal disasters like shipwrecks became a recurrent theme in literature. The unpredictability and terror of the sea were often depicted as a reminder of the power of nature and the vulnerability of human life. Writers like Charles Kingsley, in his novel *The Water-Babies* (1863), use the symbolism of water and storms to represent transformation, innocence, and the harshness of natural law. The storm in Kingsley's novel, though presented in a fantastical context, still highlights the danger that nature presents to human life, particularly in relation to the innocence and purity of childhood. In addition to their physical effects, natural disasters often serve as a mirror for the social and moral conditions of a given time. In literature, disasters like floods, storms, and earthquakes are sometimes used to critique the moral failings of society, highlighting the ways in which human actions can bring about or exacerbate disaster. The reflection on disaster in literature is thus not only about nature's power but also about the human response to it and the societal implications of living in an unpredictable world.

In the early 19th century, the increasing industrialization of England brought about a series of social and environmental issues, many of which were compounded by natural disasters. Writers like Mary Shelley, in *Frankenstein* (1818), used the metaphor of uncontrollable natural forces to critique the unchecked ambition of humans. In her novel, the protagonist Victor Frankenstein attempts to control and manipulate nature through his scientific experiments, only to unleash a disaster that destroys both his life and the lives of those around him. Shelley's depiction of nature as a force that, once disturbed, leads to destruction is a reflection of the anxieties surrounding industrialization and the belief that humans were overstepping their natural boundaries. Likewise, in works such as *North and South* (1854-55) by Elizabeth Gaskell, industrial progress and the resulting social inequalities are often juxtaposed with the disruptive forces of nature. Gaskell's portrayal of the industrial town of Milton, with its factories and pollution, underscores the negative effects of progress, while nature—represented by the changing seasons, storms, and the countryside—becomes a symbol of both escape and moral renewal.

Samuel Pepys holds a place in English literature alongside figures like Chaucer and Shakespeare. His diary remains one of the most valuable historical sources of the period, documenting major events such as the Great Plague, the Great Fire, the Restoration of Charles II, and the Anglo-Dutch Wars. Despite its prominence, early editions of *The Diary of Samuel Pepys* suffered from inaccuracies and omissions. It was only when Robert Latham and William Matthews meticulously transcribed the 300-year-old original manuscript that a comprehensive, fully annotated edition was produced. Pepys' detailed accounts, rich in gossip and observation, offer a remarkably vivid portrayal of 17th-century life (Pepys, 2000). Samuel Pepys and John Evelyn, two of the most renowned English diarists, were also close friends and significant figures in Restoration society. Their diaries provide an intimate window into one of England's most dramatic historical periods (Willes, 2017).

In Daniel Defoe's era, hygiene was poorly understood, and epidemics were perceived as divine punishments, often foreshadowed by celestial omens. The Great Plague of 1665, for instance, was believed to be a form of divine judgment, though Defoe himself attributed its spread to an infection originating in Holland. The idea that the plague was influenced by environmental factors in London was beyond contemporary scientific understanding. Defoe's writings depict the unsanitary conditions of 17th-century London—overcrowded slums, wooden houses surrounded by filth, and the presence of black rats, which contributed to the spread of disease. Historical records reveal that plagues were recurrent in London, with outbreaks occurring regularly from the 15th century to the late 17th century. Plague is an acute infectious disease caused by the bacterium Yersinia pestis, which remains endemic in rodent populations across parts of the Americas, Africa, and Central Asia. Epidemic outbreaks occur when fleas from infected rats transmit the disease to humans, particularly the Oriental rat flea and the human flea. The disease can also spread through person-to-person transmission in its pneumonic form (Frith, 2012). Even after the Great Fire of 1666, which is often thought to have eradicated the infection, plague persisted in the city until 1679. Defoe's account provides a historical perspective on the long-standing presence of epidemics, highlighting how living conditions and urban structures played a role in their spread (Bell, 1924).

To summarize, the natural disasters, though not as frequent in England as in other parts of the world, have nevertheless played a significant role in shaping English literature. From the Great Flood of 1607 to the storms and shipwrecks that have ravaged the English coast, natural catastrophes have provided rich material for writers to explore the human condition, social change, and the relationship between humans and nature. Literature, in turn, has allowed for the preservation of memories of these disasters, reflecting both the trauma and the resilience that they bring. Whether depicting the physical destruction of floods and earthquakes or the moral and spiritual consequences of living in a world prone to natural catastrophes,

English literature has consistently engaged with the themes of chaos, renewal, and the enduring struggle for survival in the face of uncontrollable forces.

THE SOCIETAL IMPACT OF NATURAL DISASTERS IN ENGLAND

The natural disasters, though relatively infrequent in England compared to other regions of the world, have had profound and lasting impacts on the country's society, economy, and cultural psyche. Whether through floods, storms, earthquakes, or coastal erosion, these catastrophic events have reshaped communities, influenced government policies, and left an indelible mark on the national consciousness. While England is not as prone to large-scale natural disasters as other parts of the world, it is still susceptible to a variety of environmental hazards. The country's geographic location, with its proximity to the Atlantic Ocean, makes it vulnerable to coastal erosion, storms, and flooding. Additionally, the country's infrastructure and urban centers are not immune to the effects of extreme weather, including heavy rainfall and flash floods. Earthquakes, although rare, also occasionally make their presence felt. Despite the relative scarcity of extreme natural events, their societal impacts can be profound, especially when they occur in densely populated or historically significant areas.

Flooding has been one of the most recurrent natural disasters in England's history, often causing both physical and psychological devastation. Historical floods, such as the Great Flood of 1607, which affected parts of southwest England, Wales, and the Bristol Channel, and more recent flooding in the 21st century, have had severe societal consequences. These events disrupted daily life, caused the destruction of property and infrastructure, and often led to the displacement of entire communities. The 1953 North Sea flood, for example, had a devastating impact on the East Coast of England, claiming 307 lives and displacing thousands of people. The flood caused widespread damage to homes, farmland, and transportation networks, disrupting local economies and altering the fabric of affected communities.

The societal impact of such flooding disasters is not only seen in terms of physical destruction but also in the long-term psychological toll on affected individuals. Communities that experience recurrent flooding, particularly in flood-prone areas like the Somerset Levels or the Thames Estuary, often face heightened levels of anxiety and stress. The fear of flooding, coupled with the trauma of experiencing a disaster, can have long-lasting effects on mental health. In the aftermath of the 2007 floods, for example, studies found that individuals in affected areas experienced increased rates of anxiety, depression, and post-traumatic stress disorder.

The economic costs of flooding are also substantial. The immediate damage to homes, businesses, and infrastructure can be crippling, while the longer-term recovery efforts require significant investment. For communities that rely heavily on agriculture, flooding can destroy crops and farmland, leading to loss of livelihood. Local economies often struggle to recover, particularly in areas where flooding has become a frequent occurrence. Moreover, the financial burden of rebuilding infrastructure and providing relief to those affected places a strain on government resources, leading to calls for policy reforms and changes in disaster management strategies.

Coastal erosion, another natural disaster that affects England, has similarly had significant societal consequences. The erosion of coastlines due to rising sea levels and increasingly severe storms has led to the loss of both land and property, particularly in areas like East Anglia and the Isle of Wight. The forced displacement of communities in these areas has created social challenges, with many residents facing relocation and the loss of their homes. The social fabric of these coastal communities is often disrupted as people are forced to move away from their traditional homes, resulting in the fragmentation of local cultures and identities.

Storms have been another source of significant societal impact in England. Due to its proximity to the Atlantic Ocean, the country is regularly hit by storms that bring strong winds, heavy rain, and even flooding. These storms often lead to the destruction of infrastructure, such as power lines, roads, and communication systems, making it difficult for affected communities to recover quickly. The Great Storm of 1703, for instance, was one of the most devastating storms in English history, causing widespread damage across the country and killing over 8,000 people. The storm also caused severe damage to the royal navy, leading to a temporary disruption in Britain's military capabilities. More recently, storms like the 1987 "Great Storm" and the 2013-2014 winter storms have had significant impacts on communities, particularly those located along the coast. These storms have caused extensive property damage, including the destruction of homes, businesses, and historic landmarks. The psychological and emotional toll on affected individuals has been considerable, with many people losing their homes and possessions. Furthermore, the economic cost of repairing and rebuilding infrastructure after such storms is often staggering, with significant public and private resources required to restore affected areas. While the immediate impact of storms can be devastating, they also serve as a reminder of the resilience of communities. In the aftermath of the 1987 storm, for instance, there was a concerted effort to rebuild and strengthen communities, particularly through improvements to disaster preparedness and response systems. Over time, storms have fostered a greater sense of solidarity and resilience, with affected communities often coming together to support one another in the recovery process.

Although earthquakes are relatively rare in England, the country has experienced several notable tremors, the most significant of which occurred in 1580. The 1580 earthquake, which struck the area around London, caused widespread panic, and although there was relatively little physical damage, the psychological effects were far-reaching. Earthquakes, while not as frequent in England as in other parts of the world, have had lasting social implications, particularly in terms of how they are perceived by the public. Earthquakes often evoke fear and anxiety, particularly when they strike densely populated urban areas like London, where the possibility of widespread destruction is greater. In the modern era, earthquakes continue to be a source of concern, particularly in the context of ongoing discussions about climate change and environmental instability. Though the likelihood of a major earthquake in England is low, the potential for significant societal impact remains high, particularly in terms of infrastructure damage and the potential disruption of essential services.

In the face of natural disasters, the role of government and public policy is critical. England's government has long been involved in disaster management, with policies evolving over time to better address the challenges posed by natural disasters. In the aftermath of major floods, for example, the government has often enacted reforms to improve flood defenses and enhance emergency response systems. The 1953 North Sea flood, for instance, led to the construction of the Thames Barrier, a major piece of infrastructure designed to protect London from flooding. Similarly, the flooding of 2007 prompted the creation of the Flood and Water Management Act of 2010, which aimed to improve flood resilience across the country. In addition to physical infrastructure, government responses to natural disasters have increasingly focused on social support systems. In the aftermath of major storms and floods, local governments have been tasked with providing relief to affected communities, offering financial assistance, temporary housing, and mental health services. Over time, these efforts have led to a growing recognition of the need for comprehensive disaster management strategies that incorporate both physical and social resilience.

Natural disasters have also played a significant role in shaping the cultural and social responses to adversity. English literature, from the works of Shakespeare to contemporary writers, has long reflected the societal impact of natural disasters. In Shakespeare's *The Tempest*, for example, the storm that leads to the shipwreck symbolizes not only chaos but also the possibility of reconciliation and change. Writers have often used natural disasters as a metaphor for social upheaval or personal transformation, highlighting the ways in which such events can force individuals and communities to confront their vulnerabilities and strengths.nIn the 19th century, authors like Charles Dickens and Elizabeth Gaskell wrote extensively about the effects of industrialization and social inequality, often using natural disasters as symbols of the moral decay of society. Gaskell's *Mary Barton* (1848), for instance, explores the social consequences of economic hardship, with the city's industrial landscape serving as a backdrop for the struggles faced by working-class families. Similarly, in works like *Bleak House*, Dickens critiques the harsh realities of life for the poor, using metaphorical "storms" to illustrate the forces that shape society.

The societal impact of natural disasters in England is multifaceted, touching on everything from physical destruction and economic disruption to emotional trauma and social change. Floods, storms, and other natural events have historically shaped English society, prompting policy reforms, rebuilding efforts, and shifts in cultural consciousness. While the frequency and intensity of such disasters may not be as extreme as in other parts of the world, their effects on local communities and the nation as a whole have been profound. Over time, England has learned to adapt to these challenges, fostering resilience in both its infrastructure and its social fabric. In literature, natural disasters continue to serve as powerful symbols of human vulnerability, strength, and the ever-present need for renewal in the face of nature's unpredictability.

RESILIENCE AND RENEWAL AFTER NATURAL DISASTERS IN ENGLAND

The natural disasters, though relatively infrequent in England compared to other parts of the world, have had a significant and lasting impact on the country's communities, economies, and societal structures. From flooding to coastal erosion, England's response to these calamities has evolved over time, showing remarkable resilience and renewal. Before exploring the specific examples of resilience and renewal in post-disaster England, it is important to define the terms. Resilience refers to the capacity of individuals, communities, and systems to recover from adversity, adapt to challenges, and continue functioning despite disruptions. Renewal, on the other hand, denotes the process of rebuilding or revitalizing areas, institutions, or cultural practices after experiencing significant disruption, often leading to the transformation and improvement of the affected system. In the aftermath of natural disasters, resilience and renewal are often intertwined. Resilience allows communities and individuals to endure the immediate consequences of the disaster, while renewal focuses on the longer-term rebuilding process. Together, they represent the dynamic relationship between overcoming immediate challenges and fostering sustainable growth and improvement in the aftermath of a crisis.

Flooding is one of the most common and destructive natural disasters in England, often leading to widespread damage, displacement, and loss of life. England's resilience in the face of flooding has been shaped by both government intervention and community efforts, with the ongoing threat of flooding often serving as a catalyst for renewal and reform. A significant example of England's response to flooding occurred after the devastating 1953 North Sea flood, one of the deadliest natural disasters in British history. The flood killed over 300 people, displaced thousands, and caused extensive damage to homes, infrastructure, and farmland. The aftermath of the flood prompted significant changes in flood management policy, most notably the construction of the Thames Barrier, a massive flood defense structure that protects London from rising tides and storm surges. The barrier, completed in 1982, is a symbol of the country's resilience in the face of natural disaster, providing both physical protection and a sense of security to millions of people living in flood-prone areas. In addition to governmental action, local communities have displayed remarkable resilience in the face of flooding. For instance, the 2007 floods in the UK, which affected large parts of England, prompted widespread efforts at community-level recovery. Volunteer groups and charity organizations, such as the Red Cross and the Salvation Army, played a crucial role in providing immediate relief to those affected, offering food, shelter, and emotional support. In the long term, community organizations helped coordinate rebuilding efforts, assisting people in repairing homes and restoring local infrastructure.

Moreover, these disasters have spurred discussions about sustainable urban planning and environmental conservation. The 2007 floods led to reforms in flood defense mechanisms, including the Flood and Water Management Act of 2010, which focused on better flood risk management and prevention strategies. This act incorporated lessons learned from past floods and reinforced the need for both local and national efforts to mitigate future flood risks. Thus, while floods have brought devastation to communities, they have also led to the renewal of policies and the development of strategies designed to prevent or minimize future damage.

England's coastline has long been a site of both natural beauty and vulnerability. Coastal erosion, particularly in areas like East Anglia and the Isle of Wight, has been a recurrent issue for centuries. As rising sea levels and increasingly powerful storms threaten the coast, the country's response to these challenges has been one of resilience and renewal. The phenomenon of coastal erosion has had a profound impact on local communities, as homes and businesses near the coast face the risk of being swallowed by the sea. For example, in the village of Happisburgh in Norfolk, residents have watched as the coastline steadily eroded, causing homes to fall into the sea. In such instances, communities have demonstrated resilience by adapting to the changing landscape, often by relocating homes and businesses inland or adopting coastal defense measures such as seawalls and breakwaters. However, the process of renewal in these coastal areas is not always easy or straightforward. Many communities face economic hardship due to the loss of land, property, and access to coastal resources. In some cases, communities have had to come together to advocate for stronger government intervention and investment in coastal protection. For example, in the 1990s, the residents of Happisburgh campaigned for government funding to help prevent further erosion, ultimately securing a commitment for the construction of flood defenses and beach nourishment programs. These efforts at resilience and renewal have not only involved physical infrastructure but also social and cultural renewal. As communities face the prospect of losing their homes to the sea, many have worked to preserve local heritage, promoting cultural tourism and historical landmarks as a way to sustain their economies. In this sense, coastal erosion has prompted a cultural renewal, as communities reimagine their identities in the face of environmental change. These efforts reflect a broader trend in England's response to natural disasters, where communities do not simply return to their pre-disaster state but also strive for innovation and renewal in the wake of loss.

Storms and extreme weather events, including powerful hurricanes, blizzards, and heatwaves, have also shaped England's resilience and renewal strategies. One of the most notable examples of England's response to a major storm is the 1987 "Great Storm," which caused widespread damage across the country. With winds reaching up to 115 mph, the storm uprooted trees, damaged buildings, and disrupted transportation networks. In the wake of the disaster, the British public displayed remarkable resilience, with local communities banding together to clear debris and repair damage. Furthermore, the storm prompted significant improvements in weather forecasting, emergency response systems, and disaster preparedness. The renewal of infrastructure after extreme weather events like the 1987 storm has been a key aspect of England's response. The storm highlighted the vulnerability of the country's infrastructure, particularly in relation to power lines and transportation systems. As a result, significant investments were made in reinforcing infrastructure to better withstand future storms. The storm also contributed to the development of more robust disaster management systems, including the establishment of the Met Office's Severe Weather Warning service, which has improved the country's ability to predict and respond to extreme weather events. In addition to infrastructure renewal, storms have often fostered a sense of social solidarity and community spirit. After the 1987 storm, residents of affected areas worked together to support one another, offering shelter, food, and assistance in rebuilding. The collective effort to recover from the storm strengthened community bonds and contributed to a broader sense of national unity. This social resilience, in turn, has inspired subsequent generations to approach future natural disasters with a sense of solidarity and collective action.

THE ROLE OF GOVERNMENT AND POLICY IN RESILIENCE AND RENEWAL

The role of government policy in facilitating resilience and renewal after natural disasters in England cannot be overstated. The government has played a central role in providing resources for recovery, as well as in the creation of policies designed to mitigate the impact of future disasters. Over time, the UK has developed a robust system for disaster management, which includes not only emergency response efforts but also long-term strategies for rebuilding and renewal. For example, the 2007 floods led to the creation of the Flood and Water Management Act of 2010, which established new guidelines for flood risk management and emphasized the importance of community-based approaches to disaster response. This act also facilitated the development of flood risk maps, which allow local authorities to identify high-risk areas and take appropriate action. The government's approach to disaster management has evolved to reflect a growing recognition of the importance of prevention, resilience, and community involvement.

The renewal process also involves supporting affected communities in rebuilding their lives. This includes providing financial aid, temporary housing, and mental health support to help individuals cope with the trauma of disaster. Additionally, the government has worked with charities, such as the Red Cross and the Salvation Army, to provide essential services during the recovery process. Beyond physical and infrastructural renewal, natural disasters in England have often led to cultural and psychological renewal. After a disaster, communities are faced with the task of healing both physically and emotionally. The psychological toll of experiencing a natural disaster can be significant, with many individuals suffering from trauma, anxiety, and depression. However, in the aftermath of such events, there is often a process of collective healing, as communities come together to support one another and rebuild their lives.

Cultural renewal after natural disasters often involves the preservation of local traditions, histories, and identities. In many cases, communities use the recovery process as an opportunity to reaffirm their cultural heritage, creating new cultural institutions or reviving old ones. For example, after the 1953 North Sea flood, many affected communities worked to preserve their local histories, using museums and oral histories to document the impact of the disaster. This cultural renewal has helped foster a sense of pride and resilience among affected communities, contributing to their ability to recover in the long term.

The resilience and renewal of England in the face of natural disasters have been shaped by a combination of government policy, community efforts, and cultural adaptation. From flooding and coastal erosion to storms and extreme weather, the country has demonstrated a remarkable ability to bounce back from adversity. While the physical destruction wrought by natural disasters is often devastating, the process of recovery has led to the development of stronger infrastructure, more effective disaster management policies, and a renewed sense of community solidarity. Ultimately, England's response to natural disasters has not only been about rebuilding what was lost but also about creating a more resilient, sustainable, and connected society that can endure the challenges of the future.

DISCUSSION, FINDINGS, AND RESULTS

The societal impact of the natural disasters in England is multifaceted and deeply intertwined with various historical, socio-political, and cultural factors. Through this research, the central themes of resilience, recovery, and renewal have emerged as essential components in understanding how England has responded to natural disasters over time. This section presents the key findings of the study, discussing both the direct and indirect effects of natural disasters on communities, as well as how these events have shaped broader societal structures. The results reflect a nuanced understanding of how resilience is cultivated and renewed in the face of devastation, and how disasters serve as catalysts for long-term change.

Beyond the immediate physical and infrastructural consequences of the natural disasters, one of the most significant findings of this study is the profound psychological and social impact these events have on affected individuals and communities. The resilience of a community in the aftermath of a disaster is not only dependent on physical reconstruction but also on the ability of individuals to recover emotionally and socially. The emotional toll of disasters is often underestimated, yet it is a crucial factor in determining how effectively individuals and communities can rebuild their lives.

Personal testimonies from survivors of the 1953 North Sea Floods, for example, reveal that many individuals continued to suffer from feelings of loss, grief, and anxiety long after the flood waters had receded. Similarly, individuals affected by the 2007 UK Floods reported long-term psychological effects, including post-traumatic stress disorder, depression, and anxiety. The trauma experienced by survivors can have a ripple effect, influencing relationships, work life, and community involvement for years to come. The study highlights the importance of integrating mental health services into disaster response efforts to address these psychological impacts. Local authorities, non-governmental organizations, and mental health professionals have a key role to play in providing counseling, emotional support, and coping strategies to those affected by natural disasters.

Moreover, the social resilience of communities plays an equally important role in disaster recovery. Communities that are able to come together in the face of disaster tend to recover more quickly, as social networks provide crucial emotional support, practical assistance, and a sense of belonging. The research found that in both the 1953 floods and the 2007 floods, community networks, such as local volunteer organizations, neighborhood groups, and faith-based institutions, played a vital role in helping residents cope with the aftermath of the disaster. These social bonds not only provide immediate relief but also foster long-term recovery by rebuilding trust, solidarity, and a sense of shared purpose.

One notable example of this social resilience can be seen in the aftermath of the 1987 Great Storm, which caused widespread damage in southern England. Despite the destruction, the response from local communities was swift, with residents working together to clear debris, provide shelter, and offer assistance to those in need. This collective effort contributed to the recovery of affected areas, demonstrating the strength of social capital in the disaster recovery process.

The way in which natural disasters are represented in the media also plays a significant role in shaping public perceptions of resilience and recovery. Media coverage of disasters serves to inform the public about the scope of the damage, the ongoing recovery efforts, and the actions being taken by the government and emergency services. However, media representations can also influence how people perceive the severity of the disaster, the adequacy of response efforts, and the resilience of affected communities. In the case of the 2007 UK Floods, for example, the media played a crucial role in raising awareness of the extent of the damage and the challenges faced by affected communities. News outlets provided extensive coverage of the flooding, highlighting the personal stories of those affected and the efforts of local authorities to provide assistance. This media coverage contributed to a sense of urgency and public concern, prompting both government action and public support for disaster recovery efforts. However, media portrayals can also be sensationalized, focusing on dramatic images and personal tragedies, which may inadvertently contribute to feelings of fear, helplessness, and anxiety among the public.

The literary representations of disaster also serve to shape cultural attitudes towards resilience and renewal. From works like John Galsworthy's *The Forsyte Saga*, which explores the psychological effects of loss and displacement, to Ian McEwan's *Atonement*, which deals with the moral and emotional consequences of traumatic events, literature provides a lens through which society can reflect on its experiences of disaster and recovery. These cultural narratives contribute to a broader understanding of how resilience is forged, both at the individual and collective levels, and how society can emerge from disaster with renewed strength and purpose. The Great Plague of 1665 and the Great Fire of London in 1666 were two of the most catastrophic events in England's history, profoundly reshaping social structures, urban planning, and cultural memory. These crises exposed vulnerabilities within society but also revealed remarkable resilience, adaptation, and innovation in response to adversity. By analyzing primary sources such as Samuel Pepys' diaries and Daniel Defoe's *A Journal of the Plague Year*, this study explores how these events influenced literature, public health policies, and collective consciousness.

For centuries, plague was the most feared disease in Europe, known for its devastating mortality rate and repugnant symptoms. The last and deadliest outbreak in London occurred in 1665, when, as Bishop Gilbert Burnet recorded, the disease depopulated the city, disrupted trade, and claimed nearly 100,000 lives. In just eight months, around one-fifth of the city's population perished. However, the epidemic was not confined to London; it spread across East Anglia, southern England, and as far north as Tyneside and Wearside. Towns like Colchester, Winchester, Southampton, and Norwich, as well as the village of Eyam in Derbyshire—famous for its self-imposed quarantine—experienced even higher proportional death tolls than the capital. Daniel Defoe would later refer to 1665 as "this calamitous Year" (Porter, 2009).

The Great Plague of 1665 led to widespread fear, isolation, and breakdowns in social structures. The enforcement of quarantine measures, such as marking infected houses with red crosses and isolating the sick, resulted in both compliance and resistance. Many wealthier individuals, including King Charles II and the royal court, fled London, while the poor were left to struggle with inadequate medical care. However, the plague also fostered solidarity and acts of compassion within communities. Some citizens provided food and aid to quarantined families, while religious institutions played a crucial role in offering spiritual support. Samuel Pepys' diary entries depict how Londoners sought solace in religious faith, viewing the plague as divine retribution while also engaging in desperate attempts to find remedies, from herbal treatments to the controversial use of bloodletting.

Historian A. Lloyd Moote and microbiologist Dorothy C. Moote offer a compelling and well-researched account of the plague year. Their work provides a panoramic yet personal view, taking readers from the homes of the elite to the crowded slums and the countryside where many sought refuge. They reveal that, despite the horrors, the city did not descend into anarchy. Physicians, apothecaries, clergy, and public officials stayed behind to help the afflicted, using legal measures to manage the crisis. Commerce persisted, even as many businesses closed. The authors focus on the experiences of nine individuals, including an apothecary serving a poor district, a rector from an affluent parish, a wealthy silk merchant and alderman, a country gentleman, and the famous diarist Samuel Pepys. Through personal letters and diaries, the book provides fresh perspectives on the crisis, exploring how different communities perceived the disease, how authorities and religious institutions responded, and the moral and economic dilemmas faced by those deciding whether to flee or stay. By emphasizing the human experience of the epidemic, the authors offer a powerful retelling of how Londoners endured and resisted a catastrophe of unimaginable proportions (Moote & Moote, 2004). Samuel Pepys' diary offers one of the most vivid accounts of the fire, describing the horror of burning embers falling from the sky. His observations, both personal and public, highlight how language shapes historical narratives. Pepys documented his emotional turmoil after witnessing the destruction, making his diary a uniquely compelling trauma narrative. His writing captures both his private distress and the broader social response, providing invaluable insights into how Londoners coped with the catastrophe (Trigg, 2016). The 17th century, particularly the Restoration period from 1660 to 1700, was one of the most fascinating eras in British history. It was an age of great transformation, marked by figures such as Samuel Pepys, Christopher Wren, and Isaac Newton. With the English Civil Wars over, society entered a new phase of scientific inquiry, artistic flourishing, and shifting social norms. Ian Mortimer's *Time Traveller's Guide to Restoration Britain* brings this period to life, addressing questions about daily life, food, fashion, and commerce. The book highlights how the 17th century represented a turning point, as superstition gradually gave way to rationalism and professionalism. By examining Restoration Britain, readers gain a deeper appreciation of historical customs while also seeing their influence on the modern world (Mortimer, 2017).

The Great Plague of London was one of the most devastating epidemics in the city's history, beginning in the winter of 1664-65. As Christmas approached, a bitter cold gripped London, and a bright comet streaked across the sky, seen as a harbinger of ominous events. In the impoverished district of St. Giles-in-the-Fields, just outside the city walls, Goodwoman Phillips succumbed to the plague, and her house was sealed with the inscription "Lord Have Mercy On Us" painted on the door in red. Within a year, nearly 100,000 people—about a third of those who remained in the city—would perish due to the outbreak. The epidemic severely impacted London's economy and society, yet the city continued to function, with life carrying on despite the crisis.

In 1666, just a year after the plague, London faced another catastrophe—the Great Fire. One of the most destructive events in the city's history, the fire reshaped London's architecture and had lasting economic consequences (Field, 2017). Such disasters, whether personal or communal, often lead people to seek deeper spiritual meaning. Many observed that God's presence is most acutely felt in moments of suffering. This was true for the Great Fire, which impacted all aspects of English society. Since the Church of England played a central role in national life, the disaster was met with a religious response. Within a month, a special church service was created and mandated across the country, later becoming part of the Book of Common Prayer and undergoing revision 30 years later. These services provide insight into the theological and cultural mindset of the time. The fire began on the night of September 2, 1666, in the bakehouse of Thomas Faryner, the king's baker, on Pudding Lane. At the time, London remained a largely medieval city, with tightly packed wooden buildings and narrow streets, making it highly susceptible to fire (Gilpin, 2013). The Great Fire of London in 1666 caused immediate devastation but ultimately paved the way for urban resilience and modernization. Initially, the lack of an organized firefighting system led to chaotic responses, with citizens using buckets of water and tearing down buildings to halt the spread. However, in the aftermath, the disaster led to the implementation of new fire prevention measures, including the use of brick and stone buildings instead of timber and the creation of organized fire brigades. Fires had long been a major threat in early modern Britain, where open flames were essential for heating, lighting, and cooking. However, it was only in the 18th century that fire insurance became widely available. Although major fires in the 17th century underscored the economic devastation they could cause, it wasn't until industrialization shifted economic structures that fire insurance became a viable concept (Pearson, 2017). The Great Fire ultimately contributed to London's development as a financial hub. While fire insurance did not emerge immediately, financial innovation grew in the aftermath of the disaster (Rawlings, 2016). Following the Great Fire, London's recovery was swift, though economic activity shifted toward Westminster by 1690. Markets expanded beyond the city center, while financial services remained concentrated within London. Analysis of Hearth Tax records indicates significant changes in wealth distribution, with wealthier residents returning to fire-affected areas, reshaping London's housing and social structure (Ager et al., 2024).

The economic impact of the plague and fire was profound. During the plague, businesses and trade suffered severe setbacks due to labor shortages and the fear of contagion. Many artisans and merchants saw their livelihoods collapse as people avoided markets, inns, and public spaces. However, these disruptions also created opportunities for upward mobility for some survivors. With labor in high demand after the plague, wages rose, leading to improved conditions for many workers. The fire's aftermath similarly reshaped economic activity. While thousands lost their homes and businesses, the rebuilding efforts stimulated employment and investment. The rebuilding of London, spearheaded by architects such as Sir Christopher Wren, provided jobs for masons, carpenters, and builders, driving an economic recovery. The fire also led to long-term financial reforms, including the introduction of new insurance policies to protect against future disasters.

Both events left lasting psychological scars on survivors, shaping cultural narratives of suffering, endurance, and renewal. Literature and personal accounts, such as Pepys' diary, highlight the fear and uncertainty that plagued the population. In *A Journal of the Plague Year*, Daniel Defoe explores themes of human endurance, fear, and survival strategies, capturing the psychological toll of living through a pandemic. The cultural memory of these events contributed to shaping England's identity. The Great Plague reinforced the need for public health measures, while the Great Fire symbolized rebirth and resilience. The fire's aftermath, in particular, became a symbol of London's strength, as the city emerged with improved urban planning, sanitation, and infrastructure.

Quarantine measures and public health responses during the 1665 plague set precedents for modern disease control strategies. The government enforced isolation policies, though enforcement was inconsistent. Religious and superstitious beliefs influenced people's reactions to the plague, with many viewing it as divine punishment, while others sought medical explanations. The pandemic led to demographic shifts, with urban populations declining temporarily due to high mortality rates and migration to rural areas. The fire exposed structural weaknesses in London's urban planning, leading to the reconstruction of the city with fire-resistant materials and wider streets to prevent future fires. The event accelerated the professionalization of firefighting, resulting in the establishment of organized fire brigades. The disaster stimulated economic growth by creating job opportunities in construction, architecture, and urban planning. Survivors of both disasters exhibited psychological resilience, relying on religious faith, community support, and literary expressions to process trauma.

Literature played a crucial role in preserving memories of these events, with Defoe's fictionalized account offering insights into the fears, behaviors, and survival mechanisms of the time. The disasters influenced government policies on disaster preparedness, leading to reforms that shaped modern emergency response strategies. The study's results provide significant insights into how historical crises shape social resilience, economic policies, and cultural narratives. The plague led to increased awareness of quarantine measures and sanitation, which influenced later public health policies. The notion of disease control through isolation and hygiene became more widely accepted, laying the foundation for future epidemic responses. The Great Fire's aftermath transformed urban planning practices, leading to safer building designs, better sanitation, and improved infrastructure. London's reconstruction served as a model for other cities facing disaster recovery. The emotional and psychological effects of these disasters continued to influence English literature, shaping themes of loss, renewal, and human resilience. Writers like Defoe used the plague as a backdrop to explore societal responses to the crisis, while the fire's narrative of destruction and rebirth became a symbol of London's endurance.

Throughout history, England has experienced a variety of disasters, each leaving its mark on the fabric of society, culture, and collective memory. The 1665 Great Plague and the 1666 Great Fire of London, while undoubtedly monumental, are only two examples in a longer history of transformative events that forced communities to reckon with disaster, adapt, and rebuild. From political upheavals to economic crises and natural disasters, these events, often reflected in literature, reveal much about the nature of resilience and renewal. This study examines the societal and literary responses to such crises, highlighting how disasters shape public health, urban planning, cultural identity, and psychological resilience, and offering lessons for modern challenges.

The Great Plague of 1665 and the Great Fire of 1666 are often remembered for their immediate and devastating impacts, yet they also serve as reminders of the resilience inherent in human societies. As seen in the diaries of Samuel Pepys and Daniel Defoe's *A Journal of the Plague Year*, these disasters tested the limits of London's infrastructure, exposing significant weaknesses, particularly in public health and urban safety. The plague disrupted daily life, as people fled the city in fear, while others sought solace in religious practices, remedies, and increasingly desperate measures of survival. Despite the overwhelming sense of dread and loss, communities came together in solidarity, helping those affected and finding new ways of organizing themselves.

The Great Fire, though initially a massive loss, was a catalyst for profound urban transformation. As the flames consumed large parts of London, the city was forced to confront the need for reformed urban planning and fire prevention. The reconstruction efforts led to wider streets, fire-resistant buildings, and a more organized city structure, setting the stage for London's rise as a modern urban center. However, these two events are not isolated cases in England's history of resilience. Over centuries, England has faced other monumental crises—from the Black Death of the 14th century to the London Blitz of World War II—each testing societal endurance. The recurring theme throughout these disasters is not just the suffering but the renewal that follows. Communities have historically shown remarkable resilience, not simply through survival, but through their ability to adapt, innovate, and rebuild.

The role of literature in reflecting and shaping societal responses to disaster cannot be overstated. The Great Plague and Great Fire of 1666, for example, have been immortalized in the works of Defoe and Pepys, but they are by no means unique in this regard. English literature, across periods, has consistently explored the human condition in the wake of disaster—whether war, disease, political collapse, or natural catastrophe. Through these narratives, literature does more than record events; it provides a space for reflection, processing trauma, and, ultimately, reconstructing meaning. The psychological impact of these crises is a central theme in much of the literature. As seen in Defoe's portrayal of the plague, individuals' responses to crises ranged from fear and superstition to moral reckoning and perseverance. Pepys' personal reflections on the plague and fire capture the internal turmoil of living in constant fear and uncertainty, yet also offer glimpses of hope, resilience, and the potential for renewal. Literature thus serves as a vehicle for cultural memory, ensuring that these lessons of survival, fear, and recovery are passed down through generations. The impact of disasters, however, is not always immediate in literature. Later works, like those written after the Blitz, often reflect on past traumas with the benefit of time and reflection. Writers such as Virginia Woolf and George Orwell explored the psychological and societal consequences of disaster, often placing emphasis on the reconstruction of identity and social order in the aftermath of chaos. These literary responses are not limited to specific periods of crisis; rather, they offer enduring insights into how societies cope with both collective and individual trauma.

While the immediate effects of the Great Plague and Great Fire may have been devastating, the long-term transformations they sparked had profound effects on public health, urban development, and disaster management. The Great Plague led to the introduction of quarantine measures and an increasing focus on sanitation. These measures were, in many ways, precursors to modern public health strategies used to combat epidemics. Similarly, the fire prompted the establishment of building codes, the professionalization of firefighting, and a renewed focus on urban planning. Disasters like these also prompted pivotal economic reforms. The rise of insurance after the fire, for instance, marked a significant shift in how individuals and businesses approached risk, creating a financial infrastructure that would later be key to recovery during future crises. The reconstruction of London, driven by both necessity and opportunity, became a model for urban renewal worldwide. The economic recovery after these events not only stabilized the city but also accelerated the growth of commerce and trade, ensuring that London would remain a leading economic power.

Another significant finding is the long-term impact of natural disasters on infrastructure and policy. The physical devastation caused by disasters often serves as a catalyst for policy change and infrastructure improvement. The 1953 North Sea Floods, for instance, led to the creation of the Thames Barrier, a massive flood defense structure designed to protect London from future flooding events. Similarly, the 2007 UK Floods prompted a reevaluation of flood risk management, leading to the introduction of new regulations and strategies aimed at reducing flood risks and enhancing the country's flood resilience. These examples illustrate the significance of viewing the natural disasters not only as catastrophic events but also as opportunities for innovation and improvement. The research suggests that the recovery process following a disaster provides an impetus for rethinking urban planning, infrastructure development, and environmental management. By integrating lessons learned from past disasters into future policy and planning, England can enhance its capacity to withstand and recover from future natural disasters.

In conclusion, the societal impact of natural disasters in England is profound and multifaceted, with significant implications for resilience, recovery, and renewal. The findings of this study underscore the complex interplay between individual, community, governmental, and infrastructural responses to disasters. While significant progress has been made in disaster preparedness, response, and recovery, the research suggests that continued investment in policy development, infrastructure, mental health support, and social resilience is crucial for enhancing the country's long-term resilience to natural disasters. By fostering a culture of resilience and learning from past experiences, England can better navigate the challenges posed by future disasters and emerge stronger in the face of adversity.

This chapter has explored the societal impact of natural disasters in England, focusing particularly on the themes of resilience and renewal. Throughout history, the natural disasters have significantly shaped both the physical landscape and the social fabric of communities. From the catastrophic 1953 North Sea Floods to more recent events such as the 2007 UK Floods and the 1987 Great Storm, England has experienced a range of natural disasters that have tested the resilience of its people, infrastructure, and systems. The findings of this study emphasize how the country's response to such disasters has evolved over time, shaped by both local and national efforts toward recovery, adaptation, and renewal.

One of the key insights from this chapter is the prominence of resilience in the aftermath of natural disasters. Resilience, in this context, refers not only to the ability to withstand the immediate impacts of a disaster but also to the capacity to recover, adapt, and emerge stronger in the long term. In the case of England, this resilience is evident at multiple levels—within individual communities, local governments, and national policies. Communities affected by disasters often demonstrate a remarkable ability to rebuild, both physically and emotionally. The shared experience of loss, hardship, and recovery can foster a sense of solidarity, encouraging people to come together in the face of adversity. This sense of collective strength and social cohesion is central to the recovery process, particularly when individuals and communities feel supported by both their peers and local authorities. Moreover, the renewal process following a natural disaster often leads to long-term improvements in infrastructure, disaster management practices, and policies. For example, the response to the 1953 North Sea Floods resulted in the creation of the Thames Barrier and other flood protection measures that continue to serve as critical safeguards against future flooding events. Similarly, the 2007 UK Floods prompted a reevaluation of flood risk management, leading to significant changes in policies related to water management and the prevention of flooding. These examples underscore the importance of learning from past disasters and implementing improvements that contribute to both immediate recovery and long-term resilience.

A key finding of this chapter is the critical role of government and policy in shaping the response to natural disasters. The UK government has implemented a range of policies and initiatives aimed at improving disaster preparedness and recovery. The Flood and Water Management Act of 2010 is one such example, which reflects a growing recognition of the need to address flood risks proactively and to develop systems that enable more effective response and recovery. The involvement of local governments and agencies, alongside national policy frameworks, ensures a coordinated and effective response, which is essential for managing the complex challenges posed by natural disasters. However, while policies and legislation are important, their success often depends on the extent to which they can be implemented effectively at the local level, where the direct impact of disasters is felt most acutely.

Furthermore, the chapter highlights the significant role of media representations and literature in shaping public perceptions of natural disasters and recovery. The way in which disasters are portrayed in the media can influence how people understand their risks, prepare for potential events, and respond in the aftermath. The media also plays a crucial role in providing information, raising awareness, and promoting a sense of urgency during disasters, while also contributing to the public discourse surrounding resilience and renewal. In literature, the themes of disaster, recovery, and renewal have long been explored, with works of fiction reflecting the societal anxieties and hopes that arise during and after catastrophic events. These cultural representations contribute to a broader understanding of how societies cope with the trauma of disaster and rebuild in the wake of destruction.

Looking to the future, the lessons of resilience and renewal from past disasters remain highly relevant. The experiences of London during the plague and fire can offer valuable insights for modern cities facing challenges such as pandemics, climate change, or terrorist attacks. The importance of disaster preparedness, community solidarity, and urban planning cannot be understated. Likewise, the psychological resilience demonstrated during these crises provides a model for how societies can bounce back from contemporary challenges, emphasizing the significance of mental health and community support systems in navigating future catastrophes. The study of England's past disasters reveals that resilience is not limited to the immediate aftermath of a crisis, but is woven into the fabric of society through its ability to adapt and transform. Whether it's through literary reflections or public policy changes, the ability to recover and rebuild is a testament to the strength of human ingenuity and communal effort. The literary responses to these events, and the cultural narratives that emerge, help preserve the lessons of the past, ensuring that future generations are equipped with the tools needed to face new crises.

CONCLUSION

The historical and literary examination of the 1665 Great Plague and the 1666 Great Fire of London reveals their profound and lasting impact on English society. These disasters not only disrupted social and economic structures but also fostered resilience, adaptation, and innovation. The Great Plague exposed vulnerabilities in public health, yet it also strengthened community bonds and collective responses.

Similarly, the Great Fire, while devastating, became a catalyst for urban renewal, modern sanitation improvements, and more structured disaster management strategies. Through the analysis of primary sources such as Samuel Pepys' diaries and Daniel Defoe's *A Journal of the Plague Year*, this chapter underscores the significance of literature in documenting and shaping societal memory. The narratives of these crises illustrate both the fragility and strength of human societies in the face of adversity. Moreover, the research highlights the long-term influence of these disasters on cultural resilience, governance, and urban development.

In conclusion, the chapter underscores the importance of resilience and renewal in the aftermath of natural disasters in England. While challenges remain, the country's response to past disasters demonstrates a growing capacity for recovery, adaptation, and long-term renewal. The resilience shown by individuals, communities, and governments reflects the strength of British society in the face of adversity, and the continued focus on improving disaster preparedness and recovery will be crucial for addressing future challenges. By fostering a culture of resilience, learning from past experiences, and ensuring that policy and infrastructure evolve in response to new threats, England can continue to build a more sustainable, resilient society that is better equipped to face the natural disasters of the future.

By drawing parallels between historical strategies—such as quarantine measures and early firefighting efforts—and contemporary disaster preparedness, this study contributes to discussions on crisis management and urban planning. Understanding how past societies navigated catastrophic events offers valuable lessons for modern resilience-building, emphasizing the importance of strategic planning, public health policies, and community solidarity in facing future challenges. The Great Plague, Great Fire, and other historical disasters stand as monuments to resilience, not just as historical events but as symbols of human capacity for renewal. Through literature, public health reforms, urban development, and cultural change, these events continue to influence how we view the world in the face of adversity. The resilience and renewal they catalyzed show that, even in the darkest of times, there is always the potential for recovery, transformation, and a brighter future.

REFERENCES

Ager, P., Pedersen, M. U., Sharp, P., & Tsoukli, X. (2024). *When London burned to sticks: The economic impact of the Great Fire of 1666* (No. 261). EHES Working Paper.

Alexander, D. (2018). *Natural disasters*. Routledge. DOI: 10.4324/9781315859149

Barker, S. (2020). Translating Tempests and Tremblements: Natural Disasters, News, and the Nation in Early Modern England and France. *Renaissance and Reformation*, *43*(2), 183–212. DOI: 10.33137/rr.v43i2.34796

Bell, W. G. (1924). The great plague in London in 1665.

Berlemann, M., Regner, J., & Tutt, J. (2020). Natural disasters and self-reported wellbeing: empirical evidence for rainfall extremes in the United Kingdom. In *Handbook on Wellbeing, Happiness and the Environment* (pp. 127-143). Edward Elgar Publishing. DOI: 10.4337/9781788119344.00015

Dyer, C. (2020). Recovering from catastrophe: How medieval society in England coped with disasters. In *Waiting for the End of the World?* (pp. 218–238). Routledge. DOI: 10.4324/9781003023449-14

Dynes, R. R., & Quarantelli, E. L. (1975). Community conflict: Its absence and its presence in natural disasters.

Field, J. F. (2017). *London, Londoners and the Great Fire of 1666: disaster and recovery*. Routledge. DOI: 10.4324/9781315099323

Frith, J. (2012). The history of plague-part 1: The three great pandemics. *Journal of Military and Veterans' Health*, *20*(2), 11–16.

Gilpin, J. (2013). God's terrible voice: Liturgical response to the Great Fire of London. *Anglican and Episcopal History*, *82*(3), 318–334.

Keim, M. E. (2015). 2. The public health impacts of natural disasters. In the Handbook *of public health in natural disasters* (pp. 33-58). Wageningen.

Kempe, M. (2003). Noah's Flood: The Genesis story and natural disasters in early modern times. *Environment and History*, *9*(2), 151–171. DOI: 10.3197/096734003129342809

Kondratyev, K. Y., Varotsos, C. A., & Krapivin, V. F. (2006). Statistics of natural disasters. *Natural Disasters as Interactive Components of Global Ecodynamics*, 1-59.

Lewis, J., & Kelman, I. (2025). Development and Disasters.

Mendia-Landa, P. (2013). Myths and Legends on Natural Disasters: Making Sense of Our World. *Yale-New Haven Teachers Institute, 7.*

Moote, A. L., & Moote, D. C. (2004). *The great plague: the story of London's most deadly year*. JHU Press. DOI: 10.56021/9780801877834

Mortimer, I. (2017). *The Time Traveller's Guide to Restoration Britain: Life in the Age of Samuel Pepys, Isaac Newton and The Great Fire of London*. Random House.

Pearson, R. (2017). *Insuring the Industrial Revolution: Fire Insurance in Great Britain, 1700–1850*. Routledge. DOI: 10.4324/9781315252193

Pepys, S. (2000). *The Diary of Samuel Pepys, Vol. 1: 1660* (Vol. 1). Univ of California Press.

Pepys, S. (2015). *The Great Fire of London*.

Porter, S. (2009). *The great plague*. Amberley Publishing.

Rawlings, P. (2016). The Great Fire of London and the Origins of Fire Insurance: a brief note. *Queen Mary School of Law Legal Studies Research Paper*, (246).

Trigg, S. (2016). Samuel Pepys and the great fire of London: Trauma and Emotion, Private and Public. *Disaster, Death and the Emotions in the Shadow of the Apocalypse, 1400–1700*, 341-356.

Tunstall, S. M., Johnson, C. L., & Penning-Rowsell, E. C. (2004, February). Flood hazard management in England and Wales: from land drainage to flood risk management. In *World Congress on Natural Disaster Mitigation* (Vol. 14, No. 2, pp. 447-454).

Willes, M. (2017). *The Curious World of Samuel Pepys and John Evelyn*. Yale University Press. DOI: 10.2307/j.ctvh4zfnh

Compilation of References

Abrash Walton, A., Marr, J., Cahillane, M. J., & Bush, K. (2021). Building community resilience to disasters: A review of interventions to improve and measure public health outcomes in the Northeastern United States. *Sustainability (Basel), 13*(21), 11829. DOI: 10.3390/su132111699

Adger, W. N., Boyd, E., Fábos, A., Fransen, S., Jolivet, D., Neville, G., Safra de Campos, R., & Vijge, M. J. (2020). Migration transforms the conditions for the achievement of the Sustainable Development Goals. *The Lancet. Planetary Health, 4*(10), e440–e442. PMID: 31777332

Adger, W. N., Dessai, S., Goulden, M., Hulme, M., Lorenzoni, I., Nelson, D. R., Naess, L. O., Wolf, J., & Wreford, A. (2009). Are there social limits to adaptation to climate change? *Climatic Change, 93*(3), 335–354. DOI: 10.1007/s10584-008-9520-z

Afifi, R. A., Abdulrahim, S., & Betancourt, T. (2020). Implementing community-based participatory research with communities affected by humanitarian crises: The potential to recalibrate equity and power in vulnerable contexts. *American Journal of Public Health, 110*(10), 1493–1495. DOI: 10.2105/AJPH.2020.305868 PMID: 32903072

Ager, P., Pedersen, M. U., Sharp, P., & Tsoukli, X. (2024). *When London burned to sticks: The economic impact of the Great Fire of 1666* (No. 261). EHES Working Paper.

Agrawal, A. (1995). Dismantling the divide between indigenous and scientific knowledge. *Development and Change, 26*(3), 413–439. DOI: 10.1111/j.1467-7660.1995.tb00560.x

Agrawal, A., Brown, D. G., & Sullivan, J. A. (2019). Are global land grabs ticking socio-ecological bombs or just inefficient investments? *One Earth, 1*(2), 159–162. DOI: 10.1016/j.oneear.2019.10.004

Ahrens, J., & Rudolph, P. M. (2006). The importance of governance in risk reduction and disaster management. *Journal of Contingencies and Crisis Management, 14*(4), 207–220. DOI: 10.1111/j.1468-5973.2006.00497.x

Aitsi-Selmi, A., Murray, V., Wannous, C., Dickinson, C., Johnston, D., Kawasaki, A., Stevance, A. S., & Yeung, T. (2016). Reflections on a science and technology agenda for 21st century disaster risk reduction. *International Journal of Disaster Risk Science, 7*(1), 1–29. DOI: 10.1007/s13753-016-0081-x

Akao, K. I., & Sakamoto, H. (2018). A theory of disasters and long-run growth. *Journal of Economic Dynamics & Control, 95*, 89–109. DOI: 10.1016/j.jedc.2018.08.006

Alarcon, G. M. (2011). A meta-analysis of burnout with job demands, resources, and attitudes. *Journal of Vocational Behavior, 79*(2), 549–562. DOI: 10.1016/j.jvb.2011.03.007

Albuquerque, P. H., & Rajhi, T. (2019). Banking stability, natural disasters, and state fragility: Evidence from developing countries. *Research in International Business and Finance, 50*, 430–443. DOI: 10.1016/j.ribaf.2019.06.001

Alcocer, S. M., Murià-Vila, D., Fernández-Sola, L. R., Ordaz, M., & Arce, J. C. (2020). Observed damage in public school buildings during the 2017 Mexico earthquakes. *Earthquake Spectra, 36*(2), 110–129. DOI: 10.1177/8755293020926183

Aldrich, D. P. (2012). *Building resilience: Social capital in post-disaster recovery*. University of Chicago Press. DOI: 10.7208/chicago/9780226012896.001.0001

Aldrich, D. P., & Meyer, M. A. (2015). Social capital and community resilience. *The American Behavioral Scientist, 59*(2), 254–269. DOI: 10.1177/0002764214550299

Aldunce, P., Beilin, R., Howden, M., & Handmer, J. (2016). Stakeholder participation in building resilience to disasters in a changing climate. *Environmental Hazards, 15*(1), 58–73. DOI: 10.1080/17477891.2015.1134427

Alexander, D. (2002). *Principles of emergency planning and management*. Terra Publishing.

Alexander, D. (2018). *Natural disasters*. Routledge. DOI: 10.4324/9781315859149

Ali, S., Liu, Y., Ishaq, M., Shah, T., Ilyas, A., & Din, I. U. (2021). Climate change and its impact on the yield of major food crops: Evidence from Pakistan. *Foods, 10*(6), 1285. DOI: 10.3390/foods10061285 PMID: 34199791

Allen, M. R., Barros, V. R., Broome, J., Cramer, W., Christ, R., Church, J. A., Clarke, L., . . . Dubash, N. K. (2014). Climate change 2014: Synthesis report. Contribution of Working Groups I, II and III to the fifth assessment report of the Intergovernmental Panel on Climate Change. Intergovernmental Panel on Climate Change. https://www.ipcc.ch/report/ar5/syr/

Allen, J., Jimmieson, N. L., Bordia, P., & Irmer, B. E. (2007). Uncertainty during organizational change: Managing perceptions through communication. *Journal of Change Management*, *7*(2), 187–210. DOI: 10.1080/14697010701563379

Allen, K. M. (2006). Community-based disaster preparedness and climate adaptation: Local capacity-building in the Philippines. *Disasters*, *30*(1), 81–101. DOI: 10.1111/j.1467-9523.2006.00308.x PMID: 16512863

Alliance, C. H. S., Groupe, U. R. D., & the The Sphere Project. (2014). Core Humanitarian Standard on Quality and Accountability. CHS Alliance. https://corehumanitarianstandard.org/files/files/Core%20Humanitarian%20Standard%20-%20English.pdf

Alston, J. M., MacEwan, J. P., & Sambucci, O. (2018). Price controls and market efficiency: Evidence from California's Proposition 65. *Applied Economic Perspectives and Policy*, *40*(3), 373–395. DOI: 10.1093/aepp/ppy008

Alston, M. (2009). Drought policy in Australia: Gender mainstreaming or gender blindness. *Gender, Place and Culture*, *16*(2), 139–154. DOI: 10.1080/09663690902795738

Altay, N., & Labonte, M. (2014). Challenges in humanitarian information management and exchange: Evidence from Haiti. *Disasters*, *38*(s1), S50–S72. DOI: 10.1111/disa.12052 PMID: 24601932

AltlasRiesgo. (2021). Atlas de riesgos-Municipio de Heroica Ciudad de Huajuapan de León, Oaxaca [Risk Atlas-Municipality of heroica City of Huajuapan de león, Oaxaca]. Secretaria de Desarrollo Agrario, Territorial y Urbano (SEDATU).

American Psychiatric Association. (2013). *Diagnostic and Statistical Manual of Mental Disorders* (5th ed.). American Psychiatric Publishing.

American Red Cross. (2021). *Team Rubicon*. https://www.redcross.org

Amri, A., Lassa, J. A., Tebe, Y., Hanifa, N. R., Kumar, J., & Sagala, S. (2022). Pathways to disaster risk reduction education integration in schools: Insights from SPAB evaluation in Indonesia. *International Journal of Disaster Risk Reduction*, *73*, 102851. DOI: 10.1016/j.ijdrr.2022.102860

Antronico, L., Coscarelli, R., Gariano, S. L., & Salvati, P. (2023). Perception of climate change and geo-hydrological risk among high-school students: A local-scale study in Italy. *International Journal of Disaster Risk Reduction, 90*, 103663. DOI: 10.1016/j.ijdrr.2023.103663

Arcaya, M. C., Raker, E. J., & Waters, M. C. (2020). The social consequences of disasters: Individual and community change. *Annual Review of Sociology, 46*(1), 671–691. DOI: 10.1146/annurev-soc-121919-054827

Argyroudis, S. A., Mitoulis, S. A., Hofer, L., Zanini, M. A., Tubaldi, E., & Frangopol, D. M. (2020). Resilience assessment framework for critical infrastructure in a multi-hazard environment. *The Science of the Total Environment, 714*, 136854. DOI: 10.1016/j.scitotenv.2020.136854 PMID: 32018987

Ariyabandu, M. (2006). Gender issues in recovery from the December 2004 Indian Ocean tsunami: The case of Sri Lanka. *Earthquake Spectra, 22*(3), 759–775. DOI: 10.1193/1.2204838

Arora-Jonsson, S. (2011). Virtue and vulnerability: Discourses on women, gender and climate change. *Global Environmental Change, 21*(2), 744–751. DOI: 10.1016/j.gloenvcha.2011.01.005

Arslan, M. (2023). Okul müdürlerinin 6 Şubat Kahramanmaraş Depremi sonrası okullarda karşılaştıkları sorunlar ve bu sorunlara ilişkin çözüm önerileri. *International Journal of Social Humanities Sciences Research, 10*(97), 1550–1559.

Asgary, A., Anjum, M. I., & Azimi, N. (2012). Disaster recovery and business continuity after the 2010 flood in Pakistan: Case of small businesses. *International Journal of Disaster Risk Reduction, 2*, 46–56. DOI: 10.1016/j.ijdrr.2012.08.001

Asim, M., Mekkodathil, A., Sathian, B., Elayedath, R., Kumar, R. N., Simkhada, P., & van Teijlingen, E. (2019). Post-traumatic stress disorder among the flood-affected population in the Indian subcontinent. *Nepal Journal of Epidemiology, 9*(1), 755–758. DOI: 10.3126/nje.v9i1.24003 PMID: 31210997

Auzzir, Z. A., Haigh, R. P., & Amaratunga, D. (2014). Public-private partnerships (PPP) in disaster management in developing countries: A conceptual framework. *Procedia Economics and Finance, 18*, 807–814. DOI: 10.1016/S2212-5671(14)01006-5

Ayan, V. M., & Keten, E. T. (2023). Medyanın yapısal sorunları bağlamında afet haberciliği: 6 Şubat 2023 Kahramanmaraş deprem haberlerinin gazetelerde çerçevelenmesi. *Niğde Ömer Halisdemir Üniversitesi İletişim Fakültesi Akademik Dergisi, 2*(2), 116–132.

Ayers, J., & Forsyth, T. (2009). Community-based adaptation to climate change. *Environment, 51*(4), 22–31. DOI: 10.3200/ENV.51.4.22-31

Aysan, Y., & Lavell, A. (2014). Disaster risk management: From reaction to prevention. In López-Carresi, A., Fordham, M., Wisner, B., Kelman, I., & Gaillard, J. C. (Eds.), *Disaster management: International lessons in risk reduction, response and recovery* (pp. 13–28). Routledge.

Ayyub, B. M. (2014). Systems resilience for multihazard environments: Definition, metrics, and valuation for decision making. *Risk Analysis, 34*(2), 340–355. DOI: 10.1111/risa.12093 PMID: 23875704

Azad, P., Muhammed Salim, A. P., & Anoop Das, K. S. (2024). *Socioeconomic and ecological impact of Wayanad landslides 2024: Lessons for future resilience in Kerala.*

Baig, N., Khan, S., & Khan, M. A. (2018). Natural disasters and economic growth: Evidence from Pakistan. *Pakistan Journal of Applied Economics, 28*(2), 189–207.

Balcik, B., Beamon, B. M., Krejci, C. C., Muramatsu, K. M., & Ramirez, M. (2010). Coordination in humanitarian relief chains: Practices, challenges and opportunities. *International Journal of Production Economics, 126*(1), 22–34. DOI: 10.1016/j.ijpe.2009.09.008

Ballesteros, M. M., & Domingo, S. N. (2015). Building Philippine SMEs resilience to natural disasters. Philippine Institute for Development Studies Discussion Paper Series, 2015-20.

Bandecchi, A. E., Pazzi, V., Morelli, S., Valori, L., & Casagli, N. (2019). Geo-hydrological and seismic risk awareness at school: Emergency preparedness and risk perception evaluation. *International Journal of Disaster Risk Reduction, 40*, 101280. DOI: 10.1016/j.ijdrr.2019.101280

Banholzer, S., Kossin, J., & Donner, S. (2014). The impact of climate change on natural disasters. In Singh, A., & Zommers, M. (Eds.), *Reducing disaster: Early warning systems for climate change* (pp. 21–49). Springer., DOI: 10.1007/978-94-017-8598-3_2

Barker, S. (2020). Translating Tempests and Tremblements: Natural Disasters, News, and the Nation in Early Modern England and France. *Renaissance and Reformation, 43*(2), 183–212. DOI: 10.33137/rr.v43i2.34796

Basnet, B. K. (2020). Earthquake and its impacts on education: Aftemath Nepal quake 2015. *The European Educational Researcher, 3*(3), 101–118. DOI: 10.31757/euer.332

Bastagli, F., Hagen-Zanker, J., Harman, L., Barca, V., Sturge, G., Schmidt, T., & Pellerano, L. (2016). *Cash transfers: What does the evidence say? A rigorous review of programme impact and of the role of design and implementation features*. Overseas Development Institute.

Battisti, M., & Deakins, D. (2017). The relationship between dynamic capabilities, the firm's resource base and performance in a post-disaster environment. *International Small Business Journal, 35*(1), 78–98. DOI: 10.1177/0266242615611471

Baytiyeh, H., & Ocal, A. (2016). High school students' perception of earthquake disaster: A comparative study of Lebanon and Turkey. *International Journal of Disaster Risk Reduction, 18*, 56–63. DOI: 10.1016/j.ijdrr.2016.06.004

Beard, A. N., & Santos-Reyes, J. (2003). A Safety Management System Model with Application to Fire Safety Offshore. *The Geneva Papers on Risk and Insurance. Issues and Practice, 28*(3), 413–425. https://link.springer.com/content/pdf/10.1111/1468-0440.00234.pdf. DOI: 10.1111/1468-0440.00234

Beazley, R., Solórzano, A., & Barca, V. (2019). *Study on shock-responsive social protection in Latin America and the Caribbean: Summary of key findings and policy recommendations*. Oxford Policy Management.

Beazley, R., Solórzano, A., & Sossouvi, K. (2016). *Study on shock-responsive social protection in Latin America and the Caribbean: Theoretical framework and literature review*. Oxford Policy Management.

Beccari, B. (2016). A comparative analysis of disaster risk, vulnerability and resilience composite indicators. PLOS Currents Disasters, 8.

Becken, S., & Hughey, K. F. D. (2013). Linking tourism into emergency management structures to enhance disaster risk reduction. *Tourism Management, 36*, 77–85. DOI: 10.1016/j.tourman.2012.11.006

Becker, B. E., & Hustled, M. A. (2006). *Strategic Human Resources Management: A Key to Competitiveness*. In The Academy of Management Perspectives.

Becker, B., & Huselid, M. (2006). Strategic human resources management: Where do we go from here? *Journal of Management, 32*(6), 898–925. DOI: 10.1177/0149206306293668

Beck, M. W., Losada, I. J., Menéndez, P., Reguero, B. G., Díaz-Simal, P., & Fernández, F. (2018). The global flood protection savings provided by coral reefs. *Nature Communications, 9*(1), 2186. DOI: 10.1038/s41467-018-04568-z PMID: 29895942

Bell, W. G. (1924). The great plague in London in 1665.

Benali, N., Abdelkafi, I., & Feki, R. (2017). The impact of natural disasters on public debt in middle-income countries. *International Journal of Economics and Finance, 9*(5), 179–186. DOI: 10.5539/ijef.v9n5p179

Benali, N., Abdelkafi, I., & Feki, R. (2018). Natural-disaster shocks and government's behavior: Evidence from middle-income countries. *International Journal of Disaster Risk Reduction, 27*, 1–6. DOI: 10.1016/j.ijdrr.2016.12.014

Bendell, J., Sutherland, N., & Little, R. (2015). Beyond unsustainable leadership: Critical social theory for sustainable leadership. Sustainability Accounting. *Management and Policy Journal, 8*(4), 418–444.

Bender, M. A., Knutson, T. R., Tuleya, R. E., Sirutis, J. J., Vecchi, G. A., Garner, S. T., & Held, I. M. (2010). Modeled impact of anthropogenic warming on the frequency of intense Atlantic hurricanes. *Science, 327*(5964), 454–458. DOI: 10.1126/science.1180568

Bendimerad, F. (2003). Disaster Risk Reduction and Sustainable Development. World Bank Seminar on The Role of Local Governments in Reducing the Risk of Disasters, Held in Istanbul, Turkey, 28 April–2 May 2003, 57–75.

Béné, C., Godfrey Wood, R., Newsham, A., & Davies, M. (2012). Resilience: New utopia or new tyranny? Reflection about the potentials and limits of the concept of resilience in relation to vulnerability reduction programmes. *IDS Working Paper, 2012*(405), 1–61. DOI: 10.1111/j.2040-0209.2012.00405.x

Benson, C. (2016). Promoting sustainable development through disaster risk management (ADB Sustainable Development Working Paper Series No. 41). Asian Development Bank.

Benson, C., & Clay, E. J. (2004). *Understanding the economic and financial impacts of natural disasters*. The World Bank. DOI: 10.1596/0-8213-5685-2

Bergamo, P. A. S., Streng, E. S., de Carvalho, M. A., Rosenkranz, J., & Ghorbani, Y. (2022). Simulation-based training and learning: A review on technology-enhanced education for the minerals industry. *Minerals Engineering, 175*, 107297. DOI: 10.1016/j.mineng.2021.107272

Berke, P. R., Kartez, J., & Wenger, D. (1993). Recovery after disaster: Achieving sustainable development, mitigation and equity. *Disasters, 17*(2), 93–109. DOI: 10.1111/j.1467-7717.1993.tb01137.x PMID: 20958760

Berkes, F. (2018). *Sacred ecology* (4th ed.). Routledge.

Berkes, F., & Folke, C. (2002). *Back to the future: Ecosystem dynamics and local knowledge*. Cambridge University Press.

Berlemann, M., Regner, J., & Tutt, J. (2020). Natural disasters and self-reported wellbeing: empirical evidence for rainfall extremes in the United Kingdom. In *Handbook on Wellbeing, Happiness and the Environment* (pp. 127-143). Edward Elgar Publishing. DOI: 10.4337/9781788119344.00015

Berlemann, M., & Steinhardt, M. F. (2017). Climate change, natural disasters, and migration—A survey of the empirical evidence. *CESifo Economic Studies*, *63*(4), 353–385. DOI: 10.1093/cesifo/ifx019

Berlin, J. (2016). *Deloitte Global Human Capital Trends Report*. Deloitte.

Berry, P., Bowen, K., & Kjellstrom, T. (2022). Climate change and health equity: Actions to protect vulnerable populations. *Annual Review of Public Health*, *43*, 185–206.

Bevan, D., & Adam, C. (2016). Financing the reconstruction of public capital after a natural disaster. World Bank Policy Research Working Paper, (7718).

Bhavnani, R. (2006). *Natural disaster conflicts*. Harvard University.

Biggs, D., Hall, C. M., & Stoeckl, N. (2015). The resilience of formal and informal tourism enterprises to disasters: Reef tourism in Phuket, Thailand. *Journal of Sustainable Tourism*, *23*(7), 1032–1056. DOI: 10.1080/09669582.2015.1038429

Bigith, G., Sooraj, Rajat Nainwal, & Kumara, H. S. (2024). *Assessing the impact and response to the disaster in Mananthavady town: A case study. IPO Conference Series: Earth and Environment Science*. DOI: 10.1088/1755-1315/1409/1/012023

Bilham, R. (2004). Earthquakes in India and the Himalaya: Tectonics, geodesy and history. *Annals of Geophysics*, *47*(2/3), 839–858.

Birkmann, J., Feldmeyer, D., McMillan, J. M., Solecki, W., Totin, E., Roberts, D., & Sengupta, S. (2021). Regional clusters of vulnerability show the need for transboundary cooperation. *Environmental Research Letters*, *16*(9), 094013. DOI: 10.1088/1748-9326/ac1f43

Birkmann, J., & Mechler, R. (2015). Advancing climate adaptation and risk management: New insights, concepts and approaches. *Natural Hazards*, *78*(1), 1–7. DOI: 10.1007/s11069-015-1856-7

Birkmann, J., & Teichman, K. (2010). Integrating disaster risk reduction and climate change adaptation: Key challenges—scales, knowledge, and norms. *Sustainability Science*, *5*(2), 171–184. DOI: 10.1007/s11625-010-0108-y

Bisri, M. B. F., & Beniya, S. (2016). Analyzing the national disaster response framework and inter-organizational network of the 2015 Nepal earthquake. *Procedia Engineering, 159*, 93–100. DOI: 10.1016/j.proeng.2016.08.059

Blaikie, P. (2006). Risk and disaster reduction: A challenge for the social sciences. *International Journal of Risk Assessment and Management, 6*(2-3), 137–145.

Bloodhart, B., Maibach, E., Myers, T., & Zhao, X. (2015). Local climate experts: The influence of local TV weather information on climate change perceptions. *PLoS One, 10*(11), e0141526. Advance online publication. DOI: 10.1371/journal.pone.0141526

Bodas, M., Peleg, K., Stolero, N., & Adini, B. (2022). Risk Perception of Natural and Human-Made Disasters—Cross Sectional Study in Eight Countries in Europe and Beyond. *Frontiers in Public Health, 10*, 825985. DOI: 10.3389/fpubh.2022.825985 PMID: 35252099

Bogg, T., & Roberts, B. W. (2004). Conscientiousness and health-related behaviors: A meta-analysis of the leading behavioral contributors to mortality. *Psychological Bulletin, 130*(6), 887–919. DOI: 10.1037/0033-2909.130.6.887 PMID: 15535742

Borawska, A. (2020). Application of behavioral economics insights to increase effectiveness of public awareness campaigns. In *Quantitative methods in contemporary economics* (pp. 45–62). Springer., DOI: 10.1007/978-3-030-30251-1_5

Botzen, W. J. W., Duijndam, S. J., & van Beukering, P. (2021). Lessons for climate policy from behavioral biases towards COVID-19 and climate change risks. *World Development, 137*, 105214. Advance online publication. DOI: 10.1016/j.worlddev.2020.105214

Botzen, W. W., Deschenes, O., & Sanders, M. (2019). The economic impacts of natural disasters: A review of models and empirical studies. *Review of Environmental Economics and Policy, 13*(2), 167–188. DOI: 10.1093/reep/rez004

Boudreau, J. W., & Ramstad, P. M. (2007). *Beyond HR: The New Science of Human Capital*. Harvard Business Review Press.

Boudreaux, C. J., Escaleras, M., & Skidmore, M. (2019). Natural disasters and entrepreneurship: Evidence from global data. *Journal of Natural Disaster Science, 40*(1), 15–27. DOI: 10.2328/jnds.40.15

Bouwer, L. M. (2019). Observed and projected impacts from extreme weather events: Implications for loss and damage. In Mechler, R., Bouwer, L. M., Schinko, T., Surminski, S., & Linnerooth-Bayer, J. (Eds.), *Loss and damage from climate change: Concepts, methods and policy options* (pp. 63–82). Springer. DOI: 10.1007/978-3-319-72026-5_3

Bowen, K. J., Ebi, K. L., & Friel, S. (2022). Climate change adaptation and mitigation: Next steps for cross-sectoral action to protect global health. *The Lancet. Planetary Health, 6*(5), e386–e396. PMID: 35550712

Bowen, T., del Ninno, C., Andrews, C., Coll-Black, S., Gentilini, U., Johnson, K., Kawasoe, Y., Kryeziu, A., Maher, B., & Williams, A. (2020). *Adaptive social protection: Building resilience to shocks*. World Bank. DOI: 10.1596/978-1-4648-1575-1

Bristow, G., & Healy, A. (2018). Innovation and regional economic resilience: An exploratory analysis. *The Annals of Regional Science, 60*(2), 265–284. DOI: 10.1007/s00168-017-0841-6

Brooks, N., Anderson, S., & Burton, I. (2023). Adaptive management in climate change adaptation: Concepts and methods. *Climate and Development, 15*(2), 145–162.

Brown, C., Stevenson, J., Giovinazzi, S., Seville, E., & Vargo, J. (2018). Factors influencing SMEs' resilience to natural hazards: A framework and literature review. *International Journal of Disaster Risk Reduction, 28*, 872–883.

Brown, L., Green, R., & Taylor, S. (2021). Mental Health and Natural Disasters: A Comprehensive Review. *Journal of Traumatic Stress, 34*(2), 123–135.

Burby, R. J., Deyle, R. E., Godschalk, D. R., & Olshansky, R. B. (2000). Creating hazard resilient communities through land-use planning. *Natural Hazards Review, 1*(2), 99–106. DOI: 10.1061/(ASCE)1527-6988(2000)1:2(99)

Burke, M., Hsiang, S. M., & Miguel, E. (2015). Global non-linear effect of temperature on economic production. *Nature, 527*(7577), 235–239. DOI: 10.1038/nature15725 PMID: 26503051

Cabello, V. M., Véliz, K. D., Moncada-Arce, A. M., Irarrázaval García-Huidobro, M., & Juillerat, F. (2021). Disaster risk reduction education: Tensions and connections with sustainable development goals. *Sustainability (Basel), 13*(24), 10886. DOI: 10.3390/su131910933

Cal Fire. (2020). *2020 Fire Season*. California Department of Forestry and Fire Protection.

Campbell, J. R. (2009). Islandness: Vulnerability and resilience in Oceania. *Shima. The International Journal of Research into Island Cultures*, *3*(1), 85–97.

Cangialosi, J. P., Latto, A. S., & Berg, R. (2018). Hurricane Irma (AL112017): 30 August-12 September 2017. National Hurricane Center Tropical Cyclone Report. NOAA/NWS.

Cannon, T., & Müller-Mahn, D. (2010). Vulnerability, resilience and development discourses in context of climate change. *Natural Hazards*, *55*(3), 621–635. DOI: 10.1007/s11069-010-9499-4

Casady, C. B., & Baxter, D. (2020). Pandemics, public-private partnerships (PPPs), and force majeure | COVID-19 expectations and implications. *Construction Management and Economics*, *38*(12), 1077–1085. DOI: 10.1080/01446193.2020.1817516

Cascio, W. F. (2018). Leveraging Analysis for Human Capital Analysis for competitive advantage. *Organizational Dynamics*, *47*(2), 98–106.

Casco, W. F. (2014). *Managing Human Resources: Productivity, Quality of Work Life, Profits*. McGraw-Hill.

Cavallo, E., Galiani, S., Noy, I., & Pantano, J. (2014). Catastrophic natural disasters and economic growth. *The Review of Economics and Statistics*, *95*(5), 1549–1561. DOI: 10.1162/REST_a_00413

Cavallo, E., & Noy, I. (2011). Natural disasters and the economy—A survey. *International Review of Environmental and Resource Economics*, *5*(1), 63–102. DOI: 10.1561/101.00000039

Census Bureau. (2019a). Household income [Data set]. United States Census Bureau. https://www.census.gov/topics/income-poverty/income/data/tables.html

Census Bureau. (2019b). County population by characteristics 2010-2019 [Data set]. United States Census Bureau. https://www.census.gov/data/tables/time-series/demo/popest/2010s-counties-detail.html

Centre for Research on the Epidemiology of Disasters (CRED). (2019). EM-DAT: The International Disaster Database. Université Catholique de Louvain. Retrieved from https://www.emdat.be

Centre for Research on the Epidemiology of Disasters. (2021). EM-DAT: The international disaster database. Université Catholique de Louvain. https://www.emdat.be/database

Chambers, R. (1994). The origins and practice of participatory rural appraisal. *World Development*, *22*(7), 953–969. DOI: 10.1016/0305-750X(94)90141-4

Chamlee-Wright, E., & Storr, V. H. (2010). The role of social entrepreneurship in post-Katrina community recovery. *International Journal of Innovative Research and Development, 2*(1-2), 149–164. DOI: 10.1504/IJIRD.2010.029860

Chandra, A., & Fealey, T. (2009). Business incubation in the United States, China and Brazil: A comparison of role of government, incubator funding and financial services. *International Journal of Entrepreneurship, 13*, 67–86.

Chandrasekhar, D. (2012). Digging deeper: Participation and non-participation in post-disaster community recovery. *Community Development (Columbus, Ohio), 43*(5), 614–629. DOI: 10.1080/15575330.2012.730538

Chang, S. E., McDaniels, T., Fox, J., Dhariwal, R., & Longstaff, H. (2014). Toward disaster-resilient cities: Characterizing resilience of infrastructure systems with expert judgments. *Risk Analysis, 34*(3), 416–434. DOI: 10.1111/risa.12133 PMID: 24152135

Chantarat, S., Mude, A. G., Barrett, C. B., & Carter, M. R. (2013). Designing index-based livestock insurance for managing asset risk in northern Kenya. *The Journal of Risk and Insurance, 80*(1), 205–237. DOI: 10.1111/j.1539-6975.2012.01463.x

Chanza, N., & de Wit, A. (2016). Enhancing climate governance through indigenous knowledge: Case of the Shona in Zimbabwe. *Climate Policy, 16*(4), 543–562. DOI: 10.1080/14693062.2015.1033674

Chapagain, T., & Raizada, M. N. (2017). Impacts of natural disasters on smallholder farmers: Gaps and recommendations. *Agriculture & Food Security, 6*(1), 39. DOI: 10.1186/s40066-017-0116-6

Chen, J., Chen, T. H. Y., Vertinsky, I., Yumagulova, L., & Park, C. (2013). Public–private partnerships for the development of disaster resilient communities. *Journal of Contingencies and Crisis Management, 21*(3), 130–143. DOI: 10.1111/1468-5973.12021

Chirisa, I. (2021). Education 5.0 in Zimbabwe: Mapping the history and trajectory of University Education. *Zambia Journal of Science and Technology, 16*(1), 22–35.

Christopher, M., & Peck, H. (2004). Building the resilient supply chain. *International Journal of Logistics Management, 15*(2), 1–14. DOI: 10.1108/09574090410700275

Christoplos, I. (2006). The elusive 'window of opportunity' for risk reduction in post-disaster recovery. In ProVention Consortium Forum (pp. 2-3).

Chun, H., Chi, S., & Hwang, B. (2017). A spatial disaster assessment model of social resilience based on geographically weighted regression. *Sustainability (Basel), 9*(12), 2222. DOI: 10.3390/su9122222

Churchill, C. (2006). *Protecting the poor: A microinsurance compendium.* International Labour Organization.

Clarke, D. J., & Wren-Lewis, L. (2013). Learning from lemons: The role of government in index insurance for individuals. FERDI Policy Brief, B76. https://ferdi.fr/en/publications/learning-from-lemons-the-role-of-government-in-index-insurance-for-individuals

Clarke, D. J., & Dercon, S. (2016). *Dull disasters? How planning ahead will make a difference.* Oxford University Press. DOI: 10.1093/acprof:oso/9780198785576.001.0001

Clinton, W. J. (2006). *Lessons learned from tsunami recovery: Key propositions for building back better.* United Nations Secretary-General's Special Envoy for Tsunami Recovery.

Coaffee, J., Therrien, M. C., Cheyne, A., & Wood, J. M. (2018). Urban resilience implementation: A policy challenge and research agenda for the 21st century. *Journal of Contingencies and Crisis Management, 26*(3), 403–410. DOI: 10.1111/1468-5973.12233

Cochrane, H. (2004). Economic loss estimation models: Time for a change? In Okuyama, Y., & Chang, S. E. (Eds.), *Modeling spatial and economic impacts of disasters* (pp. 229–248). Springer., DOI: 10.1007/978-3-540-24787-6_11

Comfort, L. K. (2007). Crisis management in hindsight: Cognition, communication, coordination, and control. *Public Administration Review, 67*(s1), 189–197. DOI: 10.1111/j.1540-6210.2007.00827.x

Comfort, L. K., Boin, A., & Demchak, C. C. (2010). *Designing resilience: Preparing for extreme events.* University of Pittsburgh Press. DOI: 10.2307/j.ctt5hjq0c

Coppola, D. P. (2015). *Introduction to international disaster management.* Kindle edition, B&H Publishing, UK.

Coppola, D. P. (2015). *Introduction to international disaster management.* Butterworth-Heinemann.

Coughlan de Perez, E., van den Hurk, B., van Aalst, M. K., Jongman, B., Klose, T., & Suarez, P. (2015). Forecast-based financing: An approach for catalyzing humanitarian action based on extreme weather and climate forecasts. *Natural Hazards and Earth System Sciences, 15*(4), 895–904. DOI: 10.5194/nhess-15-895-2015

CRED, & UNISDR. (2018). Economic losses, poverty & disasters: 1998-2017. Centre for Research on the Epidemiology of Disasters (CRED) and United Nations Office for Disaster Risk Reduction (UNISDR).

Creswell, J. W., & Miller, D. L. (2000). Determining validity in qualitative inquiry. *Theory into Practice*, *39*(3), 124–130. DOI: 10.1207/s15430421tip3903_2

Cruz, G. (2023). 100 municipios de la Mixteca con sequía severa y extrema: Conagua [100 municipalities of the Mixteca with severe and extreme Drought: Conagua]. https://www.xeouradio.com/2023/11/10/100-municipios-de-la-mixteca-con-sequia-severa-y-extrema-conagua/

Cuaresma, J. C. (2010). Natural disasters and human capital accumulation. *The World Bank Economic Review*, *24*(2), 280–302. DOI: 10.1093/wber/lhq008

Cummins, J. D., & Weiss, M. A. (2009). Convergence of insurance and financial markets: Hybrid and securitized risk-transfer solutions. *The Journal of Risk and Insurance*, *76*(3), 493–545. DOI: 10.1111/j.1539-6975.2009.01311.x

Cutter, S. L. (2016). Resilience to what? Resilience for whom? *The Geographical Journal*, *182*(2), 110–113. DOI: 10.1111/geoj.12174

Cutter, S. L., Barnes, L., Berry, M., Burton, C., Evans, E., Tate, E., & Webb, J. (2008). A place-based model for understanding community resilience to natural disasters. *Global Environmental Change*, *18*(4), 598–606. DOI: 10.1016/j.gloenvcha.2008.07.013

Cutter, S. L., Boruff, B. J., & Shirley, W. L. (2003). Social vulnerability to environmental hazards. *Social Science Quarterly*, *84*(2), 242–261. DOI: 10.1111/1540-6237.8402002

Cvetkovic, V., Ocal, A., & Ivanov, A. (2019). Young adults' fear of disasters: A case study of residents from Turkey, Serbia and Macedonia. *International Journal of Disaster Risk Reduction*, *350*, 101095. DOI: 10.1016/j.ijdrr.2019.101095

Dasgupta, P., Morton, J. F., & Dodman, D. (2020). *Community-based adaptation to climate change: Scaling it up*. Routledge.

Davenport, F. V., Burke, M., & Diffenbaugh, N. S. (2021). Contribution of historical precipitation change to US flood damages. *Proceedings of the National Academy of Sciences of the United States of America*, *118*(4), e2017524118. Advance online publication. DOI: 10.1073/pnas.2017524118

Davis, L. W., & Gertler, P. J. (2015). Contribution of air conditioning adoption to future energy use under global warming. *Proceedings of the National Academy of Sciences of the United States of America, 112*(19), 5962–5967. DOI: 10.1073/pnas.1423558112

del Ninno, C., Subbarao, K., & Milazzo, A. (2009). How to make public works work: A review of the experiences. World Bank Social Protection Discussion Paper, 0905. https://openknowledge.worldbank.org/handle/10986/11718

Deloitte. (2017). *Global Human Capital Analysis trends 2017: Rewriting the rules for the digital age*. Deloitte University Press.

Devereux, S. (2016). Social protection for enhanced food security in sub-Saharan Africa. *Food Policy, 60*, 52–62. DOI: 10.1016/j.foodpol.2015.03.009

Dhar, T. K., & Khirfan, L. (2017). A multi-scale analysis of urban resilience planning: The case of the London climate change strategy. *Habitat International, 67*, 108–119. DOI: 10.1016/j.habitatint.2017.07.004

DHS. (2002). *Homeland Security Act of 2002*. U.S. Department of Homeland Security.

Di Pietro, G. (2018). The academic impact of natural disasters: Evidence from L'Aquila earthquake. *Education Economics, 26*(1), 62–77. DOI: 10.1080/09645292.2017.1394984

Dias, N. T., Keraminiyage, K., & DeSilva, K. K. (2016). Long-term satisfaction of post disaster resettled communities: The case of post tsunami – Sri Lanka. *Disaster Prevention and Management, 25*(5), 581–594. DOI: 10.1108/DPM-11-2015-0264

Dixit, A. M., Shrestha, S., Guragain, R., Pandey, B., Sen Oli, K., Adhikari, S., Acharya, S., Jimee, G., Upadhyay, B., Sangachhe, S., Shrestha, N., Pradhan, S., Dhungel, R., Khatiwada, P., Baskota, A., Poudel, A., Tandingan, M., Mathema, N., Parajuli, B., & Marasini, N. (2018). *Risk Management*. Response, Relief, Recovery, Reconstruction, and Future Disaster Risk Reduction., DOI: 10.1016/B978-0-12-812808-4.00005-5

Djalante, R., Holley, C., Thomalla, F., & Carnegie, M. (2013). Pathways for adaptive and integrated disaster resilience. *Natural Hazards, 69*(3), 2105–2135. DOI: 10.1007/s11069-013-0797-5

Dodman, D., & Mitlin, D. (2022). Challenges in implementing community-based adaptation. *Nature Climate Change, 12*(2), 118–124.

Doocy, S., Daniels, A., Packer, C., Dick, A., & Kirsch, T. D. (2013). The human impact of floods: A historical review of events and systemic response. *Prehospital and Disaster Medicine*, *28*(1), 1–9. PMID: 23158909

Doocy, S., & Tappis, H. (2017). Cash-based approaches in humanitarian emergencies: A systematic review. *Campbell Systematic Reviews*, *13*(1), 1–200. DOI: 10.4073/csr.2017.17

Doytch, N., & Klein, Y. (2018). Natural disasters and energy consumption: A panel analysis. *Energy Economics*, *76*, 1–13. DOI: 10.1016/j.eneco.2018.09.014

Drolet, J., Dominelli, L., Alston, M., Ersing, R., Mathbor, G., & Wu, H. (2015). Women rebuilding lives post-disaster: Innovative community practices for building resilience and promoting sustainable development. *Gender and Development*, *23*(3), 433–448. DOI: 10.1080/13552074.2015.1096040

Dufty, N. (2020). *Disaster education, communication and engagement*. Wiley., DOI: 10.1002/9781119569817

Dündar, H. Ç. (2023). 6 Şubat Kahramanmaraş depremlerinin iş dünyasına etkileri. *Çevre Şehir ve İklim Dergisi, 2*(4), 262-281.

Dwiningrum, S. I. A., Sitompul, N., Nisa, A. A., & Sumunar, D. R. S. (2022). School resilience policy development and strategic steps to disaster-resilient communities. *IOP Conference Series. Earth and Environmental Science*, *1039*(1), 012006. DOI: 10.1088/1755-1315/989/1/012017

Dyer, C. (2020). Recovering from catastrophe: How medieval society in England coped with disasters. In *Waiting for the End of the World?* (pp. 218–238). Routledge. DOI: 10.4324/9781003023449-14

Dynes, R. R., & Quarantelli, E. L. (1975). Community conflict: Its absence and its presence in natural disasters.

Dynes, R. R. (2006). Social capital: Dealing with community emergencies. *Homeland Security Affairs*, *2*(2).

Działek, J., Biernacki, W., & Bokwa, A. (2013). Challenges to social capacity building in flood-affected areas of southern Poland. *Natural Hazards and Earth System Sciences*, *13*(10), 2555–2566. DOI: 10.5194/nhess-13-2555-2013

Dzigbede, K. D., Gehl, S. B., & Willoughby, K. (2020). Disaster resiliency of U.S. local governments: Insights to strengthen local response and recovery from the COVID-19 pandemic. *Public Administration Review*, *80*(4), 634–643. DOI: 10.1111/puar.13249 PMID: 32836460

Earthquake Engineering Research Institute (EERI). (2006). The Kashmir Earthquake of October 8, 2005: Impacts in Pakistan. EERI Special Earthquake Report. https://www.eeri.org/lfe/pdf/kashmir_eeri_report_2006.pdf

Ebi, K. L., Berry, P., & Hayes, K. (2023). Vulnerability assessments for climate change and health. *Annual Review of Public Health*, *44*, 167–188.

Ebi, K. L., & Hess, J. J. (2023). Health risks of climate change: Implications for adaptation and resilience. *Annual Review of Environment and Resources*, *48*, 331–355.

Ebi, K. L., Hess, J. J., & Watkiss, P. (2021). Health risks and costs of climate variability and change. In *Climate Change 2021: Impacts, Adaptation and Vulnerability. Contribution of Working Group II to the Sixth Assessment Report of the Intergovernmental Panel on Climate Change*. Cambridge University Press.

Ebi, K. L., Hess, J. J., & Watkiss, P. (2021). Health risks and costs of climate variability and change. In *The Economics of Climate Change* (pp. 123–152). Cambridge University Press.

Ebi, K. L., & Otmani del Barrio, M. (2023). Community engagement in climate change and health adaptation: A systematic review. *The Lancet. Planetary Health*, *7*(3), e220–e231.

Eckhardt, D., Leiras, A., & Thomé, A. M. T. (2019). Systematic literature review of methodologies for assessing the costs of disasters. *International Journal of Disaster Risk Reduction*, *33*, 398–416. DOI: 10.1016/j.ijdrr.2018.10.010

Eguchi, R. T., Huyck, C. K., Ghosh, S., & Adams, B. J. (2010). The application of remote sensing technologies for disaster management. *The International Archives of the Photogrammetry, Remote Sensing and Spatial Information Sciences*, *38*(8), 125–130. https://www.isprs.org/proceedings/XXXVIII/part8/pdf/W08_04_20100519115847.pdf

Eling, M., Pradhan, S., & Schmit, J. T. (2014). The determinants of microinsurance demand. *The Geneva Papers on Risk and Insurance. Issues and Practice*, *39*(2), 224–263. DOI: 10.1057/gpp.2014.5

El-Khani, A., Calam, R., Sabine-Rakotomalala, L. C., & Maalouf, W. (2023). Urgent help needed for children affected by the earthquake in Syria and Turkiye. *Lancet*, *10*(4), 246–247. Advance online publication. DOI: 10.1016/S2215-0366(23)00060-3 PMID: 36870354

Elkhidir, E., Mannakkara, S., Henning, T. F. P., & Wilkinson, S. (2023). A pathway towards resilient cities: National resilience knowledge networks. *Cities (London, England)*, *134*, 104194. DOI: 10.1016/j.cities.2023.104243

Enarson, E. (2012). *Women and disasters: A gender analysis*. International Federation of Red Cross and Red Crescent Societies.

Enarson, E., & Chakrabarti, P. D. (Eds.). (2009). *Women, gender and disaster: Global issues and initiatives*. SAGE Publications India. DOI: 10.4135/9788132108078

Enarson, E., & Fordham, M. (2001). From women's needs to women's rights in disasters. *Environmental Hazards, 3*(3-4), 133–136. DOI: 10.1016/S1464-2867(02)00006-2

Engel, E., Fischer, R. D., & Galetovic, A. (2014). *The economics of public-private partnerships: A basic guide*. Cambridge University Press. DOI: 10.1017/CBO9781139565615

Ensor, J., Berger, R., & Huq, S. (2021). Community-based adaptation and culture in theory and practice. *Progress in Human Geography, 45*(4), 813–832.

Ensor, J., Berger, R., & Huq, S. (2023). Community-based adaptation: Lessons from practice. *Progress in Human Geography, 47*(2), 289–307.

Erdelj, M., Król, M., & Natalizio, E. (2017). Wireless sensor networks and multi-UAV systems for natural disaster management. *Computer Networks, 124*, 72–86. DOI: 10.1016/j.comnet.2017.05.021

Erdoğan, B. (2023). Depremin sosyolojisi: 6 Şubat felaketinin toplumsal ve kültürel boyutları. *TRT Akademi, 8*(18), 718–725. DOI: 10.37679/trta.1306900

Fabian, M., Lessmann, C., & Sofke, T. (2019). Natural disasters and regional development: Evidence from earthquakes and night-time lights. *Journal of Economic Geography, 19*(4), 797–819. DOI: 10.1093/jeg/lby036

Fang, J., Lau, C. K. M., Lu, Z., Wu, W., & Zhu, L. (2019). Natural disasters, climate change, and their impact on inclusive wealth in G20 countries. *Environmental Science and Pollution Research International, 26*(2), 1455–1463. DOI: 10.1007/s11356-018-3634-2 PMID: 30426378

Fankhauser, S., & McDermott, T. K. (2016). Climate-resilient development: An introduction. In Fankhauser, S., & McDermott, T. K. (Eds.), *The economics of climate-resilient development* (pp. 1–14). Edward Elgar Publishing. DOI: 10.4337/9781785360312.00009

Fan, L. (2013). *Disaster as opportunity? Building back better in Aceh*. Overseas Development Institute.

Farquharson, E., Torres de Mästle, C., Yescombe, E. R., & Encinas, J. (2011). *How to engage with the private sector in public-private partnerships in emerging markets*. The World Bank. DOI: 10.1596/978-0-8213-7863-2

Federal Emergency Management Agency (FEMA). (2018). *Federal Emergency Management Agency strategic plan 2018-2022*. Federal Emergency Management Agency.

Federal Emergency Management Agency (FEMA). (2020). *Citizen Corps Program*. https://www.fema.gov

Federal Emergency Management Agency (FEMA). (2021). *Community Rating System (CRS)*. https://www.fema.gov

Federal Emergency Management Agency (FEMA). (2021). *Great ShakeOut Earthquake Drill*. https://www.shakeout.org

Federal Emergency Management Agency. (2019). FEMA data feeds. https://www.fema.gov/data-feeds

Feldman, L., Maibach, E. W., Roser-Renouf, C., & Leiserowitz, A. (2012). Climate on cable: The nature and impact of global warming coverage on Fox News, CNN, and MSNBC. *The International Journal of Press/Politics*, *17*(1), 3–31. DOI: 10.1177/1940161211425410

Fengler, W., Ihsan, A., & Kaiser, K. (2008). *Managing post-disaster reconstruction finance: International experience in public financial management*. The World Bank.

Fernandes, S., & Thampi, K. (2017). *Traditional Knowledge and Disaster Resilience: Lessons from Kerala*. Sage Publications.

Fernandez, J., Black, M., Jones, L., Wilson, L., Salvador-Carulla, T., Astell-Burt, D., & Black, D. (2015). Flooding and mental health: A systematic mapping review. *PLoS One*, *10*(4), e0119929. DOI: 10.1371/journal.pone.0119929 PMID: 25860572

Fernandez, L. S., Byard, D., Lin, C. C., Benson, S., & Barbera, J. A. (2015). Frail elderly as disaster victims: Emergency management strategies. *Prehospital and Disaster Medicine*, *17*(2), 67–74. DOI: 10.1017/S1049023X00000200 PMID: 12500729

Field, J. F. (2017). *London, Londoners and the Great Fire of 1666: disaster and recovery*. Routledge. DOI: 10.4324/9781315099323

Finnis, K. K., Johnston, D. M., Ronan, K. R., & White, J. D. (2010). Hazard perceptions and preparedness of Taranaki youth. *Disaster Prevention and Management*, *19*(2), 175–184. DOI: 10.1108/09653561011037986

Finucane, M. L., & Holup, J. L. (2006). Risk as value: Combining affect and analysis in risk judgements. *Journal of Risk Research*, *9*(2), 141–164. DOI: 10.1080/13669870500166930

Finucane, M. L., May, L. W., & Chang, J. (2021). *A scoping literature review on indicators and metrics for assessing racial equity in disaster preparation, response, and recovery*. RAND Corporation., DOI: 10.7249/RRA1092-1

Fletcher-Miles, H., Gammon, J., Williams, S., & Hunt, J. (2020). A scoping review to assess the impact of public education campaigns to affect behavior change pertaining to antimicrobial resistance. *American Journal of Infection Control*, *48*(8), 933–940. DOI: 10.1016/j.ajic.2019.07.011 PMID: 31444097

Folke, C., Carpenter, S. R., Walker, B., Scheffer, M., Chapin, T., & Rockström, J. (2010). Resilience thinking: Integrating resilience, adaptability and transformability. *Ecology and Society*, *15*(4), 20. DOI: 10.5751/ES-03610-150420

Food and Agriculture Organization of the United Nations. (2015). The impact of disasters on agriculture and food security. FAO. https://www.fao.org/3/a-i5128e.pdf

Ford, J. D., King, N., & Galappaththi, E. K. (2020). Strengthening the resilience of Indigenous health systems to climate change. *The Lancet. Planetary Health*, *4*(6), e237–e238.

Ford, J. D., Sherman, M., & Berrang-Ford, L. (2021). Building capacity for climate change adaptation in the health sector. *The Lancet. Planetary Health*, *5*(4), e237–e246. PMID: 33684341

Frantzeskaki, N. (2019). Seven lessons for planning nature-based solutions in cities. *Environmental Science & Policy*, *93*, 101–111. DOI: 10.1016/j.envsci.2018.12.033

Frith, J. (2012). The history of plague-part 1: The three great pandemics. *Journal of Military and Veterans' Health*, *20*(2), 11–16.

Funk, C., Peterson, P., Landsfeld, M., Pedreros, D., Verdin, J., Shukla, S., & Michaelsen, J. (2019). The climate hazards infrared precipitation with stations—A new environmental record for monitoring extremes. *Scientific Data*, *6*(1), 180066. DOI: 10.1038/sdata.2018.66 PMID: 26646728

Fu, Q., & Zhang, X. (2024). Promoting community resilience through disaster education: Review of community-based interventions with a focus on teacher resilience and well-being. *PLoS One*, *19*(1), e0296454. DOI: 10.1371/journal.pone.0296393 PMID: 38166092

Gaillard, J. C., & Mercer, J. (2013). From knowledge to action: Bridging gaps in disaster risk reduction. *Progress in Human Geography*, *37*(1), 93–114. DOI: 10.1177/0309132512446717

Gaillard, J. C., & Mercer, J. (2022). Understanding and addressing root causes of vulnerability in disaster risk reduction. *Progress in Disaster Science*, *10*, 100203. DOI: 10.1016/j.pdisas.2022.100203

Gallina, V., Torresan, S., Critto, A., Sperotto, A., Glade, T., & Marcomini, A. (2016). A review of multi-risk methodologies for natural hazards: Consequences and challenges for a climate change impact assessment. *Journal of Environmental Management*, *168*, 123–132. DOI: 10.1016/j.jenvman.2015.11.011 PMID: 26704454

Gallup. (2018). State of the American Workplace: Employee Engagement Insights for U.S. Business Leaders.

Gassebner, M., Keck, A., & Teh, R. (2010). Shaken, not stirred: The impact of disasters on international trade. *Review of International Economics*, *18*(2), 351–368. DOI: 10.1111/j.1467-9396.2010.00868.x

George, B., Kumar, R., & Banerjee, S. (2022). Vulnerability of women in the face of climate change: A study of Wayanad District of Kerala, India. *Journal of Chinese Economic and Foreign Trade Studies*, *15*(3), 279–297. DOI: 10.1108/JCEFTS-09-2021-0058

GFDRR. (2015). *Guide to developing disaster recovery frameworks: Sendai conference version*. Global Facility for Disaster Reduction and Recovery.

GFDRR. (2018). *Bringing resilience to scale: A work plan for the Global Facility for Disaster Reduction and Recovery, 2018-2021*. Global Facility for Disaster Reduction and Recovery.

Ghesquiere, F., & Mahul, O. (2010). Financial protection of the state against natural disasters: A primer. World Bank Policy Research Working Paper, (5429).

Gibson, M., & Mullins, J. T. (2020). Climate risk and beliefs in New York floodplains. *Journal of the Association of Environmental and Resource Economists*, *7*(6), 1069–1111. DOI: 10.1086/710666

Gill, J. C., Taylor, F. E., & Duncan, M. J. (2021). Invited perspectives: Building sustainable and resilient communities—Recommended actions for natural hazard scientists. *Natural Hazards and Earth System Sciences*, *21*(5), 1479–1490. DOI: 10.5194/nhess-21-1479-2021

Gilpin, J. (2013). God's terrible voice: Liturgical response to the Great Fire of London. *Anglican and Episcopal History*, *82*(3), 318–334.

Glavovic, B. C. (2008). *Waves of adversity, layers of resilience: Exploring the sustainable development prospect of Ohura and the Whangaehu catchment*. Massey University.

Goldmann, S., & Galea, S. (2014). Mental health consequences of disasters. *Annual Review of Public Health*, *35*(1), 169–183. DOI: 10.1146/annurev-publhealth-032013-182435 PMID: 24159920

Gosselin, P., Bélanger, D., & Lapaige, V. (2022). Implementing a community-based heat action plan in Quebec, Canada: Lessons learned and future directions. *Climate Risk Management*, *35*, 100403.

Government of Kerala. (2024). *Report on drought and landslide vulnerability in Wayanad district*.

Grafakos, S., Viero, G., Reckien, D., Trigg, K., Viguie, V., Sudmant, A., Graves, C., Foley, A., Heidrich, O., Mirailles, J. M., Carter, J., Chang, L. H., Nador, C., Liseri, M., Chelleri, L., Orru, H., Orru, K., Aelenei, R., Bilska, A., & Dawson, R. (2020). Integration of mitigation and adaptation in urban climate change action plans in Europe: A systematic assessment. *Renewable & Sustainable Energy Reviews*, *121*, 109623. DOI: 10.1016/j.rser.2019.109623

Greer, A., & Trainor, J. (2016). Designing for disaster: Lessons learned about disaster research from Superstorm Sandy. *Journal of Extreme Events*, *3*(02), 1671001.

Gros, C., Bailey, M., Schwager, S., Hassan, A., Zingg, R., Uddin, M. M., Shahjahan, M., Islam, H., Lux, S., Jaime, C., & Coughlan de Perez, E. (2019). Household-level effects of providing forecast-based cash in anticipation of extreme weather events: Quasi-experimental evidence from humanitarian interventions in the 2017 floods in Bangladesh. *International Journal of Disaster Risk Reduction*, *41*, 101275. DOI: 10.1016/j.ijdrr.2019.101275

Grosh, M., Del Ninno, C., Tesliuc, E., & Ouerghi, A. (2008). *For protection and promotion: The design and implementation of effective safety nets*. The World Bank. DOI: 10.1596/978-0-8213-7581-5

Guha-Sapir, D., Hoyois, P., Wallemacq, P., & Below, R. (2016). *Annual disaster statistical review 2016: The numbers and trends. Centre for Research on the Epidemiology of Disasters*. CRED.

Gumbo, D., & Zirebwa, K. (2017). *Indigenous knowledge systems and climate change adaptation in Zimbabwe: Challenges and opportunities*. IGI Global.

Guo, J., Liu, H., Wu, X., Gu, J., Song, S., & Tang, Y. (2015). Natural disasters, economic growth and Sustainable development in China-An empirical study using provincial panel data. *Sustainability (Basel)*, *7*(12), 16783–16800. DOI: 10.3390/su71215847

Gupta, R., & Nair, S. (2021). Social Support Networks in Disaster Recovery: Evidence from India. *Journal of Disaster Studies*, *15*(3), 45–60.

Hackett Group. (2019). Human Capital Management: Benchmarking for Business Success.

Haddow, G. D., Bullock, J. A., & Coppola, D. P. (2017). *Introduction to emergency management*. Butterworth-Heinemann.

Haibo, C., Zhang, K., & Su, Q. (2019). Sustainable development and disaster risk reduction: A case study of China. *Sustainability*, *11*(22), 6478. DOI: 10.3390/su11226478

Haines, A., & Ebi, K. (2019). The imperative for climate action to protect health. *The New England Journal of Medicine*, *380*(3), 263–273. DOI: 10.1056/NEJMra1807873 PMID: 30650330

Hallegatte, S. (2015). The indirect cost of natural disasters and an economic definition of macroeconomic resilience. World Bank Policy Research Working Paper, 7357. https://openknowledge.worldbank.org/handle/10986/22238

Hallegatte, S., Rentschler, J., & Walsh, B. (2020). Building back better: Achieving resilience through stronger, faster, and more inclusive post-disaster reconstruction. World Bank. https://openknowledge.worldbank.org/handle/10986/33727

Hallegatte, S., Vogt-Schilb, A., Bangalore, M., & Rozenberg, J. (2017). Unbreakable: Building the resilience of the poor in the face of natural disasters. World Bank. https://openknowledge.worldbank.org/handle/10986/25335

Hallegatte, S. (2014). *Economic resilience: Definition and measurement*. The World Bank. DOI: 10.1596/1813-9450-6852

Hallegatte, S., Bangalore, M., Bonzanigo, L., Fay, M., Kane, T., Narloch, U., Rozenberg, J., Treguer, D., & Vogt-Schilb, A. (2016). *Shock waves: Managing the impacts of climate change on poverty*. The World Bank.

Hallegatte, S., Henriet, F., & Corfee-Morlot, J. (2011). The economics of climate change impacts and policy benefits at city scale: A conceptual framework. *Climatic Change*, *104*(1), 51–87. DOI: 10.1007/s10584-010-9976-5

Hallegatte, S., Rentschler, J., & Rozenberg, J. (2019). *Lifelines: The resilient infrastructure opportunity*. The World Bank. DOI: 10.1596/978-1-4648-1430-3

Hallegatte, S., & Rozenberg, J. (2017). Climate change through a poverty lens. *Nature Climate Change*, *7*(4), 250–256. DOI: 10.1038/nclimate3253

Hameed, S., Sadiq, A., & Din, A. (2019). The Increased Vulnerability of Refugee Population to Mental Health Disorders. *Kansas Journal of Medicine*, *11*(1), 1–12. DOI: 10.17161/kjm.v11i1.8680 PMID: 29844851

Hamidi, A. R., Wang, J., & Guo, S. (2020). Natural disasters and economic growth: Empirical evidence from Asia. *Natural Hazards*, *101*(3), 707–727. DOI: 10.1007/s11069-020-03886-7

Haque, C. E., Choudhury, M. U., & Sikder, M. S. H. (2012). "Community-based adaptation" to climate change and cyclonic hazards in coastal Bangladesh. *Journal of Environmental Management*, *111*, 248–258. DOI: 10.1016/j.jenvman.2012.06.017

Haraguchi, M., & Lall, U. (2015). Flood risks and impacts: A case study of Thailand's floods in 2011 and research questions for supply chain decision making. *International Journal of Disaster Risk Reduction*, *14*(3), 256–272. DOI: 10.1016/j.ijdrr.2014.09.005

Harrison, C. G., & Williams, P. R. (2016). A systems approach to natural disaster resilience. *Simulation Modelling Practice and Theory*, *65*, 11–31. DOI: 10.1016/j.simpat.2016.02.008 PMID: 32288694

Harvey, P., Proudlock, K., Clay, E., Riley, B., & Jaspars, S. (2018). A review of the linkages between cash transfers and food security outcomes in humanitarian contexts. Overseas Development Institute. https://odi.org/en/publications/a-review-of-the-linkages-between-cash-transfers-and-food-security-outcomes-in-humanitarian-contexts/

Hayashi, T. (2012). Japan's post-disaster economic reconstruction: From Kobe to Tohoku. *Asian Economic Journal*, *26*(3), 189–210. DOI: 10.1111/j.1467-8381.2012.02082.x

Heanoy, E. Z., & Brown, N. R. (2024). Impact of natural disasters on mental health: Evidence and implications. *Health Care*, *12*(18), 1812. PMID: 39337153

Heltberg, R., Hossain, N., Reva, A., & Turk, C. (2012). Anatomy of coping: Evidence from people living through the crises of 2008–11. World Bank Policy Research Working Paper, 5957. https://openknowledge.worldbank.org/handle/10986/1990

Hennighausen, H., & Suter, J. F. (2020). Flood risk perception in the housing market and the impact of a major flood event. *Land Economics*, *96*(3), 366–383. DOI: 10.3368/le.96.3.366

Herrnstadt, E., & Muehlegger, E. (2014). Weather, salience of climate change and congressional voting. *Journal of Environmental Economics and Management*, *68*(3), 435–448. DOI: 10.1016/j.jeem.2014.08.002

Hidalgo, D., & Baez, A. A. (2019). Natural disasters: Preparedness and response in the Americas. *Prehospital and Disaster Medicine*, *34*(S1), s10–s11. DOI: 10.1017/S1049023X19000283

Hill, A. C., & Martinez-Diaz, L. (2020). *Building a resilient tomorrow: How to prepare for the coming climate disruption*. Oxford University Press. DOI: 10.1093/oso/9780190909345.001.0001

Holguín-Veras, J., Pérez, N., Jaller, M., Van Wassenhove, L. N., & Aros-Vera, F. (2012). On the appropriate objective function for post-disaster humanitarian logistics models. *Transportation Research Part E, Logistics and Transportation Review*, *48*(1), 8–22. DOI: 10.1016/j.tre.2011.08.002

How, V., Azmi, E. S., & Rahman, H. A. (2020). The way forward: Opportunities and challenges of sustainable school disaster education in Malaysia. *International Journal of Disaster Risk Reduction*, *51*, 101836. DOI: 10.1016/j.ijdrr.2020.101836

Hunt, S. T., Baltes, B. B., & DeHart, W. B. (2018). Analysis for Human Capital Analysis: A review of literature and implications for research. *Human Resource Management Review*, *28*(3), 347–358.

Hutton, C., Kienberger, S., Amoako Johnson, F., Allan, A., Giannini, V., Allen, R. (2011). Vulnerability to climate change: People, place and exposure to hazard. Advances in Science and Research, 7, 37.45.

Hynes, W., Trump, B., Love, P., & Linkov, I. (2020). Bouncing forward: A resilience approach to dealing with COVID-19 and future systemic shocks. *Environment Systems & Decisions*, *40*(2), 174–184. DOI: 10.1007/s10669-020-09776-x PMID: 32837818

IDMC. (2023). Children and youth in internal displacements. Internal Displacement Monitoring Centre (iDMC). https://www.internal-displacement.org/global-report/grid2022/#download

Ingram, J. C., Franco, G., Rio, C. R., & Khazai, B. (2006). Post-disaster recovery dilemmas: Challenges in balancing short-term and long-term needs for vulnerability reduction. *Environmental Science & Policy*, *9*(7-8), 607–613. DOI: 10.1016/j.envsci.2006.07.006

Inter-Agency Standing Committee. (2015). Reference module for cluster coordination at the country level. IASC. https://interagencystandingcommittee.org/system/files/iasc_reference_module_for_cluster_coordination_at_country_level_march_2015.pdf

Intergovernmental Panel on Climate Change. (2012). Managing the risks of extreme events and disasters to advance climate change adaptation: Special report of the Intergovernmental Panel on Climate Change. Cambridge University Press. https://www.ipcc.ch/report/managing-the-risks-of-extreme-events-and-disasters-to-advance-climate-change-adaptation/

IPCC. (2014). *Impacts, adaptation, and vulnerability. Part A: Global and sectoral aspects. Contribution of Working Group II to the Fifth Assessment Report of the Intergovernmental Panel on Climate Change*. Intergovernmental Panel on Climate Change. *Climatic Change*, 2014.

IPCC. (2018). *Global warming of 1.5°C. An IPCC Special Report on the impacts of global warming of 1.5°C above pre-industrial levels and related global greenhouse gas emission pathways, in the context of strengthening the global response to the threat of climate change, sustainable development, and efforts to eradicate poverty*. Intergovernmental Panel on Climate Change.

IPCC. (2022). *Climate Change 2022: Impacts, Adaptation and Vulnerability*. Cambridge University Press.

IPCC. (2022). *Climate Change 2022: Impacts, Adaptation and Vulnerability. Contribution of Working Group II to the Sixth Assessment Report of the Intergovernmental Panel on Climate Change*. Cambridge University Press.

Ishiwatari, M. (2012). Government roles in community-based disaster risk reduction. In Shaw, R. (Ed.), *Community-based disaster risk reduction* (pp. 19–33). Emerald Group Publishing., DOI: 10.1108/S2040-7262(2012)0000010008

Ishiwatari, M., & Surjan, A. (2019). Challenges in coordinating large-scale post-disaster redevelopment projects: Lessons from Asia. *International Journal of Disaster Risk Reduction, 39*, 101246. DOI: 10.1016/j.ijdrr.2019.101246

ISO. (2018). *ISO 31000:2018 Risk management - Guidelines*. International Organization for Standardization.

Jain, S., & Singh, S. (2020). Impact of Human Capital on Organizational Performance: A Review of Literature. Journal of Human Resource Management.

Jakhar, R., Samek, L., & Styszko, K. (2023). A comprehensive study of the impact of waste fires on the environment and health. *Sustainability (Basel), 15*(19), 14241. Advance online publication. DOI: 10.3390/su151914241

Jang, J. H., & Ha, K. M. (2021). Inclusion of children with disabilities in disaster management. *Children (Basel, Switzerland), 8*(7), 581. DOI: 10.3390/children8070581 PMID: 34356560

Jenkins, P., & Phillips, B. (2008). Battered women, catastrophe, and the context of safety after Hurricane Katrina. *NWSA Journal, 20*(3), 49–68. DOI: 10.1353/ff.2008.a256898

Jensen, N. D., Barrett, C. B., & Mude, A. G. (2016). Index insurance quality and the demand for insurance: Evidence from a field experiment in Kenya. *American Journal of Agricultural Economics, 98*(5), 1431–1450. DOI: 10.1093/ajae/aaw046

Jensen, N. D., Barrett, C. B., & Mude, A. G. (2017). Cash transfers and index insurance: A comparative impact analysis from northern Kenya. *Journal of Development Economics, 129*, 14–28. DOI: 10.1016/j.jdeveco.2017.08.002

Johannessen, A., Rosemarin, A., Thomalla, F., Gerger Swartling, Å., Stenström, T. A., & Vulturius, G. (2014). Strategies for building resilience to hazards in water, sanitation and hygiene (WASH) systems: The role of public private partnerships. *International Journal of Disaster Risk Reduction, 10*, 102–115. DOI: 10.1016/j.ijdrr.2014.07.002

Jomo, K. S., Chowdhury, A., Sharma, K., & Platz, D. (2016). Public-private partnerships and the 2030 Agenda for Sustainable Development: Fit for purpose? United Nations Department of Economic and Social Affairs Working Paper, 148. https://www.un.org/esa/desa/papers/2016/wp148_2016.pdf

Jones, L., Jaspars, S., Pavanello, S., Ludi, E., Slater, R., Grist, N., & Mtisi, S. (2010). Responding to a changing climate: Exploring how disaster risk reduction, social protection and livelihoods approaches promote features of adaptive capacity. Overseas Development Institute. https://odi.org/en/publications/responding-to-a-changing-climate-exploring-how-disaster-risk-reduction-social-protection-and-livelihoods-approaches-promote-features-of-adaptive-capacity/

Jones, A., & Patel, K. (2019). Anxiety and Fear in Disaster Survivors: A Meta-Analysis. *Psychological Trauma: Theory, Research, Practice, and Policy, 11*(4), 456–464.

Jones, P., & Kumar, R. (2019). Economic Diversification as a Strategy for Disaster Resilience. *The International Journal of Environmental Studies, 76*(4), 512–528.

Jose, P. J. (2024, August 17). *Wayanad landslides: When nature rewrote the map. Frontline Magazine*. Retrieved from https://frontline.thehindu.com/environment/wayanad-mundakkai-chooralmala-landslides-army-bodies-kerala-climate-change-impact-gadgil-western ghats/article68528102.ece

Joyette, V. T., Nurse, L. A., & Pulwarty, R. S. (2015). Disaster risk insurance and catastrophe models in the Caribbean. *The Geneva Papers on Risk and Insurance. Issues and Practice*, *40*(3), 553–576. DOI: 10.1057/gpp.2015.8 PMID: 25546529

Joy, J. L. K., Ramachandran, M., & George, S. (2021). Learned helplessness, psychological wellbeing, and pro-environment care behaviour among victims of frequent floods in Kerala. *Journal of Neurosciences in Rural Practice*, *12*(1), 137–144. DOI: 10.1055/s-0040-1721566 PMID: 33531773

Juillard, H., Mohiddin, L., Péchayre, M., & Smith, G. (2020). The influence of market support interventions on household food security: An evidence synthesis. Overseas Development Institute. https://odi.org/en/publications/the-influence-of-market-support-interventions-on-household-food-security-an-evidence-synthesis/

Juran, L. (2012). The Gendered Nature of Disasters: Women Survivors in Post-Tsunami Tamil Nadu. *Indian Journal of Gender Studies*, *19*(1), 1–29. DOI: 10.1177/097152151101900101

Kabisch, N., Frantzeskaki, N., Pauleit, S., Naumann, S., Davis, M., Artmann, M., . . . Zaunberger, K. (2017). Nature-based solutions to climate change adaptation in urban areas: Linkages between science, policy and practice. In N. Kabisch, H. Korn, J. Stadler, & A. Bonn (Eds.), Nature-based solutions to climate change adaptation in urban areas (pp. 1–11). Springer. DOI: 10.1007/978-3-319-56091-5_1

Kahneman, D., & Frederick, S. (2002). Representativeness revisited: Attribute substitution in intuitive judgment. In Gilovich, T., Griffin, D., & Kahneman, D. (Eds.), *Heuristics and biases: The psychology of intuitive judgment* (pp. 49–81). Cambridge University Press. DOI: 10.1017/CBO9780511808098.004

Kahn, M. E., & Kotchen, M. J. (2011). Business cycle effects on concern about climate change: The chilling effect of recession. *Climate Change Economics (Singapore)*, *2*(3), 257–273. DOI: 10.1142/S2010007811000292

Kaku, K. (2019). Satellite remote sensing for disaster management support: A holistic and staged approach based on case studies in Sentinel Asia. *International Journal of Disaster Risk Reduction*, *33*, 417–432. DOI: 10.1016/j.ijdrr.2018.09.015

Kanamori, H., & Brodsky, E. (2004). The physics of earthquakes. *Reports on Progress in Physics, 54*(8), 1429–1496. Advance online publication. DOI: 10.1088/0034-4885/67/8/R03

Kankanamge, N., Yigitcanlar, T., & Goonetilleke, A. (2020). How engaging are disaster management related social media channels? The case of Australian state emergency organisations. *International Journal of Disaster Risk Reduction, 48*, 101571. DOI: 10.1016/j.ijdrr.2020.101571

Kao, C. L., Chien, L. C., Wang, M. C., Tang, J. S., Huang, P.-C., Chuang, C.-C., & Shih, C.-L. (2023). The development of new remote technologies in disaster medicine education: A scoping review. *Frontiers in Public Health, 11*, 1099467. DOI: 10.3389/fpubh.2023.1029558 PMID: 37033011

Kapucu, N. (2008). Collaborative emergency management: Better community organising, better public preparedness and response. *Disasters, 32*(2), 239–262. DOI: 10.1111/j.1467-7717.2008.01037.x PMID: 18380853

Kapucu, N., & Garayev, V. (2011). Collaborative decision-making in emergency and disaster management. *International Journal of Public Administration, 34*(6), 366–375. DOI: 10.1080/01900692.2011.561477

Kapur, B. K. (2016). Economic consequences of natural disasters: A review. *Journal of Economic Surveys, 30*(4), 664–687. DOI: 10.1111/joes.12123

Kataoka, S. H., Langley, A. K., Wong, M., Baweja, S., & Stein, B. D. (2009). Responding to students with PTSD in schools. *Child and Adolescent Psychiatric Clinics of North America, 18*(3), 563–579.

Katz, J. M. (2013). *The big truck that went by: How the world came to save Haiti and left behind a disaster*. St. Martin's Press.

Kaufhold, M. A., Reuter, C., & Ludwig, T. (2023). Big data and multi-platform social media services in disaster management. In *International handbook of disaster research* (pp. 1–18). Springer., DOI: 10.1007/978-3-030-91691-6_103

Kearns, P. (2012). Managing the Human Capital: Performance Management in the 21st Century. *Journal of Management Development.*

Keerthiratne, S. (2017). Economic impact of natural disasters. University of Oxford, Department of Economics Working Paper. https://ora.ox.ac.uk/objects/uuid:12345678-1234-5678-1234-567890123456

Keerthiratne, S., & Tol, R. S. J. (2018). Impact of natural disasters on income inequality in Sri Lanka. *World Development, 105*, 217–230. DOI: 10.1016/j.worlddev.2018.01.001

Keim, M. E. (2015). 2. The public health impacts of natural disasters. In the Handbook *of public health in natural disasters* (pp. 33-58). Wageningen.

Kellens, W., Zaalberg, R., Neutens, T., Vanneuville, W., & De Maeyer, P. (2011). An analysis of the public perception of flood risk on the Belgian coast. *Risk Analysis, 31*(7), 1055–1068. DOI: 10.1111/j.1539-6924.2010.01571.x PMID: 21231949

Kellett, J., & Peters, K. (2014). Dare to prepare: Taking risk seriously. Overseas Development Institute. https://odi.org/en/publications/dare-to-prepare-taking-risk-seriously/

Kelly, A. E., & Goulden, M. L. (2008). Rapid shifts in plant distribution with recent climate change. *Proceedings of the National Academy of Sciences of the United States of America, 105*(33), 11823–11826. DOI: 10.1073/pnas.0802891105

Kelman, I., Gaillard, J.-C., & Mercer, J. (2015). Climate change's role in disaster risk reduction's future: Beyond vulnerability and resilience. *International Journal of Disaster Risk Science, 6*(1), 21–27. DOI: 10.1007/s13753-015-0038-5

Kempe, M. (2003). Noah's Flood: The Genesis story and natural disasters in early modern times. *Environment and History, 9*(2), 151–171. DOI: 10.3197/096734003129342809

Kemp, S., Helton, W. S., Richardson, J. J., Blampied, N. M., & Grimshaw, M. (2011). Sleeplessness, stress, cognitive disruption and academic performance following the September 4, 2010, Christchurch earthquake Australas. *Disaster Trauma Studies*, 2, 11–18.

Kennedy, J., Ashmore, J., Babister, E., & Kelman, I. (2008). The meaning of 'build back better': Evidence from post-tsunami Aceh and Sri Lanka. *Journal of Contingencies and Crisis Management, 16*(1), 24–36. DOI: 10.1111/j.1468-5973.2008.00529.x

Kerala Government. (n.d.). *Pariraksha: Comprehensive mental health and psychosocial project*. Retrieved from https://www.kerala.gov.in/pariraksha

Kerala State Disaster Management Authority. (n.d.). *DISHA Helpline (1056): 24/7 post-traumatic psychological support*. Retrieved from https://disha1056.com/

Kerala State Disaster Management Authority. (n.d.). *Psychosocial Support Teams (PSS) and School Mental Health Program*. https://sdma.kerala.gov.in/psychosocial-support/

Kessler, R. C., Sonnega, A., Bromet, E., Hughes, M., & Nelson, C. B. (2013). Posttraumatic stress disorder in the National Comorbidity Survey. In *Fear and anxiety* (pp. 22–34). Routledge.

Kim, J., Yu, I., Lee, J., & Kim, H. (2019). Probabilistic risk assessment for urban infrastructure resilience: A framework for multi-hazard environments. *Sustainable Cities and Society*, *51*, 101662. DOI: 10.1016/j.scs.2019.101662

Kirkland, T. M., Keene, D. E., & Gómez, W. (2021). Indigenous knowledge integration in community-based adaptation. *Current Opinion in Environmental Sustainability*, *51*, 55–60.

Klomp, J., & Valckx, K. (2014). Natural disasters and economic growth: A meta-analysis. *Global Environmental Change*, *26*, 183–195. DOI: 10.1016/j.gloenvcha.2014.02.006

Kocabıyık, O. O. (2016). Olgubilim ve gömülü kuram: Bazı özellikler açısından karşılaştırma. *Trakya Üniversitesi Eğitim Fakültesi Dergisi*, *6*(1), 55–66.

Köknel, Ö. (1987). *Zorlanan insan*. Altın kitaplar Yayınevi, İstanbul.

Kondratyev, K. Y., Varotsos, C. A., & Krapivin, V. F. (2006). Statistics of natural disasters. *Natural Disasters as Interactive Components of Global Ecodynamics*, 1-59.

Koshy, M., Aranya, R., & Refstie, H. (2022). Handling compounded uncertainty in spatial planning and humanitarian action in unexpected floods in Wayanad, Kerala: Towards a contextualized contingency planning approach. *Planning Theory & Practice*, *23*(1), 33–51. DOI: 10.1080/14649357.2022.2143548

Koshy, M., & Smith, D. (2024). Community resilience implications for institutional response under uncertainty. *Sustainable Development (Bradford)*, *32*(2), 1412–1427. DOI: 10.1002/sd.2678

Koubi, V., Spilker, G., Schaffer, L., & Bernauer, T. (2016). Environmental stressors and migration: Evidence from Vietnam. *World Development*, *79*, 197–210. DOI: 10.1016/j.worlddev.2015.11.016

Kousky, C., & Shabman, L. (2015). A proposed design for community-based flood insurance. Resources for the Future Discussion Paper, 15-02. https://www.rff.org/publications/working-papers/a-proposed-design-for-community-based-flood-insurance/

Kousky, C. (2014). Informing climate adaptation: A review of the economic costs of natural disasters. *Energy Economics*, *46*, 576–592. DOI: 10.1016/j.eneco.2013.09.029

Kousky, C. (2016). Impacts of natural disasters on children. *The Future of Children*, *26*(1), 73–92. DOI: 10.1353/foc.2016.0004

Kousky, C., & Kunreuther, H. (2014). Addressing affordability in the national flood insurance program. *Journal of Extreme Events*, *1*(1), 1450001. DOI: 10.1142/S2345737614500018

Kovács, G., & Spens, K. M. (2011). Trends and developments in humanitarian logistics—A gap analysis. *International Journal of Physical Distribution & Logistics Management*, *41*(1), 32–45. DOI: 10.1108/09600031111101411

Krishnamurthy, P. K. (2012). Disaster-induced migration: Assessing the impact of extreme weather events on livelihoods. *Environmental Hazards*, *11*(2), 96–111. DOI: 10.1080/17477891.2011.609879

Kudumbashree. (n.d.). *Community-based interventions for mental well-being and resilience*. Retrieved from https://www.kudumbashree.org/

Kula, Y., Cohen, O., Clempert, N., Grinstein-Cohen, O., & Slobodin, O. (2021). Educating nursing students for cultural competence in emergencies: A randomized controlled trial. *BMC Nursing*, *20*(1), 178. DOI: 10.1186/s12912-021-00704-1 PMID: 34587960

Kull, D., Mechler, R., & Hochrainer-Stigler, S. (2013). Probabilistic cost-benefit analysis of disaster risk management in a development context. *Disasters*, *37*(3), 374–400. DOI: 10.1111/disa.12002 PMID: 23551288

Kumar, A., & Singh, V. (2022). Collaborative Disaster Management: Role of NGOs and Government Agencies. *Disaster Prevention and Management*, *31*(2), 89–104.

Kun, P., Tong, X., Liu, Y., Pei, X., & Luo, H. (2013). What are the determinants of post-traumatic stress disorder: Age, gender, ethnicity or other? Evidence from 2008 Wenchuan earthquake. *Public Health*, *127*(7), 644–652. DOI: 10.1016/j.puhe.2013.04.018 PMID: 23806187

Kuntz, J. R. C. (2015). Protracted effect: Surveying teachers experiences in the aftermath of the christchurch earthquakes. *Natural Hazards Review*, *16*(1), 401–414. DOI: 10.1061/(ASCE)NH.1527-6996.0000147

La Greca, A. M., Silverman, W. K., Lai, B., & Jaccard, J. (2010). Hurricane-related exposure experiences and stressors, other life events, and social support: Concurrent and prospective impact on children's persistent posttraumatic stress symptoms. *Journal of Consulting and Clinical Psychology*, *78*(6), 794–805. DOI: 10.1037/a0020775 PMID: 20939624

Lakdawalla, D., & Zanjani, G. (2012). Catastrophe bonds, reinsurance, and the optimal collateralization of risk transfer. *The Journal of Risk and Insurance, 79*(2), 449–476. DOI: 10.1111/j.1539-6975.2011.01425.x

Lang, C., & Ryder, J. D. (2016). The effect of tropical cyclones on climate change engagement. *Climatic Change, 135*(3–4), 625–638. DOI: 10.1007/s10584-015-1590-0

Le Houérou, H. N. (1996). Climate change, drought and desertification. *Journal of Arid Environments, 34*(2), 133–185. DOI: 10.1006/jare.1996.0099

Lechowska, E. (2018). What determines flood risk perception? A review of factors of flood risk perception and relations between its basic elements. *Natural Hazards, 94*(3), 1341–1366. DOI: 10.1007/s11069-018-3480-z

Lee, C. C., & Tang, H. (2019). The impact of natural disasters on financial development: Evidence from the Philippines. *The Singapore Economic Review, 64*(5), 1207–1229. DOI: 10.1142/S0217590818500118

Lewis, J., & Kelman, I. (2025). Development and Disasters.

Libayao, H. G. C., & Navaluna, C. A.Jr. (2024). Disaster education in elementary school curriculum: Basis for framework designing. *Journal of Educational and Social Research, 14*(1), 147–158. DOI: 10.36941/jesr-2024-0012

Li, J., Meng, L., Zhang, Z., & Yang, K. (2023). Low-frequency, high impact: Discovering important rare events from UGC. *Journal of Retailing and Consumer Services, 70*, 103153. DOI: 10.1016/j.jretconser.2022.103153

Lindell, M. K., & Hwang, S. N. (2008). Households' perceived personal risk and responses in a multihazard environment. *Risk Analysis, 28*(2), 539–556. DOI: 10.1111/j.1539-6924.2008.01032.x PMID: 18419668

Lindell, M., Bostrom, A., Gotz, J. D., & Prater, C. S. (2021). Evaluating hazard awareness brochures: Assessing the textual, graphical, and numerical features of tsunami evacuation products. *International Journal of Disaster Risk Reduction, 61*, 102361. DOI: 10.1016/j.ijdrr.2021.102361

Linnenluecke, M. K., & McKnight, B. (2017). Community resilience to natural disasters: The role of disaster entrepreneurship. *Journal of Enterprising Communities: People and Places in the Global Economy, 11*(1), 166–185. DOI: 10.1108/JEC-01-2015-0005

Linnerooth-Bayer, J., & Hochrainer-Stigler, S. (2015). Financial instruments for disaster risk management and climate change adaptation. *Climatic Change, 133*(1), 85–100. DOI: 10.1007/s10584-013-1035-6

Loayza, N. V., Olaberria, E., Rigolini, J., & Christiaensen, L. (2009). Natural disasters and growth: Going beyond the averages. World Bank Policy Research Working Paper 4980. DOI: 10.1596/1813-9450-4980

Lumbroso, D., Brown, E., & Ranger, N. (2017). Stakeholders' perceptions of the overall effectiveness of early warning systems and risk assessments for weather-related hazards in Africa, the Caribbean and South Asia. *International Journal of Disaster Risk Reduction*, *21*, 419–432. DOI: 10.1016/j.ijdrr.2016.11.003

Lumbroso, D., Woolhouse, G., & Jones, L. (2017). A review of the consideration of climate change in the planning of hydropower schemes in sub-Saharan Africa. *Climatic Change*, *140*(3-4), 563–578. DOI: 10.1007/s10584-016-1857-0

Lv, Y., Sarker, N. I., & Firdaus, R. B. R. (2024). Disaster resilience in climate-vulnerable community context: Conceptual analysis. *Ecological Indicators*, *158*, 111527. DOI: 10.1016/j.ecolind.2023.111527

Lyons, M. (2009). Building back better: The large-scale impact of small-scale approaches to reconstruction. *World Development*, *37*(2), 385–398. DOI: 10.1016/j.worlddev.2008.01.006

Maas, A., Wardropper, C., Roesch-McNally, G., & Abatzoglou, J. (2020). A (mis) alignment of farmer experience and perceptions of climate change in the US inland Pacific Northwest. *Climatic Change*, *162*(3), 1011–1029. DOI: 10.1007/s10584-020-02713-6

Maat, H., Balabanova, D., Mokuwa, E., Richards, P., Mohan, V., Ssengooba, F., Twinomuhangi, R., Woldie, M., & Mayhew, S. (2021). Towards sustainable community-based systems for infectious disease and disaster response: Lessons from local initiatives in four African countries. *Sustainability (Basel)*, *13*(11), 6335. DOI: 10.3390/su131810083

Maden, S. (2023). 6 Şubat 2023'te Kahramanmaraş'ta Yaşanan Depremler Ekseninde Türkiye'de Deprem Haberciliğine Bakış: Prof. Dr. Süleyman İrvan ile Söyleşi. *Etkileşim*, *6*(11), 406–420. DOI: 10.32739/etkilesim.2023.6.11.202

Mahadev Madgule, N., Vinayaka, H., Sonkhaskar, Y. M., Patel, D. R., Karthikeyan, R., & Selvakumar, P. (2023). Mechanical properties and microstructure of activated TIG welded similar joints of Inconel alloys by desirability approaches. *Materials Today: Proceedings*, *77*, 528–533. DOI: 10.1016/j.matpr.2022.12.250

Mahul, O., & Ghesquiere, F. (2010). Financial protection of the state against natural disasters: A primer. World Bank Policy Research Working Paper, 5429. https://openknowledge.worldbank.org/handle/10986/3913

Mahul, O., & Stutley, C. J. (2010). Government support to agricultural insurance: Challenges and options for developing countries. World Bank. https://openknowledge.worldbank.org/handle/10986/2432

Manase, S. (2020). *Community-based disaster risk management in Zimbabwe: Integrating local knowledge into national strategies*. IGI Global.

Mann, M. E., & Gleick, P. H. (2015). Climate change and California drought in the 21st century. *Proceedings of the National Academy of Sciences of the United States of America, 112*(13), 3858–3859. DOI: 10.1073/pnas.1503667112

Manyena, B., O'Brien, G., O'Keefe, P., & Rose, J. (2011). Disaster resilience: A bounce back or bounce forward ability? *Local Environment, 16*(5), 417–424. DOI: 10.1080/13549839.2011.583049

Marchezini, V., Horita, F. E. A., Matsuo, P. M., Trajber, R., Trejo-Rangel, M. A., & Olivato, D. (2018). Bridging the gap between research and practice: The role of early warning systems in disaster risk reduction. *International Journal of Disaster Risk Reduction, 31*, 1082–1092. DOI: 10.1016/j.ijdrr.2018.08.011

Marchiori, L., Maystadt, J.-F., & Schumacher, I. (2012). The impact of weather anomalies on migration in sub-Saharan Africa. *Journal of Environmental Economics and Management, 63*(3), 355–374. DOI: 10.1016/j.jeem.2012.02.001

Marin, G., & Modica, M. (2017). Socio-economic exposure to natural disasters. *Environmental Impact Assessment Review, 64*, 57–66. DOI: 10.1016/j.eiar.2017.03.002

Martínez-Zarzoso, I. (2020). Searching for grouped patterns of heterogeneity in the climate–migration link. *Weather, Climate, and Society, 12*(4), 695–710. DOI: 10.1175/WCAS-D-19-0116.1

Martin, R., & Sunley, P. (2015). On the notion of regional economic resilience: Conceptualization and explanation. *Journal of Economic Geography, 15*(1), 1–42. DOI: 10.1093/jeg/lbu015

Maskrey, A. (2011). Revisiting community-based disaster risk management. *Environmental Hazards, 10*(1), 42–52. DOI: 10.3763/ehaz.2011.0005

Masten, A. S., & Narayan, A. J. (2012). Child development in the context of disaster, war, and terrorism: Pathways of risk and resilience. *Annual Review of Psychology, 63*(1), 227–257. DOI: 10.1146/annurev-psych-120710-100356 PMID: 21943168

Masten, A. S., & Osofsky, J. D. (2010). Disasters and their impact on child development: Introduction to the special section. *Child Development, 81*(4), 1029–1039. DOI: 10.1111/j.1467-8624.2010.01452.x PMID: 20636680

Mathew, M. (2019). *Environmental degradation and disaster in Wayanad: A study on the role of corporate interests and governmental measures.*

Math, S. B., Nirmala, M. C., Moirangthem, S., & Kumar, N. C. (2015). Disaster management: Mental health perspective. *Indian Journal of Psychological Medicine*, *37*(3), 261–271. DOI: 10.4103/0253-7176.162915 PMID: 26664073

Mavhura, E. (2021). Indigenous knowledge and disaster risk reduction: The role of cultural beliefs in building resilience. *Sustainability*, *13*(5), 2558. DOI: 10.3390/su13052558

Mavhura, E., Manyena, B., & Collins, A. (2020). Drought vulnerability and risk perception: A case study from Zimbabwe. *Natural Hazards*, *101*(1), 219–241. DOI: 10.1007/s11069-020-03854-5

Mayer, B., Bergstrand, J. K., & Running, K. (2020). Economic recovery and community resilience: Lessons from Hurricane Harvey. *Environmental Justice*, *13*(2), 52–57. DOI: 10.1089/env.2019.0037

Mazzucato, M., & Perez, C. (2015). Innovation as growth policy: The challenge for Europe. In Fagerberg, J., Laestadius, S., & Martin, B. R. (Eds.), *The triple challenge for Europe: Economic development, climate change, and governance* (pp. 229–264). Oxford University Press. DOI: 10.1093/acprof:oso/9780198747413.003.0009

McAdoo, B. G., Moore, A., & Baumwoll, J. (2009). Indigenous knowledge and the near field population response during the 2007 Solomon Islands tsunami. *Natural Hazards*, *48*(1), 73–82. DOI: 10.1007/s11069-008-9249-z

McCord, A. (2013). Public works and social protection in sub-Saharan Africa: Do public works work for the poor? United Nations University Press. https://www.unu.edu/publications/books/public-works-and-social-protection-sub-saharan-africa-do-public-works-work-poor

McDaniels, T., Chang, S., Cole, D., Mikawoz, J., & Longstaff, H. (2008). Fostering resilience to extreme events within infrastructure systems: Characterizing decision contexts for mitigation and adaptation. *Global Environmental Change*, *18*(2), 310–318. DOI: 10.1016/j.gloenvcha.2008.03.001

McKinsey & Company. (2021). *The Future of Work After COVID-19*. McKinsey Global Institute.

McNamara, K. E., & Buggy, L. (2017). Community-based climate change adaptation: A review of academic literature. *Local Environment*, *22*(4), 443–460. DOI: 10.1080/13549839.2016.1216954

McNamara, K. E., & Buggy, L. (2021). Community-based climate adaptation: A systematic review of the literature. *Climate and Development, 13*(2), 110–122.

McNaught, R. (2024). The application of collaborative governance in local level climate and disaster resilient development—A global review. *Environmental Science & Policy, 151*, 103632. DOI: 10.1016/j.envsci.2023.103627

Medel, K., Kousar, R., & Masood, T. (2020). A collaboration–resilience framework for disaster management supply networks: A case study of the Philippines. *Journal of Humanitarian Logistics and Supply Chain Management, 10*(4), 509–532. DOI: 10.1108/JHLSCM-09-2019-0066

Meier, P. (2015). *Digital humanitarians: How big data is changing the face of humanitarian response.* CRC Press. DOI: 10.1201/b18023

Mendeldon, E. (2011). System and method for providing alarming notification and real-time, critical emergency information to occupants in a building or emergency designed area and evacuation guidance system to and in the emergency exit route (U.S. Patent No. US20110241877A1). Google Patents. https://patents.google.com/patent/US20110241877A1/en

Mendia-Landa, P. (2013). Myths and Legends on Natural Disasters: Making Sense of Our World. *Yale-New Haven Teachers Institute, 7.*

Mercene, F. (2017). *Cyclone Nargis: A case study of disaster response.* Manila Bulletin.

Mercer, J., Kelman, I., Taranis, L., & Suchet-Pearson, S. (2010). Framework for integrating indigenous and scientific knowledge for disaster risk reduction. *Disasters, 34*(1), 214–239. DOI: 10.1111/j.1467-7717.2009.01126.x PMID: 19793324

Milla, A., Rojas, A., & Salazar, G. (2013). Traditional ecological knowledge and agricultural land-use practices in the Andes: A case study of the Quechua people of Peru. *Journal of Sustainable Agriculture, 37*(2), 221–234.

Mohan, C., & Joy, R. (n.d.). Entrepreneurs' resilience to natural disasters: Emerging entrepreneurs from flood-affected areas in Kerala. *The Empirical Economic Letters, 20*(3), 33–41.

Mohan, P. S., Strobl, E., & Watson, P. (2018). The short-term economic impact of tropical cyclones: Evidence from the Caribbean. *Natural Hazards, 91*(1), 87–104. DOI: 10.1007/s11069-017-3113-9

Mojtahedi, M., & Oo, B. L. (2017). Critical attributes for proactive engagement of stakeholders in disaster risk management. *International Journal of Disaster Risk Reduction, 21*, 35–43. DOI: 10.1016/j.ijdrr.2016.10.017

Moote, A. L., & Moote, D. C. (2004). *The great plague: the story of London's most deadly year*. JHU Press. DOI: 10.56021/9780801877834

Mortimer, I. (2017). *The Time Traveller's Guide to Restoration Britain: Life in the Age of Samuel Pepys, Isaac Newton and The Great Fire of London*. Random House.

Moser, S. C., Coffee, J., & Seville, A. (2021). *Rising to the challenge, together: A review and critical assessment of the state of the US climate adaptation field*. World Resources Institute.

Mostofi, M., & Brown, L. M. (2021). Cultural competence and disaster mental health. In *The intersection of trauma and disaster behavioral health* (pp. 285–300). Springer., DOI: 10.1007/978-3-030-51525-6_7

Mudenge, M. (2016). Indigenous knowledge systems and practices in Zimbabwe: Implications for sustainable development. *Indigenous Knowledge and Development Journal, 15*(1), 75–89.

Mukherjee, S., & Hastak, M. (2018). Economic consequences of natural disasters: A regional perspective. *International Journal of Disaster Risk Reduction, 31*, 1216–1225. DOI: 10.1016/j.ijdrr.2018.03.017

Muñoz, V. A., Carby, B., Abella, E. C., Cardona, O. D., López-Marrero, T., Marchezini, V., Meyreles, L., Olivato, D., Trajber, R., & Wisner, B. (2020). Success, innovation and challenge: School safety and disaster education in South America and the Caribbean. *International Journal of Disaster Risk Reduction, 44*, 101424. DOI: 10.1016/j.ijdrr.2019.101395

Mutch, C. (2021). The role of schools in helping communities copes with earthquake disasters: The case of the 2010–2011 New Zealand earthquakes. In Xu, J., Lu, Y., & Penning-Rowsell, E. C. (Eds.), *Earthquake disasters: Prevention and reconstruction*. Routledge. DOI: 10.4324/9781003173328-5

Naheed, S. (2021). Understanding disaster risk reduction and resilience: A conceptual framework. In *Handbook of disaster risk reduction for resilience: New perspectives and challenges* (pp. 1–20). Springer., DOI: 10.1007/978-3-030-61278-8_1

National Oceanic and Atmospheric Administration. (2019). Storm events database. https://www.ncdc.noaa.gov/stormevents/

National Research Council. (2011). *Building community disaster resilience through private-public collaboration*. National Academies Press., DOI: 10.17226/13028

Neal, D. M. (2013). Social time and disaster. *International Journal of Mass Emergencies and Disasters, 31*(2), 247–270. DOI: 10.1177/028072701303100207

Neria, Y., Nandi, A., & Galea, S. (2008). Post-traumatic stress disorder following disasters: A systematic review. *Psychological Medicine, 38*(4), 467–480. DOI: 10.1017/S0033291707001353 PMID: 17803838

Nguyen, P. T., Anderson, C., & Läderach, P. (2023). Leveraging indigenous knowledge to improve climate resilience: Insights from the Maasai pastoralists in Kenya. *Journal of Environmental Management, 324*, 116435. DOI: 10.1016/j.jenvman.2022.116435

Nikolaraizi, M., Argyropoulos, V., Papazafiri, M., & Kofidou, C. (2021). Promoting accessible and inclusive education on disaster risk reduction: The case of students with sensory disabilities. *International Journal of Inclusive Education, 25*(7), 772–786. DOI: 10.1080/13603116.2020.1862408

Norris, F. H., Stevens, S. P., Pfefferbaum, B., Wyche, K. F., & Pfefferbaum, R. L. (2008). Community resilience as a metaphor, theory, set of capacities, and strategy for disaster readiness. *American Journal of Community Psychology, 41*(1-2), 127–150. DOI: 10.1007/s10464-007-9156-6 PMID: 18157631

Norris, F., Friedman, M., Watson, P., Byrne, C., Diaz, E., & Kaniasty, K. (2002). 60,000 disaster victims speak: Part I: An empirical review of the empirical literature, 1981–2001. [Crossref. PubMed. ISI.]. *Psychiatry, 65*(3), 207–239. DOI: 10.1521/psyc.65.3.207.20173 PMID: 12405079

North, C. S., & Pfefferbaum, B. (2013). Mental health response to community disasters: A systematic review. *Journal of the American Medical Association, 310*(5), 507–518. DOI: 10.1001/jama.2013.107799 PMID: 23925621

Noy, I., & Yonson, R. (2018). Economic vulnerability and resilience to natural hazards: A survey of concepts and measurements. *Sustainability (Basel), 10*(8), 2850. DOI: 10.3390/su10082850

Nyong, A., Adesina, F., & Elasha, B. O. (2007). The value of indigenous knowledge in climate change mitigation and adaptation strategies in the African Sahel. *Mitigation and Adaptation Strategies for Global Change, 12*(5), 787–797. DOI: 10.1007/s11027-007-9099-0

Nyong, A., Naswa, P., & Tabopda, C. (2021). The importance of integrating indigenous knowledge in climate change adaptation and mitigation strategies in Africa. *Climate Change Review, 45*(3), 197–210.

O'Brien, G., O'Keefe, P., Rose, J., & Wisner, B. (2006). Climate change and disaster management. *Disasters*, *30*(1), 64–80. DOI: 10.1111/j.1467-9523.2006.00307.x PMID: 16512862

Ofli, F., Meier, P., Imran, M., Castillo, C., Tuia, D., Rey, N., Briant, J., Millet, P., Reinhard, F., Parkan, M., & Joost, S. (2016). Combining human computing and machine learning to make sense of big (aerial) data for disaster response. *Big Data*, *4*(1), 47–59. DOI: 10.1089/big.2014.0064 PMID: 27441584

Oliveira, V. H. (2019). Natural disasters and economic growth in Brazil: A sectoral analysis. *Environment and Development Economics*, *24*(5), 497–517. DOI: 10.1017/S1355770X19000217

Olshansky, R. B., Hopkins, L. D., & Johnson, L. A. (2012). Disaster and recovery: Processes compressed in time. *Natural Hazards Review*, *13*(3), 173–178. DOI: 10.1061/(ASCE)NH.1527-6996.0000077

Opitz-Stapleton, S., Nadin, R., Watson, C., & Kellett, J. (2019). Risk-informed development: From crisis to resilience. Overseas Development Institute (ODI) Report. https://odi.org/en/publications/risk-informed-development-from-crisis-to-resilience

Oven, K., & Bankoff, G. (2020). The neglected country(side): Earthquake risk perceptions and disaster risk reduction in post-Soviet rural Kazakhastan. *Journal of Rural Studies*, *80*, 171–184. DOI: 10.1016/j.jrurstud.2020.08.048

Owen, A. L., Conover, E., Videras, J., & Wu, S. (2012). Heat waves, droughts, and preferences for environmental policy. *Journal of Policy Analysis and Management*, *31*(3), 556–577. DOI: 10.1002/pam.21627

Owen, P., & Wesselbaum, D. (2020). On thresholds in the climate–migration relationship. *International Review of Applied Economics*, *34*(3), 400–412. DOI: 10.1080/02692171.2020.1749242

Öztürk, M. K. (2013). A research on earthquake experiences of primary school teacher candidates. [Hacettepe Unıversity Journal of Education]. *Hacettepe Üniversitesi Eğitim Fakültesi Dergisi*, *28*(1), 308–319.

Pachauri, R. K., Allen, M. R., Barros, V. R., Broome, J., Cramer, W., Christ, R., . . . Dubash, N. K. (2014). Climate change 2014: Synthesis report. Contribution of Working Groups I, II and III to the fifth assessment report of the Intergovernmental Panel on Climate Change. Intergovernmental Panel on Climate Change. https://www.ipcc.ch/report/ar5/syr/

Padli, J., Habibullah, M. S., & Baharom, A. H. (2018). The impact of human development on natural disaster fatalities and damage: Panel data evidence. *Ekonomska Istrazivanja*, *31*(1), 1557–1573. DOI: 10.1080/1331677X.2018.1504689

Pahl, S., Sheppard, S., Boomsma, C., & Groves, C. (2014). Perceptions of time in relation to climate change. *Wiley Interdisciplinary Reviews: Climate Change*, *5*(3), 375–388. DOI: 10.1002/wcc.272

Panteli, M., & Mancarella, P. (2015). Influence of extreme weather and climate change on the resilience of power systems: Impacts and possible mitigation strategies. *Electric Power Systems Research*, *127*, 259–270. DOI: 10.1016/j.epsr.2015.06.012

Panwar, V., & Sen, S. (2019). Economic impact of natural disasters: An empirical re-examination. *Margin*, *13*(1), 109–139. DOI: 10.1177/0973801018800087

Paramasivam, S., Mariappan, P., Petchiammal, S., & Richard Paul, S. J. (2024). Removal of chemicals from effluent using photobioreactor technology to improve environmental and health impacts. *Zeitschrift für Physikalische Chemie*, *238*(6), 1119–1132. DOI: 10.1515/zpch-2023-0450

Paramasivam, S., Palanisamy, S. K., Cinthaikinian, S., Selvamurugan, P., Vigneshkumar, P., Mariappan, R., & Selvakumar, P. K. (2024). Biosensors and its diverse applications in healthcare systems. *Zeitschrift für Physikalische Chemie*, *238*(5), 885–904. DOI: 10.1515/zpch-2023-0406

Patel, S. (2018). Stress Management Techniques in Disaster-Affected Populations. *Journal of Mental Health and Resilience*, *10*(1), 22–35.

Paton, D. (2019). Disaster risk reduction: Psychological perspectives on preparedness. *Australian Journal of Psychology*, *71*(4), 327–341. DOI: 10.1111/ajpy.12237

Paton, D., Anderson, E., Becker, J., & Petersen, J. (2015). Developing a comprehensive model of hazard preparedness: From the Christchurch earthquake. *International Journal of Disaster Risk Reduction*, *14*, 37–45. DOI: 10.1016/j.ijdrr.2014.11.011

Paton, D., & Johnston, D. (2001). Disasters and communities: Vulnerability, resilience, and preparedness. *Disaster Prevention and Management*, *10*(4), 270–277. DOI: 10.1108/EUM0000000005930

Paton, D., & Johnston, D. (2017). *Disaster resilience: An integrated approach* (2nd ed.). Charles C Thomas Publisher.

Paton, D., & Johnston, D. (2017). *Disaster resilience: an integrated approach.* Charles C Thomas Publisher.

Patton, M. Q. (2014). Nitel araştırma ve değerlendirme yöntemleri. M. Bütün ve S. B. Demir (Çev. Ed.). Ankara: Pegem Akademi.

Patz, J. A., & Thomson, M. C. (2022). Climate change and health: Moving from theory to practice. *PLoS Medicine*, *19*(4), e1003985. PMID: 30063707

Pearson, R. (2017). *Insuring the Industrial Revolution: Fire Insurance in Great Britain, 1700–1850*. Routledge. DOI: 10.4324/9781315252193

Peduzzi, P. (2019). The disaster risk, global change, and sustainability nexus. *Sustainability (Basel)*, *11*(4), 957. DOI: 10.3390/su11040957

Peek, L. (2008). Children and disasters: Understanding vulnerability, developing capacities, and promoting resilience. *Children, Youth and Environments*, *18*(1), 1–29. DOI: 10.1353/cye.2008.0052

Pepys, S. (2000). *The Diary of Samuel Pepys, Vol. 1: 1660* (Vol. 1). Univ of California Press.

Pepys, S. (2015). *The Great Fire of London*.

Phillips, J. J., & Phillips, P. P. (2005). *ROI at Work: Best Practices in the Real World*. Barrett-Koehler Publishers.

Pickering, C. J., Guy, E., Al-Baldawi, Z., McVean, L., Sargent, S., & O'Sullivan, T. (2021). "I believe this team will change how society views youth in disasters": The EnRiCH youth research team: A youth-led community-based disaster risk reduction program. *Canadian Journal of Public Health = Revue Canadienne de Santé Publique*, *112*(4), 664–672. DOI: 10.17269/s41997-021-00486-8 PMID: 34008134

Pielke, R. A.Jr, Gratz, J., Landsea, C. W., Collins, D., Saunders, M. A., & Musulin, R. (2008). Normalized hurricane damage in the United States: 1900–2005. *Natural Hazards Review*, *9*(1), 29–42. DOI: 10.1061/(ASCE)1527-6988(2008)9:1(29)

Piguet, E. (2022). Linking climate change, environmental degradation, and migration: An update after 10 years. *Wiley Interdisciplinary Reviews: Climate Change*, *13*(1), e746. Advance online publication. DOI: 10.1002/wcc.746

Piltch-Loeb, R., Bernard, D., & Vallejo, B. Q. (2022). Engaging community leaders in sharing local knowledge for emergency preparedness to leverage communication and trusted assets for vulnerable populations. *Disaster Medicine and Public Health Preparedness*, *16*(5), 1882–1889. DOI: 10.1017/dmp.2021.103 PMID: 34016210

Porter, S. (2009). *The great plague*. Amberley Publishing.

Potter, S. H., Becker, J. S., Johnston, D. M., & Rossiter, K. P. (2015). An overview of the impacts of the 2010-2011 Canterbury earthquakes. *International Journal of Disaster Risk Reduction, 14*(1), 6–14. DOI: 10.1016/j.ijdrr.2015.01.014

Poudyal, R., Loskot, P., Nepal, R., Parajuli, R., & Khadka, S. K. (2019). Mitigating the current energy crisis in Nepal with renewable energy sources. *Renewable & Sustainable Energy Reviews, 116*, 109388. DOI: 10.1016/j.rser.2019.109388

Poulakis, E. D., & O'Leary, R. (2011). Why the Performance Management System Fails: How to Fix It. *Industrial and Organizational Psychology: Perspectives on Science and Practice*.

Pranata, S., Widodo, S., & Vranada, A. (2021). How to prepare a school-based disaster preparedness in Indonesia. *Health Science Journal, 15*(7), 1–6. https://www.hsj.gr/medicine/how-to-prepare-a-schoolbased-disaster-preparedness-in-indonesia.php?aid=23674

Preston, B. L., Westaway, R. M., & Yuen, E. J. (2022). Climate change vulnerability assessment: Framework and methods. *Environmental Science & Policy, 89*, 56–69.

Putnam, R. (2000). *Bowling Alone: The Collapse and Revival of American Community*. Simon & Schuster.

Raaijmakers, R., Krywkow, J., & van der Veen, A. (2008). Flood risk perceptions and spatial multi-criteria analysis: An exploratory research for hazard mitigation. *Natural Hazards, 46*(3), 307–322. DOI: 10.1007/s11069-007-9189-z

Rahman, M. M., Nabila, I. A., Islam, F., & Tasnim, F. (2020). Challenges to implement disaster risk reduction in schools of developing country: Study on Dhaka City, Bangladesh. In *International Conference on Disaster Risk Reduction, Dhaka, Bangladesh.*

Rahman, M. S., Karim, M. R., & Islam, M. S. (2022). Building climate-resilient health facilities in coastal Bangladesh: A community-based approach. *Climate and Development, 14*(7), 645–657.

Ramakrishnan, T., Mohan Gift, M. D., Chitradevi, S., Jegan, R., Subha Hency Jose, P., Nagaraja, R., Sharma, H. N., & Selvakumar, P. (2022). Study of numerous resins used in polymer matrix composite materials. *Advances in Materials Science and Engineering, 2022*, 1088926. DOI: 10.1155/2022/1088926

Ramakrishnan, T., Rahman, S., & Ngamassi, L. (2022). Examining the factors that influence the use of social media for disaster management by underserved communities. *Jàambá, 14*(1), 1035. DOI: 10.4102/jamba.v14i1.1035

Ramli, A., Mokhtar, M., & Aziz, B. A. (2014). Revisiting the concept of development, disaster and safety management: The Quranic perspective. *International Journal of Disaster Risk Reduction*, *9*, 26–37. DOI: 10.1016/j.ijdrr.2014.03.006

Ramos-Cruz, J. M., & Ruiz-Garcia, J. (2024). Mean period prediction model for Mexican interpolate and intermediate-depth intraslab earthquake ground motions. *Soil Dynamics and Earthquake Engineering*, *178*, 108448. DOI: 10.1016/j.soildyn.2023.108448

Rao, A., Vazquez, J. A., & Kumpati, G. (2017). Next generation infrastructure materials for sustainable and resilient systems. In Rao, A., & Subramanyam, K. G. (Eds.), *Sustainability and resilience of infrastructure systems* (pp. 1–18). CRC Press.

Raschky, P. A. (2008). Institutions and the losses from natural disasters. *Natural Hazards and Earth System Sciences*, *8*(4), 627–634. DOI: 10.5194/nhess-8-627-2008

Rawlings, P. (2016). The Great Fire of London and the Origins of Fire Insurance: a brief note. *Queen Mary School of Law Legal Studies Research Paper*, (246).

Reid, H. (2016). Ecosystem- and community-based adaptation: Learning from community-based natural resource management. *Climate and Development*, *8*(1), 4–9. DOI: 10.1080/17565529.2015.1034233

Reid, H. F. (1911). *The elastic-rebound theory of earthquakes*. University of California Press.

Reid, H., Alam, S. S., & Berger, R. (2019). Community-based adaptation to climate change: Lessons from the field. In *Climate Change Adaptation in the Global South* (pp. 67–86). Springer.

Reid, H., Alam, S. S., & Berger, R. (2022). Scaling up community-based adaptation. In *Climate Change Adaptation in Practice* (pp. 123–144). Routledge.

Reliefweb. (2024). Türkiye/Syria – One in 3 children displaced by Türkiye earthquakes still homeless while needs at record levels in Syria. Türkiye/Syria – One in 3 children displaced by Türkiye earthquakes still homeless while needs at record levels in Syria - Syrian Arab Republic | ReliefWeb

Righi, E., Lauriola, P., Ghinoi, A., Giovannetti, E., & Soldati, M. (2021). Disaster risk reduction and interdisciplinary education and training. *Progress in Disaster Science*, *10*, 100054. DOI: 10.1016/j.pdisas.2021.100165

Rodgers, J., Hassan, W., Motter, C., & Thornley, J. (2021). Impacts of the 2018 M7.1 Anchorage earthquake on schools. *Earthquake Spectra*, *37*(3), 1849–1874. DOI: 10.1177/8755293020988022

Rodríguez, H., Donner, W., & Trainor, J. E. (2018). *Handbook of disaster research*. Springer. DOI: 10.1007/978-3-319-63254-4

Rohrmann, B. (2008). Risk perception, risk attitude, risk communication, risk management: A conceptual appraisal. https://tiems.info/dmdocuments/events/TIEMS_2008_Bernd_Rohrmann_Keynote.pdf

Romanello, M., Di Napoli, C., & Drummond, P. (2023). The 2023 report of The Lancet Countdown on health and climate change. *Lancet*, *402*(10414), 1785–1863. PMID: 37977174

Ronan, K. R., Johnston, D. M., Daly, M., & Fairley, R. (2001). School Children's Risk Perceptions and Preparedness: A Hazards Education Survey. *Australasian Journal of Disaster and Trauma Studies*, *1*, 1–42.

Rose, A. (2004). Defining and measuring economic resilience to disasters. *Disaster Prevention and Management*, *13*(4), 307–314. DOI: 10.1108/09653560410556528

Rossello, J., Becken, S., & Santana-Gallego, M. (2020). The effects of natural disasters on international tourism: A global analysis. *Tourism Management*, *79*, 104080. DOI: 10.1016/j.tourman.2020.104080 PMID: 32287755

Rothwell, W. J. (2010). The Strategic Development of Human Capital. In Strategic Human Resource Development.

Rothwell, W. J., & Kazanas, H. C. (2003). *Mastering the instructional design process: A systematic approach* (3rd ed.). Pfeiffer.

Rozenberg, J., & Fay, M. (Eds.). (2019). *Beyond the gap: How countries can afford the infrastructure they need while protecting the planet*. The World Bank.

Sakurai, M., & Shaw, R. (2022). The potential of digitally enabled disaster education for sustainable development goals. *Sustainability (Basel)*, *14*(8), 4269. DOI: 10.3390/su14116568

Sambuli, N., & Whitt, J. P. (2017). Technology innovation hubs and policy engagement. In Adesina, N., Agboluaje, A., Akinola, O., Anise, O., Guttmann, R., Karambiri, T., Nwagwu, K., Rudman, A., Sambuli, N., Whitt, J. P., & Yonazi, J. (Eds.), *Foresight Africa: Top priorities for the continent in 2017* (pp. 84–90). Brookings Institution.

Sandoval, A. (2009). Se desbordan los ríos Salado y Mixteco; afectados 6 municipios de Huajuapan [The Salado and Mixteco rivers overflow; 6 municipalities of Huajuapan affected]. https://www.xeouradio.com/2009/07/01/se-desbordan-los-ros-salado-y-mixteco-afectados-6-municipios-de-huajuapan/#:~:text=Un%20total%20de%206%20municipios,intensa%20lluvia%20que%20azot%C3%B3%20Huajuapan

Santos-Reyes, J. (2020). Factors motivating Mexico City residents to earthquake mass evacuation drills. *International Journal of Disaster Risk Reduction*, *49*, 101661. DOI: 10.1016/j.ijdrr.2020.101661 PMID: 32455104

Santos-Reyes, J. (2024). Awareness and risk perception of a mult-hazard megacity: The case of adolescent students. *Safety Science*, *171*, 106382. DOI: 10.1016/j.ssci.2023.106382

Santos-Reyes, J., Beard, A. N., & Smith, R. A. (2005). A Systemic Analysis of Railway Accidents, *Journal of Rail and Rapid Transit: Proc. IMechE. Proceedings of the Institution of Mechanical Engineers. Part F, Journal of Rail and Rapid Transit*, *219*(2), 47–65. DOI: 10.1243/095440905X8745

Santos-Reyes, J., & Gouzeva, T. (2023). Adolescents' Responses to the 2017 Puebla Earthquake in Mexico City. *Journal of Disaster Research*, *18*(7), 771–183332. DOI: 10.20965/jdr.2023.p0771

Sardar, A., Javed, S. A., & Amir-ud-Din, R. (2016). Natural disasters and economic growth in Pakistan: Evidence from flood-prone areas. *Pakistan Economic and Social Review*, *54*(2), 199–222. https://www.jstor.org/stable/26616704

Sawada, Y., & Takasaki, Y. (2017). Natural disaster, poverty, and development: An introduction. *World Development*, *94*, 2–15. DOI: 10.1016/j.worlddev.2016.12.035

Schultz, J., & Elliott, J. R. (2013). Natural disasters and local demographic change in the United States. *Population and Environment*, *34*(3), 293–312. DOI: 10.1007/s11111-012-0171-7

Schutte, S., Vestby, J., Carling, J., & Buhaug, H. (2021). Climatic conditions are weak predictors of asylum migration. *Nature Communications*, *12*, 2067. Advance online publication. DOI: 10.1038/s41467-021-22255-4

Seddighi, H., Sajjadi, H., Yousefzadeh, S., López López, M., Vameghi, M., Rafiey, H., & Khankeh, H. (2022). School-based education programs for preparing children for natural hazards: A systematic review. *Disaster Medicine and Public Health Preparedness*, *16*(3), 1228–1236. DOI: 10.1017/dmp.2020.479 PMID: 33818366

Senanayake, A. C., Samarakkody, A., Malalgoda, C., Amaratunga, D., Haigh, R., Liyanage, C., Hamza, M., Kaklauskas, A., & Shaw, R. (2023). Towards an inclusive disaster education: The state of online disaster education from the learner's perspective. *Sustainability (Basel)*, *15*(14), 11044. DOI: 10.3390/su151411042

Seyle, D. C., Widyatmoko, C. S., & Silver, R. C. (2013). Coping with natural disasters in Yogyakarta, Indonesia: A study of elementary school teachers. *School Psychology International*, *34*(4), 387–404. DOI: 10.1177/0143034312446889

Shah, A. A., Ye, J., Abid, M., & Ullah, R. (2019). Determinants of flood risk mitigation strategies at household level: A case of Khyber Pakhtunkhwa (KP) province, Pakistan. *Natural Hazards*, *95*(1-2), 401–423. DOI: 10.1007/s11069-018-3508-2

Shaheen, M. A. (2008). Earthquake effects on educational institutions and libraries of Azad Kashmir: An appraisal. *Library Review*, *57*(6), 449–456. DOI: 10.1108/00242530810886724

Shaji, K. A. (2024, September 6). *A month after India's deadliest landslide ever, Wayanad villages begin to recover. Down To Earth*. Retrieved from https://www.downtoearth.org.in/natural-disasters/a-month-after-indias-deadliest-landslide-ever-wayanad-villages-begin-to-recover

Sharma, S., & Kumar, V. Saruchi. (2022). Community approach toward disaster resilience. In: Siddhartha Bhattacharyya, Kumar Mondal, Bhanu Prakash (eds.), *Cognitive Data Models for Sustainable Environment- A volume in Cognitive Data Science in Sustainable Computing* (pp. 121-161). Elsevier. https://doi.org/DOI: 10.1016/B978-0-12-824038-0.00003-1

Shaw, R., & Sharma, A. (2020). *Community-Based Disaster Risk Reduction*. Springer.

Sherman, M., Berrang-Ford, L., & Ford, J. (2022). Stakeholder engagement in climate change adaptation planning: A systematic review. *Environmental Science & Policy*, *128*, 1–13.

Sherman, M., Berrang-Ford, L., & Ford, J. (2023). Monitoring and evaluation in community-based adaptation. *Global Environmental Change*, *68*, 102–114.

Shirazi, N. A., Omidvar, B., & Nohegar, A. (2023). Introduction to decision-making and disaster management systems. In Reza, M., Omidvar, B., & Makarem, M. (Eds.), *Crises in Oil, Gas and Petrochemical Industries-Volume 1: Disasters and Environmental Challengues* (pp. 3–24). Elsevier., DOI: 10.1016/B978-0-323-95154-8.00008-6

Shi, Y., & Sun, J. (2021). The economic impacts of natural disasters: A review of models and empirical studies. *Natural Hazards Review*, *22*(3), 04021015. DOI: 10.1061/(ASCE)NH.1527-6996.0000473

SHRM. (2016). *Human Capital Analytics: Driving Value from Data*. Society for Human Resource Management.

Silzer, R., & Dowell, B. E. (2010). *Strategy-driven talent management: A leadership imperative*. John Wiley & Sons.

Sims, C. A. (2003). Implications of rational inattention. *Journal of Monetary Economics*, *50*(3), 665–690. DOI: 10.1016/S0304-3932(03)00029-1

SISPLADE. (2000). Macrosismos en el estado de Oaxaca [Macrosisms in the state of Oaxaca]. http://sisplade.oaxaca.gob.mx/sisplade/PeligrosDocs/Tomo2/tomo2%20macrosismos%20en%20el%20estado%20de%20oaxaca.pdf

Sivri, S. (2023). Depremin Psikolojik Etkileri. https://www.acibadem.com.tr/hayat/depremin-psikolojiketkileri/#:~:text=Do%C4%9Fal%20bir%20felaket%20olan%20deprem,temelli%20bir%20anlam%20ta%C5%9F%C4%B1maya%20ba%C5%9Flar. (Erişim Tarihi: 13.04.2024).

Slovic, P. (2000). *The perception of risk*. Earthscan.

Smith, J. (2020). Risk Assessment and Preparedness in Disaster-Prone Regions. *Disaster Resilience Journal*, *8*(2), 112–130.

Smith, J., White, T., & Lee, M. (2020). Psychological Responses to Natural Disasters: A Cross-Cultural Study. *International Journal of Disaster Risk Reduction*, *45*, 101–112.

Smith, K. (2013). *Environmental hazards: Assessing risk and reducing disaster* (6th ed.). Routledge. DOI: 10.4324/9780203805305

Smith, K. R., Woodward, A., & Campbell-Lendrum, D. (2023). Climate change and global health equity. *The Lancet. Global Health*, *11*(6), e815–e826. PMID: 37202013

Smith, S., Farra, S. L., & Hodgson, E. (2021). Evaluation of two simulation methods for teaching a disaster skill. *Journal of Educational Technology Development and Exchange*, *14*(1), 31–44. DOI: 10.18785/jetde.1401.03 PMID: 35520385

Sodhi, M. S. (2016). Natural disasters, the economy and population vulnerability: Lessons from global data. *International Journal of Production Economics*, *176*, 69–78. DOI: 10.1016/j.ijpe.2016.03.008

Song, J., Huang, B., & Li, R. (2017). Measuring recovery to build up metrics of flood resilience based on pollutant discharge data: A case study in East China. *Water (Basel)*, *9*(8), 619. DOI: 10.3390/w9080619

Sphere India. (2021). *Preliminary Assessment Report – Wayanad Landslides, Kerala*. Retrieved from https://www.sphereindia.org.in/sites/default/files/2024-08/Wayanad%20Preliminary%20Assessment%20Report%20Aug%202024.pdf?utm_source=chatgpt.com

Sreenivasulu, A., Manimozhi, T., Satheesh Kumar, P., Selvakumar, P., & Singh, P. (2022). The model of smart sensing device for sensitive nanoclusters modification in sensing properties. In *2022 2nd International Conference on Advance Computing and Innovative Technologies in Engineering (ICACITE)* (pp. 1043–1047). IEEE. DOI: 10.1109/ICACITE53722.2022.9823846

SSN. (2023). Mapa de sismicidad anual 2023 [Annual sesimicity map 2023]. http://www.ssn.unam.mx/sismicidad/mapas-de-sismicidad-anual/

Stanke, V. Murray, R. Amlot, J. Nurse, R. Williams (2012). *The effects of flooding on mental health: outcomes and recommendations from a review of the literature,* PLoS Curr 4 e4f9f1fa9c3cae.

Stern, N. (2016). Economics: Current climate models are grossly misleading. *Nature, 530*(7591), 407–409. DOI: 10.1038/530407a PMID: 26911765

Subramanian, S. P. (2023). Evaluating the effectiveness of online environmental activism campaigns: Assessing the impact on public awareness and behavioral change. *Journal of Intelligent Connectivity and Emerging Technologies, 8*(2), 45–60. https://www.journalicet.org/index.php/jicet/article/view/123

Sudmeier-Rieux, K., Ash, N., Murti, R., & Dunning, E. (Eds.). (2019). *Ecosystems, livelihoods, and disasters: An integrated approach to disaster risk management.* Springer.

Sugumaran, K. M. D., Pandiyarajan, V., Sonkhaskar, Y. B., Selvakumar, P., & Manimozhi, M. S. (2024). Analysis of electromyography signals for control of mechanical prosthesis using machine learning techniques. In *2024 2nd International Conference on Intelligent Data Communication Technologies and Internet of Things (IDCIoT)* (pp. 1059–1067). IEEE. DOI: 10.1109/IDCIoT59759.2024.10467707

Sun, L., Deng, Y., & Qi, W. (2018). Two impact pathways from religious belief to public disaster response: Findings from a literature review. *International Journal of Disaster Risk Reduction, 27,* 588–595. DOI: 10.1016/j.ijdrr.2017.10.004

Surminski, S., & Tanner, T. (Eds.). (2016). *Realising the 'triple dividend of resilience': A new business case for disaster risk management.* Springer. DOI: 10.1007/978-3-319-40694-7

Taghizadeh-Hesary, F., Yoshino, N., & Mortha, A. (2019). Assessing the macroeconomic impacts of natural disasters in Asia and the Pacific. ADBI Working Papers 966. Asian Development Bank Institute. https://www.adb.org/publications/assessing-macroeconomic-impacts-natural-disasters-asia-pacific

Taghizadeh-Hesary, F., Sarker, T., Yoshino, N., Mortha, A., & Vo, X. V. (2021). Quality infrastructure and natural disaster resiliency: A panel analysis of Asia and the Pacific. *Economic Analysis and Policy*, *69*, 394–406. DOI: 10.1016/j.eap.2020.12.021

Tanaka, E., Shibata, H., Sugiura, K., Hoshino, N., & Nishikawa, J. (2016). Long-term psychological effects of the Wenchuan earthquake on adolescent survivors: A six-year follow-up study. *Journal of Affective Disorders*, *204*, 342–348002E. DOI: 10.1016/j.jad.2016.08.001

Tao, Y., Lin, T., Feng, X., Gao, Y., & Mashino, S. (2022). Cultural competence for disaster nursing: A scoping review of the Chinese and English literature. *International Journal of Disaster Risk Reduction*, *78*, 103147. DOI: 10.1016/j.ijdrr.2022.103188

Taye, M., Dyer, E., & Charles, K. J. (2021). Community-based surveillance and response to climate-sensitive diseases in the Somali Region of Ethiopia. *Pastoralism*, *11*(1), 1–14.

Tetik, N., & Akbulut, İ. İ. (2023). *6 şubat 2023'te yaşanan depremin ekonomik ve finansal etkileri: İhracat üzerinden bir inceleme. Kahramanmaraş merkezli depremler sonrası için akademik öneriler*. Özgür Yayınları.

The PRS Group. (2019). The International Country Risk Guide. https://www.prsgroup.com/explore-our-products/icrg/

The White House. (2006). *The Federal Response to Hurricane Katrina: Lessons Learned*. U.S. Government Printing Office.

The World Bank. (2022). World development indicators [Data set]. https://databank.worldbank.org/source/world-development-indicators

Thomson, M. C., Mason, S. J., & Connor, S. J. (2023). *Climate information for public health action*. Routledge.

Thoresen, S., Birkeland, M. S., Arnberg, F. K., Wentzel-Larsen, T., & Blix, I. (2019). Long-term mental health and social support in victims of disaster: Comparison with a general population sample. *BJPsych Open*, *5*(1), e2. DOI: 10.1192/bjo.2018.74 PMID: 30762498

Tierney, K. (2012). Disaster governance: Social, political, and economic dimensions. *Annual Review of Environment and Resources*, *37*(1), 341–363. DOI: 10.1146/annurev-environ-020911-095618

Tierney, K. (2014). *The social roots of risk: Producing disasters, promoting resilience*. Stanford University Press.

Trenberth, K. E., Dai, A., van der Schrier, G., Jones, P. D., Barichivich, J., Briffa, K. R., & Sheffield, J. (2014). Global warming and changes in drought. *Nature Climate Change*, *4*(1), 17–22. DOI: 10.1038/nclimate2067

Trigg, S. (2016). Samuel Pepys and the great fire of London: Trauma and Emotion, Private and Public. *Disaster, Death and the Emotions in the Shadow of the Apocalypse, 1400–1700*, 341-356.

Trinh, T. A., Feeny, S., & Posso, A. (2021). The impact of natural disasters on migration: Evidence from Vietnam. *Economic Analysis and Policy*, *70*, 445–463. DOI: 10.1016/j.eap.2021.03.009

Tselios, V., & Tompkins, E. L. (2017). Local government, political decentralisation and resilience to natural hazard-induced disasters. *Environmental Hazards*, *16*(3), 228–252. DOI: 10.1080/17477891.2016.1277967

Tuladhar, G., Yatabe, R., Dahal, R. K., & Bhandary, N. P. (2015). Disaster risk reduction knowledge of local people in Nepal. *Geoenvironmental Disasters*, *2*(1), 2–12. DOI: 10.1186/s40677-014-0011-4

Tunstall, S. M., Johnson, C. L., & Penning-Rowsell, E. C. (2004, February). Flood hazard management in England and Wales: from land drainage to flood risk management. In *World Congress on Natural Disaster Mitigation* (Vol. 14, No. 2, pp. 447-454).

Turco, M., Llasat, M. C., von Hardenberg, J., & Provenzale, A. (2014). Climate change impacts on wildfires in a Mediterranean environment. *Climatic Change*, *125*(3–4), 369–380. DOI: 10.1007/s10584-014-1183-3

Turkey, N. A., & Jawahar Saud, S. (2024, June). Assessment of Governance Gaps in Landslide Recovery and Post Recovery Phase Using Analytical Hierarchy Process (AHP)—A Case Study from Wayanad District, Kerala, India. *In International Conference on Structural Engineering and Construction Management* (pp. 483-499). Cham: Springer Nature Switzerland. DOI: 10.1007/978-3-031-70431-4_36

Twigg, J. (2015). *Disaster risk reduction*. Overseas Development Institute.

Ulrich, D., & Brockbank, W. (2005). The HR Value Proposition. Harvard Business School Press.

UNDRR. (2019). *Global assessment report on disaster risk reduction 2019*. United Nations Office for Disaster Risk Reduction.

UNICEF. (2018). El impacto de los terremotos en México en los niños, niñas y adolescentes [The impacto of the earthquakes in Mexico on children and adolescents]. https://www.unicef.org/mexico/informes/impacto-de-los-terremotos-en-m%C3%A9xico-en-los-ni%C3%B1os-ni%C3%B1as-y-adolescentes

UNICEF. (2023). 100 days after earthquakes hit Turkiye and Syria, millions of children continue to face desperate conditions. https://reliefweb.int/report/syrian-arab-republic/100-days-after-earthquakes-hit-turkiye-and-syria-millions-children-continue-face-desperate-conditions-enar

UNISDR. (2009). 2009 UNISDR terminology on disaster risk reduction. United nations International Strategy for Disaster Reduction (UNISDR), Geneva, Switzerland. 2009 UNISDR Terminology on Disaster Risk Reduction - World | ReliefWeb

UNISDR. (2015). *Sendai framework for disaster risk reduction 2015-2030.* United Nations Office for Disaster Risk Reduction.

UNISDR. (2015). *Sendai Framework for Disaster Risk Reduction 2015-2030.* United Nations Office for Disaster Risk Reduction.

United Nations (UN). (2022). *The Sustainable Development Goals Report 2022.* United Nations., DOI: 10.18356/9789210017988

United Nations Development Programme (UNDP). (2020). Human Development Report 2020: The Next Frontier – Human Development and the Anthropocene. https://hdr.undp.org/en/2020-report

United Nations International Strategy for Disaster Reduction (UNISDR). (2015). Sendai Framework for Disaster Risk Reduction 2015-2030. United Nations. https://www.undrr.org/publication/sendai-framework-disaster-risk-reduction-2015-2030

United Nations International Strategy for Disaster Reduction (UNISDR). (2019). Global Assessment Report on Disaster Risk Reduction 2019. United Nations. https://gar.undrr.org

United Nations Office for Disaster Risk Reduction (UNDRR). (2019). Global Assessment Report on Disaster Risk Reduction 2019. United Nations. https://gar.undrr.org

United Nations Office for Disaster Risk Reduction (UNDRR). (2023). *Global assessment report on disaster risk reduction 2023.* UNDRR., DOI: 10.18356/9789214030488

United Nations. (2015). *Transforming our world: The 2030 Agenda for Sustainable Development.* United Nations.

UN-SDGs (United Nations-Sustainable Development Goals). (2023). Sustainable development goals. https://sdgs.un.org/goals

Vaijhala, S., & Rhodes, J. (2018). Resilience bonds: A business-model for resilient infrastructure. *Field Actions Science Reports, 18*(Special Issue), 58–63.

van der Linden, S. (2015). The social-psychological determinants of climate change risk perceptions: Towards a comprehensive model. *Journal of Environmental Psychology, 41*, 112–124. DOI: 10.1016/j.jenvp.2014.11.012

Vijayakumar, G., Rajkumar, M., Chandar, N. R., Selvakumar, P., & Duraisamy, R. (2024). Environmentally friendly TDS removal from wastewater by electrochemical ion exchange batch-type recirculation (EIR) technique. *Environmental Science. Water Research & Technology, 10*(4), 826–835. DOI: 10.1039/D3EW00793F

Wachinger, G., Renn, O., Begg, C., & Kuhlicke, C. (2013). The risk perception paradox— Implications for governance and communication of natural hazards. *Risk Analysis, 33*(6), 1049–1065. DOI: 10.1111/j.1539-6924.2012.01942.x PMID: 23278120

Wang, J., & Dong, K. (2019). What drives environmental degradation? Evidence from 14 Sub Saharan African countries. *The Science of the Total Environment, 656*, 165–173. DOI: 10.1016/j.scitotenv.2018.11.354 PMID: 30504018

Wang, Y. C., Lin, S. W., & Lee, C. H. (2020). Conducting an evaluation framework for disaster management under adaptive organization change in a school system. *Sustainability (Basel), 12*(16), 6615. DOI: 10.3390/su12166615

Watkiss, P., Hunt, A., Blyth, W., & Dyszynski, J. (2015). The use of new economic decision support tools for adaptation assessment: A review of methods and applications, towards guidance on applicability. *Climatic Change, 132*(3), 401–416. DOI: 10.1007/s10584-014-1250-9

Watt, F. (2002). *Depremler ve yanardağlar.* (Çev. Deniz Yurtören), 18. Basım. Tübitak popüler bilim kitapları: 71, Ankara.

Watts, N., Amann, M., & Arnell, N. (2021). The 2021 report of The Lancet Countdown on health and climate change. *Lancet, 398*(10311), 1619–1662. DOI: 10.1016/S0140-6736(21)01787-6 PMID: 34687662

Watts, N., Amann, M., & Arnell, N. (2022). The 2022 report of The Lancet Countdown on health and climate change. *Lancet, 400*(10363), 1619–1662. DOI: 10.1016/S0140-6736(22)01540-9 PMID: 36306815

Watts, N., Amann, M., Arnell, N., Ayeb-Karlsson, S., Belesova, K., Boykoff, M., Byass, P., Cai, W., Campbell-Lendrum, D., Capstick, S., Chambers, J., Dalin, C., Daly, M., Dasandi, N., Davies, M., Drummond, P., Dubrow, R., Ebi, K. L., Eckelman, M., & Montgomery, H. (2019). The 2019 report of The Lancet Countdown on health and climate change: Ensuring that the health of a child born today is not defined by a changing climate. *Lancet*, *394*(10211), 1836–1878. DOI: 10.1016/S0140-6736(19)32596-6 PMID: 31733928

Weerasekara, S., Wilson, C., Lee, B., & Hoang, V. N. (2021). Climate change and natural disasters: Evidence from the 2011 Thailand floods. *Environmental Development*, *39*, 100648. DOI: 10.1016/j.envdev.2021.100648

Westerling, A. L., & Bryant, B. P. (2008). Climate change and wildfire in California. *Climatic Change*, *87*(Suppl 1), 231–249. DOI: 10.1007/s10584-007-9363-z

White, J. P. (2012). The role of the military in disaster response in the U.S. *Journal of Homeland Security and Emergency Management*, *9*(1), 153–163. DOI: 10.1515/1547-7355.2119

WHO. (2015). Operational framework for building climate resilient health systems. World Health Organization. https://www.who.int/globalchange/publications/building-climate-resilient-health-systems/en/

WHO. (2020). *Operational framework for building climate resilient health systems*. World Health Organization.

WHO. (2021). *COP26 special report on climate change and health*. World Health Organization.

WHO. (2021). COP26 special report on climate change and health: The health argument for climate action. World Health Organization. https://www.who.int/publications/i/item/9789240036727

WHO. (2021). *Operational framework for building climate-resilient health systems*. World Health Organization.

WHO. (2022). *Building climate-resilient health systems through community-based adaptation*. World Health Organization.

WHO. (2023). *Building resilient health systems: A framework for climate change adaptation*. World Health Organization.

WHO. (2023). *Guidelines for climate-resilient and environmentally sustainable health care facilities*. World Health Organization.

Wilhite, D. A., Hayes, M. J., & Knutson, C. L. (2005). Drought preparedness planning: building institutional capacity. In: Wilhite, D. (ed.),2005. Drought and water crises: science, technology, and management issues, CRC Press, 93-121.

Willes, M. (2017). *The Curious World of Samuel Pepys and John Evelyn.* Yale University Press. DOI: 10.2307/j.ctvh4zfnh

Williams, P. A., Crespo, O., & Abu, M. (2021). Adapting to climate change through community-based approaches. *Regional Environmental Change, 21*(2), 1–14.

Wisner, B., Gaillard, J. C., & Kelman, I. (Eds.). (2012). *Handbook of hazards and disaster risk reduction and management.* Routledge. DOI: 10.4324/9780203844236

WMO. (2018). *Multi-hazard early warning systems: A checklist.* World Meteorological Organization.

Wooldridge, J. M. (2010). *Econometric analysis of cross section and panel data* (2nd ed.). MIT Press.

World Bank. (2017). *Sovereign climate and disaster risk pooling: World Bank technical contribution to the G20.* The World Bank.

World Health Organization. (2013). *Building back better: Sustainable mental health care after emergencies.* Retrieved from https://www.who.int/publications/i/item/9789241564571

World Health Organization. (2019). *Mental Health in Emergencies.* World Health Organization.

Wright, P. M., & Nishii, L. H. (2007). Strategic HRM and Organizational Behavior: Integrating Multiple Levels of Analysis. In Research in Personnel and Human Resources Management.

Yale Program on Climate Change Communication. (2019). Climate change opinion maps. https://climatecommunication.yale.edu/visualizations-data/ycom-us-2018/

Yari, A., Zarezadeh, Y., & Ostadtaghizadeh, A. (2019). Prevaence of fatalistic attitudes toward earthquake disaster risk management in citizens of Tehran, Iran. *International Journal of Disaster Risk Reduction, 38*, 101181. DOI: 10.1016/j.ijdrr.2019.101181

Yeon, D.-H., Chung, J.-B., & Im, D.-H. (2020). The effects of earthquake experience on disaster education for children and teens. *International Journal of Environmental Research and Public Health, 17*(15), 1–14. DOI: 10.3390/ijerph17155347 PMID: 32722236

Yıldırım, A. ve Şimşek, H. (2016). Sosyal bilimlerde nitel araştırma yöntemleri. Ankara: Seçkin Yayıncılık.

Yildiz, A., Teeuw, R., Dickinson, J., & Roberts, J. (2021). Children's perceptions of flood risk and preparedness: A study after the May 2018 flooding in Golcuk, Turkey. *Progress in Disaster Science*, *9*, 100143. DOI: 10.1016/j.pdisas.2021.100143

Yusuf, R., Fajri, I., & Gani, S. A. (2022). Disaster education in disaster-prone schools: A systematic review. *IOP Conference Series. Earth and Environmental Science*, *1039*(1), 012007. DOI: 10.1088/1755-1315/1041/1/012034

Zaval, L., Keenan, E. A., Johnson, E. J., & Weber, E. U. (2014). How warm days increase belief in global warming. *Nature Climate Change*, *4*(2), 143–147. DOI: 10.1038/nclimate2093

Zenklusen, O. (2007). *Natural disasters and economic development: A neoclassical review of theoretical perspectives and empirical evidence*. Ph.D. thesis, The University of St. Gallen, Switzerland.

Zhang, Jun MDa; Zhang, Ye MDa; Du, Changhui MSc; Zhu, Shenyue MAb; Huang, Yalin MSa; Tian, Yulian MAd; Chen, Decao MAa; Li, Haimin MSa; Gong, Yao MSa; Zhang, Mengmeng MSa; Gu, Bo MSe. (2016). Prevalence and risk factors of posttraumatic stress disorder among teachers 3 months after the Lushan earthquake: A cross-sectional study. *Medicine 95*(29), 1-6.

Zhang, M., & Zheng, W. (2024). Building disaster resilience through emergency plan updates: A case study of Ya'an, China. *International Journal of Disaster Risk Reduction*, *100*, 104173. DOI: 10.1016/j.ijdrr.2023.104173

Zuama, S. N. (2024). Disaster mitigation education: Challenges and opportunities in six areas of PAUD development. *Journal on Advanced Science. Educational Technology*, *10*(1), 23–35. https://journal.unisnu.ac.id/jaset/article/view/123

About the Contributors

Qamar Ali is an Instructor/Tutor in Economics, Virtual University of Pakistan since 15-12-2015. Addition to my teaching responsibilities, I wrote 50 research papers after joining Virtual University of Pakistan. These research papers are published in recognized national & international journals. Out of these, 26 research papers are published in impact factor journals like Journal of Cleaner Production (IF=9.227), Renewable Energy (IF=8.001), Transportation Research Part A: Policy and practice (IF=5.594), Environmental Science and Pollution Research (IF=3.056), Scientia Horticulturae (IF=3.463), Polish Journal of Environmental Studies (IF=1.200), The Journal of Animal and Plant Sciences (IF=0.422). Moreover, I am also engaged as reviewers with different research journals. I received a certificate of the Outstanding Reviewer in 2017 from Renewable and Sustainable Energy Reviews (IF=14.982), Elsevier, The Netherlands. I also received certificate of reviewing research articles from Environmental Science and Pollution Research (IF=2.800), Springer & Asian Journal of Agricultural Extension, Economics and Sociology, Science Domain International, India. The cumulative impact factor of my published work is 185.571, having 1215 citations and h-index 18. I wrote 4 Economics books, published by Dogar Publishers, Lahore

Muhammad Tariq Iqbal Khan is currently working as a Lecturer in Economics at the Government Postgraduate College, Jaranwala. He started his research career as an Internee at the Technology transfer Institute. He joined Higher Education Department Punjab on 18-08-2012, as Lecturer in Economics. He has more than 8 years of teaching and research experience. He passed B.Sc. (Hons.) Agricultural and Resources Economics in 2011 from the University of Agriculture, Faisalabad. During his education, He actively participated in university-level competitions at the University of Agriculture, Faisalabad. He five times won the Quiz Competition at the University of Agriculture, Faisalabad during graduation. He got M.Phil. Economics degree from the Government College University, Faisalabad in 2016. In 2017, He enrolled Ph.D. Economics from Government College University, Faisalabad. During his career, he has published 32 research publications in refereed International and national journals, including 18 research papers that have been published in impact factor journals. The cumulative impact factor of his research is 60.535, having 266 citations and an h-index 8. His current research focus is on disaster management, marketing efficiency, environmental economics, and agricultural economics.

***

A. Rohini working as Senior Assistant Professor in Environmental science in the department of Humanities and Sciences in CVR College of Engineering from 2007 to till date. I got my Ph.D Doctorate in Environmental science from Osmania University in 2023.Tile of the Thesis is Ecological Studies on Safilguda Lake with Reference to Water Quality.I have 18 years of experience in teaching Environmental science for B.Tech Engineering students.

Gülten Akgül works as an English language teacher and special education teacher in Kütahya, Türkiye. In addition to her teaching role, she works as a translator, philologist, and content writer. She holds a master's degree in English language and literature and has published academic works in the fields of social sciences and humanities, education, action research, measurement and assessment in language teaching, vocational English, language education, culture, literature, comparative literature, mythology, and history. She contributed as a content writer for "Diyalekt," the mobile and web language application of the Ministry of National Education, and as a theater instructor on a provincial basis and participated in some Erasmus+ projects in Romania and the Czech Republic.

Sudhanshu Chandra Assistant Professor (Academic Coordinator) at Manuu Law School, Maulana Azad National Urdu University Hyderabad A Central University Established by an Act of Parliament in 1988, Accredited with Grade A+ by NACC, (Hyderabad Telangana) -500032. He Completed his B.A.LLB(Hons)-Dr. Ram Manohar Lohiya National Law University Lucknow(U.P) LLM-National Law University Jodhpur (Rajasthan), Cleared UGC-NET, Ph.D. in Law from Gautam Buddha University Greater Noida(U.P.) Authored Three Books- The Menace and Implications of Narcotic Drug Abuse and Alcoholism in India: A Socio- Legal Study (Singhal Law Publication-2022), The Effect of Intellectual Property Rights and Conservation on Biodiversity Law in India (Associated Publishing House-2023), Custodial Violence by Police in India: Analysis of Judicial Decisions 2010-2020 (Associated Publishing House-2023).Also taken Copyright of work Custodial Violence by Police in India: Analysis of Judicial Decisions 2010-2020 Registration

No-L-124572/2023 Dated-09/06/2023.Judged Several Moot Court Competition. Published More than 20 Research Papers in International/National/Scopus/UGC-Care Listed Journals.

Kasagani Durga is a dedicated legal professional and an LLM student at NBM College of Law, under Andhra University, specializing in International Law. With a strong foundation in legal studies, she is passionate about global legal frameworks, human rights, and diplomatic relations. Durga is committed to deepening her expertise in cross-border legal issues, international treaties, and dispute resolution mechanisms. She aims to contribute to the evolving field of international law by engaging in research, advocacy, and legal practice. With a keen interest in legal reforms and justice, she strives to make a meaningful impact in the legal domain. Her academic pursuits and professional aspirations reflect her dedication to excellence in legal scholarship and international legal affairs.

Shashank G is an experienced Assistant Professor in the Department of Management Studies at ISBR Business School, Bengaluru, with 5 years of academic expertise. He specializes in teaching MBA and MCom subjects, including Marketing Management, Consumer Behavior, and Financial Markets. Additionally, he serves as a Placement Coordinator and guides student projects. Prof. Shashank has published research on green marketing and has undertaken studies on digital marketing trends in FMCG. With a strong educational background in MBA and BCom, he is skilled in MS Office and Tally ERP9. His personal attributes include confidence, communication skills, and leadership.

T.C. Manjunath was born on 6th Feb. 1967 in Bangalore. He did his BE from RV College of Engg., Bangalore Univ. in 1989 with First Class, ME in Robotics and Automation from the Govt. LD College of Engg. in 1992 with First Rank – FCD from Gujarat Univ., Ahmedabad and PhD in Systems and Control Engg. from IIT Bombay in the year 2007. He is currently working as the Dean Research, R and D and as Professor in the Dept. of Computer Science and Engineering with specialization in IoT, Cyber Security and Blockchain Tech. in Rajarajeswari College of Engg., Bengaluru, Karnataka, India. He has got a teaching, academic, research and administrative experience of more than 35+ years with 10 years as Principal, 8 yrs as Professor - HoD and 17 yrs as faculty. He has published more than 1200+ research papers in various national and international journals, conferences, with 5000+ citations, 175+ patents, 50+ book-chapters, 25 awards, published 25 text books in various field of science, engineering and technology, completed funded research projects worth Rs. 9 Crores till date apart from having guided 10 PhD scholars, 25 PG Students, more than 200 UG students towards their project works.

He is a member of 30 Professional Society Chapters and is a fellow of IE, IETE, AMIEE and a chartered engineer. He got the best research scholar award in 2006 in IIT Bombay amongst all the research scholars of the institute.

Menard Musendekwa, PhD, is a distinguished scholar in Theology, specializing in the Old Testament, and a dedicated Minister of Religion. His academic journey spans across multiple disciplines, reflecting his versatile expertise in areas such as Church History, Theology, Gender Studies, Education, and Climate Change, among others. With a rich portfolio of publications in esteemed journals and numerous contributions to book chapters, he has made significant scholarly contributions that resonate across academic circles. Dr. Musendekwa's commitment to academic excellence is further demonstrated through his active engagement in various academic conferences, where he shares insights and collaborates with fellow scholars. Currently serving as the Chair of Theology at Reformed Church University in Zimbabwe and holding the position of Vice-Principal at Murray Theological College, he brings a wealth of experience to his roles. With a profound dedication to university education, Dr. Musendekwa continues to inspire and mentor the next generation of scholars, shaping the future of theological studies and beyond.

N.B. Chandrakala is a distinguished legal scholar and educator with an extensive background in academia. With a Ph.D. in Law, she currently holds the prestigious position of Professor at the Department of Law, DR BR Ambedkar College of Law, Andhra University, Vishakhapatnam, Andhra Pradesh. Dr. Chandrakala has a remarkable teaching career spanning over three decades, with 30 years of experience at the postgraduate level. Her teaching journey began as an Assistant Professor at as the Principal of N.V.P. Law College, Madhuravada, Visakhapatnam, after Dr. B.R. Ambedkar PG Centre, Andhra University, Etcherla, Srikakulam. Later, she served as an Associate Professor at the Department of Law in Sri Padmavati Mahila Visvavidyalayam, Tirupati, before being promoted to the position of Professor, where she contributed significantly to legal education and research.

P. Selvakumar is currently an Associate Professor in the Department of Science and Humanities at Nehru Institute of Technology, Coimbatore, Tamil Nadu, India. Since 2023, he has served as the Department IQAC Coordinator at the institute. Prior to joining Nehru Institute of Technology, he held positions as an Assistant Professor in the Department of Science and Humanities various engineering colleges in Coimbatore, Tamil Nadu, India, spanning from 2011 to 2023. Dr.P. Selvakumar's academic and research interests lie primarily in the field of phytochemistry. He has made significant contributions to his field, with an impressive record that includes

30 journal publications, 12 patents, and 20 conference papers. Additionally, he has organized 5 seminars and conferences, further demonstrating his commitment to academic and scientific exchange. Throughout his career, Dr. P.Selvakumar has been recognized for his dedication and excellence in teaching, having been honored with several awards, including the Best Teacher Awards. He holds a prominent position in the academic community, contributing actively to research.

Deepika Paira is an accomplished Assistant Professor of Law, currently pursuing her doctoral research in Criminal Law from Vel Tech University. She holds an LL.B. from Law Centre-I, Delhi University, and an LL.M. in Constitutional Law from Amity School of Law. With over five years of teaching experience, she has also practiced litigation, offering her a comprehensive understanding of both academic and practical aspects of law. Deepika specializes in Constitutional Law and Criminal Law, and is known for her insightful teaching methodology. She has published numerous research papers in UGC CARE-listed journals and Scopus-indexed publications. Her active participation in Faculty Development Programs, as well as her engagement in national and international workshops, seminars, and conferences, reflects her commitment to continuous learning and academic excellence. Deepika Paira is a dedicated legal educator and researcher, contributing significantly to the field of law.

Shivansu Sachan is pursuing Ph.D. in Management, a Master's holder from Integral University, Lucknow, Bachelor's from IHM, Lucknow. Academic and professional accomplishments of 6 years with renowned organizations and institutes. My Teaching Career has been very satisfying and have continuously upgraded the skills and knowledge by attending many courses and training programs, seminars, conferences organized by different organizations and also delivered many lectures as a Hospitality sector.

Pamarthi Satyanarayana, specialize in Constitutional Law and Legal Order, focusing my research on the interface between individual rights and public policy, with a strong emphasis on civil liberties, social justice, and the right to privacy. Over the past 36 months, I have served as an Assistant Professor of Law at Vel Tech Rangarajan Dr. Sagunthala R&D Institute of Science and Technology in Chennai, Tamil Nadu. Here, I teach courses such as Constitutional Law, Legal Research and Writing, and both Civil and Criminal Law. My passion for teaching and in-depth subject knowledge have earned me recognition as a dedicated educator. In addition to my academic role, I actively engage with the community, partnering with legal clinics to provide pro bono services to underprivileged individuals. Looking forward,

I aim to expand my research into emerging legal issues in the digital age, especially in privacy rights, while guiding aspiring law students and future legal scholars.

Murat Sengöz holds a Bachelor's degree in Engineering and postgraduate degrees in Management and Finance, as well as a doctorate in Business Administration. He is the author of several academic and non-academic monographs and research papers. His researches focus on critical and reflective philosophical studies in organizational behavior, strategic management, management philosophy, and national security.

Mohit Sharma is a research scholar at the Institute of Management Studies and Research, Maharshi Dayanand University, Rohtak, Haryana, India. His research focuses on the dynamic relationship between brand image and purchase intentions, particularly in the context of brand extensions within the FMCG sector. Mohit actively engages in bibliometric research, analyzing publication trends and identifying research gaps in his area of interest. He has co-authored several scholarly works and continues to contribute to the academic community. Mohit can be reached at mohitattri.rs.imsar@mdurohtak.ac.in

Y. Ebenezer, specialize in macroeconomics and development economics, with a research focus on the interplay between macroeconomic variables and economic stability. My work emphasizes macroeconomic stability, economic growth, and sustainable development. Over the past decade, I have served as an Assistant Professor of Economics at Vel Tech Rangarajan Dr. Sagunthala R&D Institute of Science and Technology, Chennai, Tamil Nadu. In this role, I have taught a range of undergraduate courses, including basic Economic Principles, Indian Economy, and Law & Economics. My commitment to teaching and deep subject expertise has consistently earned recognition for academic excellence and student mentorship. Beyond academia, I am actively involved in community outreach, particularly in initiatives aimed at supporting underprivileged populations. Looking ahead, I intend to broaden my research into development challenges faced by emerging economies in the digital era, with a continued focus on macroeconomic dynamics. I am also dedicated to mentoring students and nurturing the next generation of economists and scholars.

Index

A

Awareness 6, 25, 29, 30, 34, 37, 38, 40, 47, 62, 73, 74, 80, 81, 83, 89, 92, 94, 96, 100, 105, 118, 123, 140, 141, 158, 159, 162, 166, 192, 195, 203, 207, 239, 240, 243, 255, 267, 270, 278, 279, 283, 293, 298, 300, 304, 307, 308, 309, 311, 334, 338, 342
awareness and risk perception 298, 311
Awareness Programs 29, 73, 94

C

Children 27, 77, 97, 100, 197, 200, 243, 244, 250, 253, 256, 257, 261, 265, 266, 273, 275, 277, 279, 288, 289, 290, 291, 292, 293, 298, 302, 306, 307, 309, 310, 312
Climate Change 1, 4, 11, 14, 16, 17, 20, 21, 22, 23, 25, 26, 27, 28, 29, 30, 32, 33, 34, 35, 36, 37, 38, 39, 40, 41, 42, 43, 44, 64, 69, 70, 94, 114, 128, 133, 134, 136, 139, 150, 151, 153, 154, 155, 156, 157, 159, 160, 161, 163, 165, 166, 167, 168, 169, 170, 171, 173, 174, 175, 176, 177, 178, 179, 180, 181, 184, 189, 193, 194, 202, 203, 208, 209, 210, 214, 217, 218, 220, 222, 223, 224, 225, 226, 227, 228, 229, 231, 232, 233, 234, 235, 239, 241, 242, 247, 248, 258, 286, 287, 292, 308, 328, 342
Climate Change Adaptation 30, 139, 150, 151, 155, 169, 177, 178, 179, 180, 194, 202, 209, 217, 225, 226, 227
Command and Control Relationship 59
Community-Based Adaptation 153, 154, 155, 156, 157, 158, 160, 163, 167, 171, 174, 177, 178, 179, 180, 224
Coordination 29, 30, 34, 37, 45, 46, 47, 48, 49, 50, 51, 52, 53, 54, 55, 56, 57, 58, 59, 60, 61, 65, 66, 67, 68, 70, 72, 87, 88, 159, 162, 193, 195, 197, 198, 199, 205, 206, 210, 211, 212, 215, 225, 237, 251, 281
Cultural Heritage 133, 332
cultural memory 313, 334, 337, 339

D

Decision Makers 103, 108, 110, 111, 120, 121, 125, 126, 127, 128
Developing Effective Disaster Education Programs 76
disaster 1, 2, 3, 4, 5, 6, 7, 8, 9, 10, 11, 12, 13, 14, 16, 17, 18, 19, 20, 22, 24, 25, 28, 30, 32, 35, 36, 37, 40, 41, 42, 45, 46, 47, 48, 49, 50, 51, 52, 53, 54, 55, 56, 57, 58, 59, 60, 61, 62, 63, 64, 65, 66, 67, 68, 69, 70, 71, 72, 73, 74, 75, 76, 77, 78, 79, 80, 81, 82, 83, 84, 85, 86, 87, 88, 89, 90, 91, 92, 93, 94, 95, 96, 97, 98, 99, 100, 101, 108, 110, 112, 114, 115, 116, 117, 118, 119, 120, 123, 127, 133, 134, 135, 136, 137, 138, 139, 140, 141, 142, 143, 144, 145, 146, 147, 148, 149, 150, 151, 152, 184, 185, 186, 187, 188, 189, 190, 192, 194, 195, 196, 197, 198, 199, 200, 201, 202, 203, 204, 205, 206, 207, 208, 209, 210, 211, 212, 214, 215, 216, 217, 218, 219, 220, 221, 222, 223, 224, 225, 226, 227, 228, 229, 230, 231, 232, 233, 234, 235, 236, 237, 238, 239, 240, 241, 242, 243, 244, 245, 246, 247, 248, 249, 250, 251, 252, 253, 254, 255, 256, 257, 258, 259, 260, 261, 263, 264, 265, 266, 267, 268, 270, 271, 272, 273, 274, 276, 278, 279, 280, 281, 282, 283, 284, 285, 286, 287, 288, 289, 290, 292, 296, 297, 306, 307, 308, 309, 310, 311, 312, 313, 314, 315, 316, 317, 318, 319, 320, 321, 322, 323, 324, 326, 327, 328, 329, 330, 331, 332, 333, 334, 336, 337, 338, 339, 340, 341, 342, 343, 344, 345
Disaster Education 73, 74, 76, 77, 78, 79,

80, 81, 82, 83, 84, 85, 86, 87, 89, 90, 91, 92, 93, 94, 95, 96, 97, 98, 100, 101, 236, 290
Disaster Preparedness 10, 40, 48, 53, 56, 62, 63, 64, 74, 77, 78, 79, 81, 82, 83, 84, 87, 88, 91, 92, 93, 94, 99, 114, 115, 119, 120, 123, 127, 133, 134, 135, 136, 139, 141, 143, 144, 145, 146, 147, 203, 214, 233, 236, 238, 239, 240, 247, 248, 251, 255, 257, 313, 327, 331, 338, 340, 341, 342, 343
Disaster Resilience 7, 8, 11, 63, 64, 87, 88, 90, 110, 139, 145, 147, 187, 221, 228, 231, 235, 236, 237, 238, 239, 240, 241, 242, 249, 251, 256, 258, 259, 261, 281, 289, 310, 311, 312, 319
Disaster Risk 2, 6, 9, 16, 17, 18, 19, 20, 22, 30, 71, 75, 96, 97, 98, 99, 100, 133, 134, 135, 136, 137, 138, 139, 140, 142, 144, 146, 147, 148, 149, 150, 151, 152, 184, 185, 186, 187, 188, 194, 196, 202, 203, 204, 205, 206, 207, 208, 209, 210, 211, 212, 214, 215, 216, 217, 218, 220, 221, 223, 224, 225, 226, 227, 228, 229, 230, 233, 238, 239, 240, 241, 242, 255, 261, 286, 287, 289, 290, 292, 296, 297, 308, 309, 310, 311, 312
Disaster Survivors 243, 244, 246, 251, 259

E

Earthquake 3, 4, 5, 14, 17, 62, 63, 70, 74, 79, 80, 92, 195, 203, 205, 217, 232, 237, 239, 240, 249, 250, 261, 264, 265, 266, 270, 271, 272, 273, 274, 275, 276, 277, 278, 279, 280, 281, 282, 283, 284, 285, 286, 287, 288, 289, 290, 291, 293, 294, 297, 298, 299, 300, 302, 304, 305, 306, 307, 308, 309, 310, 311, 312, 323, 328
Economic Impact 6, 14, 17, 18, 32, 188, 189, 247, 337, 344
Economic Resilience 19, 28, 183, 184, 185, 186, 187, 189, 193, 195, 203, 204, 205, 206, 207, 208, 209, 210, 211, 212, 213, 218, 223, 228
Education 7, 29, 31, 38, 40, 61, 62, 64, 73, 74, 75, 76, 77, 78, 79, 80, 81, 82, 83, 84, 85, 86, 87, 89, 90, 91, 92, 93, 94, 95, 96, 97, 98, 99, 100, 101, 111, 114, 115, 116, 117, 118, 119, 120, 123, 127, 138, 150, 190, 200, 203, 207, 236, 239, 240, 264, 265, 266, 267, 268, 270, 271, 278, 279, 280, 283, 284, 286, 287, 289, 290, 292, 297, 304, 310, 313
Employee Performance 106, 107
England 96, 178, 313, 317, 318, 319, 320, 321, 322, 323, 324, 325, 326, 327, 328, 329, 330, 331, 332, 333, 334, 335, 336, 337, 338, 339, 340, 341, 342, 343, 344, 345
Environmental Challenges 22, 133, 134, 137, 255
Environmental Change 18, 179, 181, 220, 226, 242, 286, 331
Environmental Sustainability 178, 202, 208, 248, 249

H

Health Equity 177, 179
HR Analysis 104, 107, 113, 126
Human Settlements 1, 2, 11, 25, 292

I

Indigenous Wisdom 133, 136, 140, 146
Infrastructure Resilience 186, 192, 193, 226

J

Joint Operational Capability 47

L

Legal Regulations 45, 48, 65, 66, 67, 68
literature 6, 14, 45, 47, 48, 97, 100, 130, 136, 137, 138, 139, 163, 178, 179, 216, 218, 221, 246, 247, 261, 268, 289, 312, 315, 316, 317, 318, 319, 320, 321, 322, 323, 324, 325, 326, 328, 329, 334, 335, 337, 338, 339, 342, 343

Local Knowledge 36, 91, 99, 133, 134, 135, 136, 137, 138, 139, 140, 141, 144, 145, 146, 147, 148, 149, 150, 154, 158, 159, 160, 163, 166, 187, 192, 236

M

Mexico 286, 291, 292, 293, 294, 298, 307, 311, 312
Multi-Stakeholder Partnerships 163, 166, 193

N

Natural Disaster 1, 3, 4, 5, 7, 12, 16, 18, 19, 28, 87, 118, 197, 200, 217, 221, 231, 232, 241, 244, 245, 247, 248, 264, 274, 276, 281, 286, 314, 315, 319, 320, 327, 330, 332, 341, 345

O

Oaxaca 291, 292, 293, 294, 296, 298, 304, 307, 308, 311

P

Participatory Approaches 31, 92, 135, 146, 147, 157, 167, 170
preparedness 10, 17, 22, 29, 30, 40, 48, 53, 54, 56, 60, 61, 62, 63, 64, 65, 71, 73, 74, 75, 76, 77, 78, 79, 80, 81, 82, 83, 84, 85, 86, 87, 88, 89, 90, 91, 92, 93, 94, 95, 99, 100, 112, 114, 115, 116, 117, 118, 119, 120, 123, 127, 128, 133, 134, 135, 136, 139, 141, 143, 144, 145, 146, 147, 173, 184, 186, 192, 194, 195, 202, 203, 204, 206, 207, 214, 233, 235, 236, 238, 239, 240, 247, 248, 251, 254, 255, 257, 261, 291, 293, 297, 298, 300, 305, 306, 307, 308, 309, 310, 312, 313, 327, 331, 338, 340, 341, 342, 343
Psychological Impact 246, 252, 319, 323, 339
Public-Private Partnerships 35, 88, 91, 162, 164, 184, 185, 189, 195, 207, 208, 211, 212, 215, 218, 221, 222, 226, 238

R

resilience 1, 7, 8, 10, 11, 16, 18, 19, 20, 22, 24, 25, 28, 30, 31, 32, 33, 34, 35, 37, 38, 39, 40, 61, 62, 63, 64, 67, 69, 70, 71, 73, 74, 75, 76, 77, 78, 80, 81, 82, 84, 86, 87, 88, 90, 91, 92, 94, 96, 97, 98, 108, 109, 110, 112, 113, 114, 115, 116, 117, 118, 119, 123, 124, 127, 128, 133, 134, 135, 136, 137, 138, 139, 140, 142, 144, 145, 146, 147, 148, 149, 150, 151, 153, 154, 155, 156, 157, 158, 160, 161, 162, 163, 165, 166, 167, 169, 170, 171, 172, 173, 174, 175, 176, 177, 178, 183, 184, 185, 186, 187, 188, 189, 190, 191, 192, 193, 194, 195, 196, 199, 200, 201, 202, 203, 204, 205, 206, 207, 208, 209, 210, 211, 212, 213, 214, 215, 216, 217, 218, 219, 220, 221, 222, 223, 224, 226, 227, 228, 229, 231, 233, 235, 236, 237, 238, 239, 240, 241, 242, 246, 248, 249, 251, 252, 254, 255, 256, 257, 258, 259, 260, 261, 264, 266, 281, 282, 283, 288, 289, 292, 293, 310, 311, 312, 313, 314, 316, 317, 318, 319, 320, 322, 325, 327, 328, 329, 330, 331, 332, 333, 334, 337, 338, 339, 340, 341, 342, 343
Resilience Building 236, 238

S

seismic risk knowledge 291, 293
Social Protection 187, 194, 197, 200, 201, 207, 216, 218, 220, 228, 241
Social Support Networks 245, 246, 254, 259, 282
Student 265, 266, 270, 271, 278, 280, 283, 284, 298
Sustainable Cities 2, 11, 187, 226
Sustainable Development 1, 2, 6, 10, 11, 17, 28, 30, 32, 75, 96, 100, 123, 128, 135, 151, 152, 184, 187, 189, 193,

194, 200, 209, 214, 216, 221, 222, 225, 226, 229, 235, 236, 240, 252, 257, 260, 286, 292, 312

T

Talent Management 131
Teacher 97, 266, 270, 271, 278, 289, 306
Technological Innovation 35, 36, 37, 57, 231
Traditional Practices 133, 134, 135, 136, 137, 138, 139, 140, 142, 147

V

Vulnerability 2, 5, 6, 7, 12, 14, 18, 19, 24, 25, 27, 29, 30, 32, 36, 70, 71, 75, 77, 90, 124, 134, 137, 138, 141, 150, 151, 158, 161, 162, 166, 169, 177, 178, 179, 185, 186, 187, 188, 190, 191, 193, 194, 196, 200, 202, 204, 205, 206, 216, 217, 218, 220, 225, 231, 234, 242, 243, 244, 245, 246, 247, 249, 253, 254, 256, 258, 259, 261, 281, 286, 287, 289, 292, 296, 314, 315, 317, 324, 329, 330, 331
Vulnerability Assessment 179

www.ingramcontent.com/pod-product-compliance
Lightning Source LLC
LaVergne TN
LVHW080343011026
849161LV00009B/2179